Light-Associated Reactions of Synthetic Polymers

A. Ravve

Light-Associated Reactions of Synthetic Polymers

 Springer

A. Ravve
Consultant
Niles, IL 60714
USA

ISBN-10: 1-4614-9800-7   ISBN-13: 978-1-4614-9800-1

Printed on acid-free paper.

9 8 7 6 5 4 3 2 1

springer.com

Contents

4. Photocrosslinkable Polymers

5. Photoresponsive Polymers

6. Photorefractive Polymers for Nonlinear Optics

Preface

This book deals only with technologically useful light-associated reactions of polymeric materials. It does not discuss, therefore, all light associated reactions of synthetic polymers. Thus, for instance, photo degradations of polymers in the environment are not included. That is a separate, very important subject presented in books dedicated to polymer degradations. Photoassociated reactions of polymers for various practical applications are in a field of polymer chemistry that I personally believe is very interesting. One need merely look at such examples as hardening of properly formulated enamels or varnishes by exposure to light of certain wave lengths. In many instances this is achieved in a fraction of a second. By comparison, hardening of enamels or varnishes by oven baking requires anywhere from ten minutes to a half an hour or even longer at elevated temperatures. Other examples include the possibility of aligning liquid crystals on flat panel displays by exposure to light of proper wave length in place of physical rubbing to force them into alignment. Interesting examples also include polymers that are capable of trapping the light energy of the sun or conducting electricity when exposed to light. Not all the light-associated reactions presented in this book have achieved commercialization. On the other hand, it appears that all those presented here have an industrial potential.

Prior to writing this book, I examined many more publications than are listed in the references. I chose only those that illustrate each subject best. This does not mean, however, that the publications that were not included are inferior in any way to those presented here.

This book consists of six chapters. The first chapter is a brief introduction to photo chemistry and physics of polymers for those not familiar with the subject. The information is given to help the reader follow the discussions in the subsequent chapters. It is not, however, a thorough discussion of the subject and the interested reader is urged to read the original sources. Chapter two is dedicated to the subject of photosensitizers and photoinitiators. They are mentioned in the discussions in the remaining four chapters. It was felt, therefore, that they belong in a separate chapter, by themselves. Chapter three discusses the chemistry of light curing of coatings and inks. Chapter four is dedicated to photocrosslinking of polymeric materials. Chapter five presents light responsive polymeric materials and some of their uses. It includes many diverse subjects. Chapter six discusses work done to develop polymeric materials for use in nonlinear optics.

This book is dedicated to all the scientists who do their research in this fascinating field.

Chapter 1

Introduction

Photoresponsive and photocrosslinkable polymeric materials are important in many industrial applications. These range from photolithography to light curable coatings and inks, to holography, and to numerous other applications. In order to adequately describe the chemistry of these materials, it is necessary to first explain the manner in which polymers interact with light. This chapter is an attempt to present some of the background and aid in understanding the discussions of light-associated reactions of polymeric materials for readers not familiar with the subjects of photochemistry and photophysics. The information presented in this introduction was obtained from the fundamental sources listed in the references. What is presented here, however, is not a thorough discussion of the subject. For a detailed discussion the reader is encouraged to go to the original sources in the references.

1.1. The Nature of Light

All electromagnetic radiations travel in vacuum at the speed of C= 2.9979x10[10] cm /sec. [1,2] Light is a form of electromagnetic radiation and, therefore, also travels at that speed. The fact that light travels at the same speed as other electromagnetic radiations leads to the assumption that light is wavelike in character. Our concept of light, however, is that it also consists of packets of energy that have wave like properties. In each packet there is a range of energies.[2] These cannot be represented by one wavelength, but rather by a whole spectrum of wavelengths. The energy of each particular wavelength in the wave-packet is a discrete unit, a *quantum.*

Electromagnetic radiation is described in terms of a transverse plane wave involving associated electric and magnetic fields. It is supposed that the electric vector **E** and magnetic vector **H** which describe the respective field strengths are aligned in planes at right angles to one another, with both planes perpendicular to the direction of propagation of the wave.[1] A convenient model for the variation of the field strength as a function of time *t* and distance *x* along the axis of propagation is given in Cartesian coordinates by the sinusoidal functions in the following equations [1]:

$$E_y = A \sin 2\pi(x / \lambda - vt)$$

$$H_z = (\epsilon/\mu)^{1/2} A \sin 2\pi(x/\lambda - v\, t)$$

In these equations E_y is the electric field strength vector lying in the xy-plane and increasing along the y-axis, H_z is the magnetic field strength vector lying in the xz-plane and increasing along the z-axis, A is the amplitude of the electric vector (the *intensity* of the wave is proportional to A^2), ϵ is the dielectric constant, and μ is the magnetic permeability of the medium through which the wave is transported. In a vacuum $\epsilon=\mu$ and they are approximately unity in air. The length of the wave, that is, the distance between adjacent maxima in the vectors measured at any instant along the direction of wave propagation (the x-axis) is λ, while v is the frequency or number of complete cycles of vector position change per second. The relationship between λ and v is [1,2]:

$$C / v = \lambda$$

where, C is the velocity of the radiation. The frequency v is independent of the medium through which the radiation passes. Wavelength λ and velocity C, on the other hand depend on ϵ and μ of the medium.

Ordinary light is not polarized. It consists of many electromagnetic vectors that are undulating in fixed, though randomly oriented with respect to each other, planes. When the light is polarized in a plane, it is believed that all the waves have their electric vectors oriented in the same direction. When the light is polarized elliptically then it is believed that two plane waves of equal wavelength and frequency and with identical directions of propagation have the electric vectors perpendicular to one another and out of phase.

The above described model is incomplete and, even called naive by some.[1] Mathematically, however, it can successfully account for many observations concerning light, and this theory has been used successfully to explain many practical phenomena associated with optics.

1.2. The Energy of Radiation

It is possible to calculate the energy associated with any particular wave length of radiation from the following relationship:

$$E = hv = hC / \lambda$$

where h is a proportionality constant, called Plank's constant, equal to 6.625×10^{-27} ergs second /quantum. The velocity of light, designated by C, in vacuum is 2.9979×10^{10} cm./ sec, and λ is the wavelength of light, expressed in centimeters. In a medium containing any matter the light will propagate at a different speed. In this case, the velocity of light, C' is expressed by an equation

$$C' = C / (\epsilon\mu)^{1/2}$$

where ε is the dielectric constant of the matter and μ represents the magnetic permeability.

The energy associated with one mole of quanta is known as an Einstein. It is equal to the energy associated with a particular wavelength multiplied by the Avogadro's number, 6.023×10^{23} (the number of molecules contained in one mole of matter). As an example we can take ultraviolet light with wavelength equal to 300 nm. The energy or an Einstein of light of 300 nm will be [1,2]:

$$E_{300} = h\ C\ /\ \lambda = 6.67 \times 10^{-27} \times 3.0 \times 10^{10} \times 6.023 \times 10^{23} \times 10^{10}\ /\ 300 \times 10^{-7} \times 4.20 =$$
$$= 95.7\ Kcal\ /\ mole$$

The above calculation shows that at the wave length of 300 nm there is enough energy in an Einstein to rupture carbon to carbon chemical bonds of organic molecules.

1.3. Reaction of Light with Organic Molecules

If monochromatic light passes through a uniform thickness of an absorbing homogeneous medium with the absorbing centers acting independently of each other, then the energy of light that is absorbed follows the Lambert-Bouguer law. According to this law of physics the light absorbed is independent of the intensity of the incident light and the intensity of radiation is reduced by the fraction that is proportional to thickness of the absorbing system. In addition, Beer's law states that absorption is proportional to the number of absorption centers. The two laws are usually combined and expressed as follows [1,3-6]:

$$dI/dl = kcI$$

where I is the intensity of the radiation, l is the length of the optical path, through the absorbing medium, c is the concentration of the absorbing centers, and k is proportionality constant. While there are no know exceptions to the Lambert-Bouguer law, there are deviations from Beer's law due to partial ionization, molecular association and complexation and fluorescence.

Portions of organic molecules or whole molecules that have π bonds can absorb light radiation, provided that it is of the right wave length. Particular groupings or arrangements of atoms in molecules give rise to characteristic absorption bands. Such groups of atoms, usually containing π bonds, are referred to as *chromophores*. Examples of such molecules with π bonds are compounds that contain carbonyl or nitro groups, and aromatic rings. A molecule that serves as an example of carbonyl arrangement, one that is often quoted, is a molecule of the formaldehyde. In this molecule, the carbon atom is linked to two hydrogens and to one oxygen by σ bonds. The hybrid sp^2 orbitals bond one electron of carbon with a second electron of oxygen in an *sp* orbital. Also, in a second *sp* orbital of a pair of unbonded *n* electrons on oxygen point away from the carbon atom. The orbitals of formaldehyde, the simplest of the carbonyl compounds were illustrated by Orchin and Jaffe, [7] as shown in Figure 1.1.

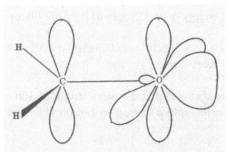

Figure 1.1. The Orbitals of Formaldehyde (from Orchin and Jaffe [18])

As described above, the molecule has a σ and π bonded skeleton, shown above. The carbon atom is attached to two hydrogen atoms by single and to oxygen atom by double bonds. This bonding of the carbon to the two hydrogens and one oxygen atoms is by means of sp^2 hybrid orbitals. The orbitals are approximately at 120 ° angles from each other. In the ground state of the molecules the pair of electrons that form a bond are paired and have opposite or antiparallel spin. In this state the formaldehyde molecule is planar. The Pauli exclusion principle states that no two electrons can have all quantum numbers identical. That means that if two electrons are in the same orbital and three of their quantum numbers are the same, the fourth quantum number, the spin quantum number, must be different. The total spin quantum number of a molecule is designated by a letter *J* and the sum of the spins of the individual electrons by a letter *S*. The spin quantum number of a molecule *J* is equal to $|2S| + 1$.

This arrangement of electrons in the p orbital can generate π bonding and π* antibonding.orbitals. Absorption of light energy by a chromophore molecule, results in formation of an excited state and an electronic transition from the ground state to an excited state. Such light may be in the ultraviolet or in the visible region of the electromagnetic spectrum, in the range of 200 mµ to approximately 780 mµ. Promotion of electrons out of the σ bonding orbitals to the excited states requires a large amount of energy and rupture of bonds in the process. On the other hand, the electronic transition to promote one of the n electrons on the oxygen atom in formaldehyde to the *antibonding* or the *non-bonding* orbital, π* level require the least amount of energy. The name, as one might deduce, is a type of orbital where the electrons make no contribution to the binding energy of the molecule. In formaldehyde this n ⟶ π* transition to the excited state gives rise to an absorption band (at about 270 mμ). This is a relatively weak band and it suggests that the transition is a forbidden one (forbidden does not mean that it never occurs, rather that it is highly improbable). It is referred to as a *symmetry forbidden transition*. The reason for it being forbidden is crudely justified by the fact that the π* is in the *xz* plane (see Fig. 1.1.) The n electrons in the p_y .orbital are in the *xz* plane and perpendicular to the π* orbital. Because, the spaces of the two orbitals overlap so poorly, the likelihood of an electronic transition from one to the other is quite low. As stated above, in the ground (normal) state of the molecules two electrons are paired. The pairing means that these electrons have opposite or anti-parallel spins. After absorbing the light energy, in the singlet excited state the two electrons maintain anti-parallel spins. The n ⟶ π* excitation, however, can lead to two excited state, a singlet (S_1) and a triplet (T_1) one with an absorption band (at about 250 mμ). Intersystem crossing to a triplet state from the singlet results in a reversal of the spin of one of the electrons and an accompanying loss of some vibrational energy. This is illustrated in Figure 1.2.

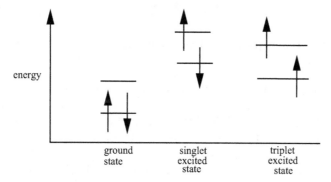

Figure 1.2. Illustration of the Singlet and the Triplet States

The intersystem crossing from the singlet to triplet states can occur with high efficiency in certain kinds of molecules, particularly in aromatic and carbonyl-containing compounds. Electron-electron repulsion in the triplet state is minimized because the electrons are farther apart in space and the energy is lower in that state than that of the corresponding excited singlet one. Solvents can exert a high influence on the n ⟶ π* transitions. While the intersystem crossing is a forbidden transition (see above) it can actually occur with high frequency in certain molecules like aromatic or carbonyl compounds.

The chemical mechanism of photoexcitation of organic molecules has been fully described in various books on photochemistry[1,3-5] It will, therefore, be discussed here only briefly. The transitions are illustrated here in a very simplified energy diagram that shows the excited singlet state and the various paths for subsequent return to the ground state in Figure 1.3.

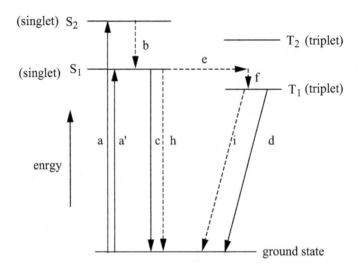

Figure 1.3. Diagram depicting excitation and relaxation pathways of an electron

.

The energy diagram (Fig. 1.3.) represents energy states of a molecule that possesses both **n** and **π*** electrons. S_1 and S_2 are the singlet excited states. T_1 and T_2 are the excited triplet states. Solid lines represent electronic transitions. They are accompanied by absorption or emissions of photons. Radiationless transitions are represented by doted lines. The above diagram shows the lowest singlet state S_1, where the electrons are spin paired and the lowest triplet state T_1, where the electrons are spin unpaired. The electron is excited by light of a particular wavelength into an upper singlet level, S_2. Relaxation follows via an internal conversion process to S_1 level. The excess energy is dissipated by vibrational interactions giving rise to evolution of heat. At the S_1 level there are three possible ways that the excited state becomes deactivated. The return to the ground state from the triplet one requires again an inversion of the spin. In Figure 1.3., **a** and **a'** represent the energies of light absorbed , **b**, **h**, and **i** the energies of internal conversion, **c** represents return to the ground state by way of fluorescence and **d** return by way of phosphorescence.

The Franck-Condon principle helps the understanding of the electronic transitions. In simple terms, what it states is that during an electronic transition the various nuclei in the molecule do not change their position or their moment.[9] What it means is that electronic transitions are much more rapid (10^{-15} sec.) than nuclear motions (10^{-12} sec.) so that immediately after the transitions the nuclei have nearly the same relative positions and velocities that they had just before the transitions. The energy of various bonding and antibonding orbitals increases for most molecules in the following order,

$$\sigma < \pi < n < \pi* < \sigma*$$

In molecules with heteroatoms, such as oxygen or nitrogen, however, the highest filled orbitals in the ground state are generally nonbonding, essentially atomic, n orbitals. This is a case with ketones and aldehydes. These molecules possess electrons that are associated with oxygen that are not involved in the bonding of the molecule, the n-electrons

n-electrons

$$\overset{H}{\underset{H}{>}}C=O:$$

Whether n electrons will be promoted to either $\sigma*$ or $\pi*$ antibonding orbitals depending upon the structure of the carbonyl compound.

As explained above, in the triplet state the spin of the excited electron becomes reversed. This results in both electrons having the same spin. From purely theoretical approach, such an electronic configuration is not allowed. Due to the fact that the excited electron cannot take up its original position in the ground state until is assumes the original spin, the triplet state is relative long lived. For instance, in benzophenone at 77 °C the lifetime can be 4.7×10^{-3} seconds. Orchin and Jaffe wrote [7] that the triplet state has a lifetime of 10^{-3} seconds. By comparison, the lifetime of a singlet state is about 10^{-8} to 10^{-7} seconds. Also, in the triplet state the molecule behaves as a free-radical and is very reactive. The carbon atom has a higher electron density in the excited state than in the ground state. This results in a higher localized site for photochemical activity at the orbital of the oxygen. Because the carbonyl oxygen in the excited state is electron deficient, it reacts similarly to an electrophilic alkoxy radical. It can, for instance, react with another molecule by abstracting hydrogen. This is discussed further in Chapter 2.

At higher frequencies (shorter wavelength) of light, if the light energy is sufficiently high, $\pi \longrightarrow \pi*$ transitions can also take place. All aromatic compounds and all conjugated diene structures possess delocalized π systems. Because there are no n electrons, all transitions in these systems are $\pi \longrightarrow \pi*$. In general, the excited states of molecule are more polar than the ground states. Polar solvents, therefore, tend to stabilize the excited state more than the ground state. As shown in Figure 1.2, the triplet state is lower in energy that the corresponding singlet state. This is due to the fact that the electron-electron repulsion is minimized, because they do not share each other's orbitals as stated by the Pauli exclusion principle Thus less energy is required for the triplet state.

The chemical reactivity of organic molecules is determined principally by the electron distribution in that molecule. When the electron distribution changes, due to absorbtion of light and subsequent transitions, photochemical reactions take place while the molecule is in an electronically excited state. The phenomenon of light absorption, formation of the excited states and subsequent reactions obey four laws of organic photochemistry, as was outlined by Turo [3]:

1. Photochemical changes take place only as a result of light being absorbed by the molecules.

2. Only one molecule is activated by one photon or by one quantum of light.

3. Each quantum or photon which is absorbed by a molecule has a given probability of populating either the singlet state or the lowest triplet state.

4. In solution the lowest excited singlet and triplet states are the starting points for the photochemical process.

The relationship between the amount of light or the number of photons absorbed and the number of molecules, that, as a result, undergo a reaction, is defined as the quantum yield, Φ. It is defined as the number of molecules involved in a particular reaction divided by the number of quanta absorbed in the process. [1,3]

Another fundamental law of photochemistry was formulated by Grotthus and Draaper.[12] It states that only the light that is absorbed by a molecule can be effective in producing photochemical changes in that molecule. There is also a fundamental law of photochemistry that states that the absorption of light by a molecule is a one-quantum process, so that the sum of the primary processes, the quantum yield must be unity. [12] Also, the law of conservation of energy requires that the sum of the primary quantum yields of all processes be equal to unity. Mathematically this can be expressed as :

$$\Phi_{total} = \sum_i \Phi_i = 1$$

Where Φ is the quantum yield. The quantum yield of photochemical reactions is important because it sheds light on the mechanisms of reactions. The number of molecules involved in a particular photoreaction can be established by an analytical kinetic process and the number of quanta absorbed can be measured with the aid of an actinometer. The quantum yield can also be expressed in general kinetic terms[1]:

$$\Phi_i = \phi_{ES}\,\phi_R\,P_i \qquad \text{or} \qquad \Phi_I = \sum_j \phi^i_{ES}\,\phi^j_R\,P^j_i$$

The above equations signify that a quantum yield of a particular photoprocess is the product of two or three distinct probabilities. These are: ϕ_{ES} , is the probability that the excited state will undergo the primary photoreaction necessary for the process. The probability that any metastable ground state intermediate will proceed to stable products is P_i and the probability that the excited state will undergo the primary photoreaction necessary of the process is ϕ_R.

The concept that matter can only acquire energy in discrete units (quanta) was introduced in 1900 by Max Planck.[2] The corollary of the quantization of energy is that matter itself must be quantized i.e., constructed of discrete levels having different potential energies. Occupying these particular levels are electrons that obviously possess the energy of the level which they occupy. In a molecule the intramolecular motions of the electrons and the associated molecular electronic levels must be taken into account. There are, in addition to electronic levels, modes of vibration and rotation that are also quantized. In other words, the absorption of a photon of light by any molecule is a reaction of light with an atom or a group of atoms that must promote transitions between quantum states. This requires two conditions. These are: (1) for a molecular state m with energy E_m there must be a state n of higher energy E_n so that $h\nu = E_n - E_m$; (2) there must be specific interaction between the radiation and the light absorbing portion of the molecule that results in a change in the dipole moment of the molecule during the transition. If we designate the wave functions of the states m and n as ψ_m and ψ_n respectfully then the transition moment integral that may not equal to zero, is:

$$\mathbf{R_{mn}} = (\psi_m / P\psi_n)$$

where P is the electric dipole operator. It has the form of $P = e\sum r_i$, where e is the electronic charge and $\mathbf{r}_i$ is the vector that corresponds to the dipole moment operator of an electron i.

The increase in the energy of a molecules as a result of absorbing a quantum of radiation can be expressed in the relationship[3-5]:

$$\Delta E = hC / \lambda$$

where λ is the wavelength of the interacting radiation. All reactions that are photochemical in nature involve electronically excited states at one time or other. Each one of these states has a definite energy, lifetime, and structure. The property of each state may differ from one to another and the excited states are different chemical entities from the ground state and behave differently. The return to the ground state from the excited state, shown in Figure 1.3. can take place by one of three processes [3-5]:

(1) The molecule returns directly to the ground state. This process is accompanied by emission of light of a different wavelength in the form of fluorescence.

(2) An intersystem conversion process takes place to the T_1 state, where the electron reverses its spin. The slower decay of excitation from the triplet state to the ground state is accompanied by emission of phosophorescence.

(3) The molecule uses the energy of excitation to undergo a chemical reaction. This dissipation of the excitation energy can also be illustrated as follows:

$$A_0 \; + \; \text{light} \longrightarrow A^* \underset{\searrow}{\overset{\nearrow}{\longrightarrow}} \begin{array}{l} A_0 \; + \text{fluorescence} \\ A_0 \; + \text{heat} \\ \text{chemical reaction} \end{array}$$

where, A_0 represents any organic molecule and A^* represents the same molecule in an excited state.

In the process of energy dissipation from the singlet and return to the ground states, the light emission by fluorescence is at a different wave-length than that of the light that was absorbed in the excitation. This is because some energy is lost in this process of the electron returning from its lowest excited state to the ground state. The energy, however, may also, depending upon the structure of the molecule, be dissipated in the form of heat, as shown above. And, also, a third form of energy dissipation can occur when the molecule undergoes a chemical reaction. Depending, again on the molecular structure, the chemical reactions can be rearrangement, isomerization, dimerization (or coupling), fragmentation, or attack on another [9-11] molecule. Some examples of such reactions are:

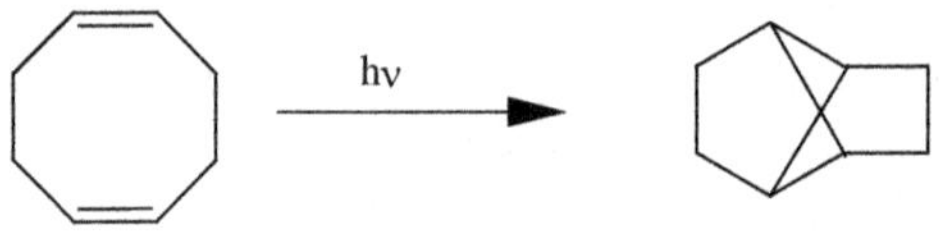

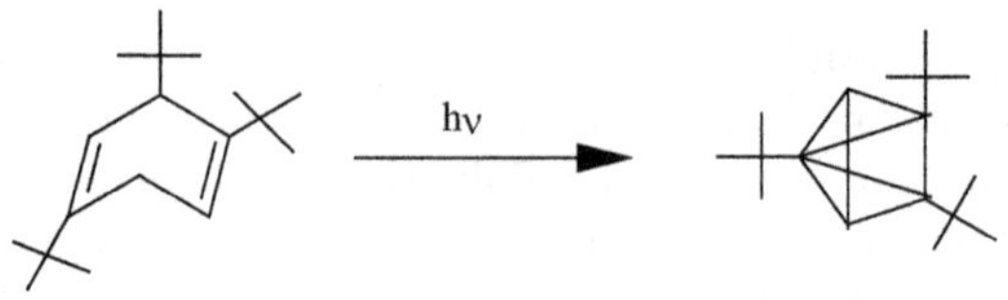

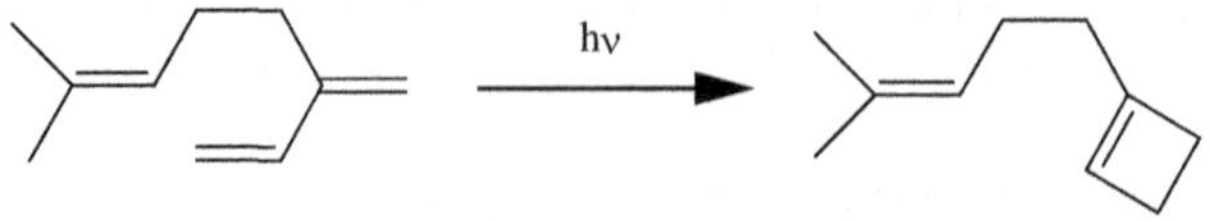

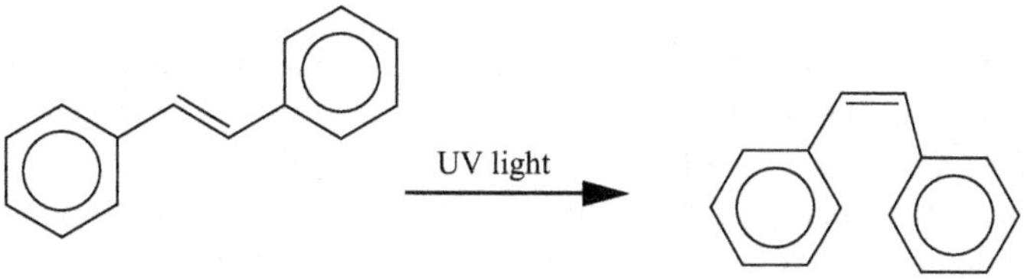

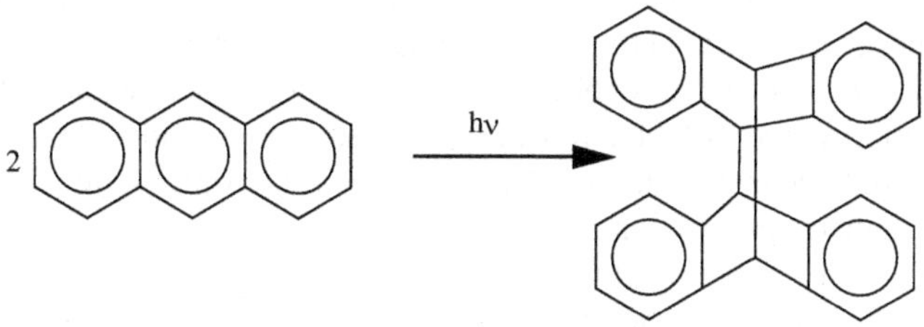

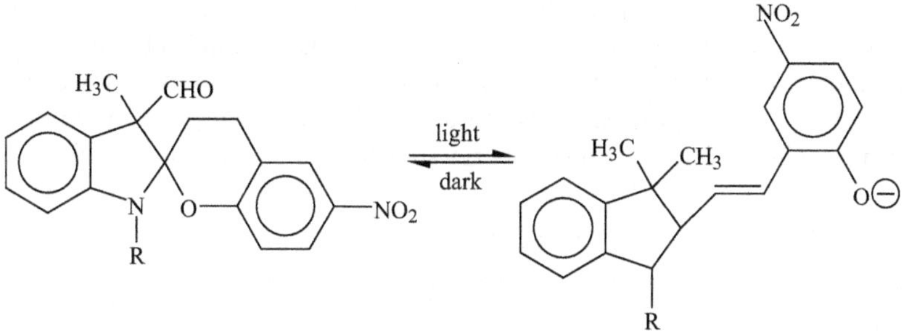

Many other examples can be found in the literature. Most familiar isomerization reaction is that of *cis*-stilbene to *trans*-stilbene, shown above. It was observed that the quantum yield of stilbene *cis-trans* isomerization decreased with an increase in viscosity of the medium. [8] In addition, it was also found that in a polymeric matrix the photoisomerization is not inhibited, provided that it occurs above the glass transition temperature of the polymer. An example of a fragmentation of a molecule is the decomposition of disulfides upon irradiated with ultraviolet light of the appropriate wavelength:

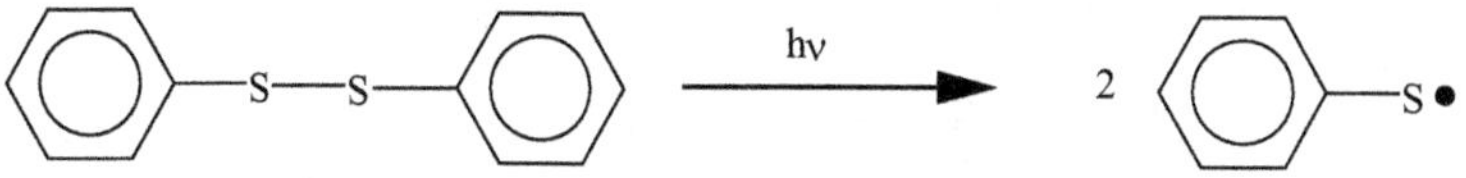

The same reaction takes place in peroxides. Ketones and aldehydes cleave by the mechanism of the Norrish reaction.

1.4. Energy Transfer Process

The energy of excitation can transfer from an excited molecule to another molecule. Thus, the term *energy transfer*[12,13] refers specifically to one-step *radiationless transfer* of electronic excitation from a *donor* molecule to another, qualified, *acceptor* molecule, from one chromophore to another one. This excludes what is referred to as, *trivial energy transfers* that result from the donor emitting light that is subsequently absorbed by an acceptor. Based on the energy and spin conservation laws, there are two a priori requirements for efficient energy transfer: (1) the process must be thermoneutral or exothermic to occur with highest efficiency, because the activation energies have to be low due to short lifetimes of electronically excited state, and (2) no net spin changes should occur. If a donor molecule was in the triplet state at the time of the energy transfer process, the acceptor molecule is then also promoted to the triplet state. Transfer of singlet to singlet energy should be possible, but it occurs less frequently, because of the shorter life times of the singlet states. [12,13]

Energy transfer, is thus the process by which excitation energy passes from one photoexcited molecule, often referred to as a *sensitizer* and in this case designated as S*, to another adjacent molecule in its ground state, often referred to as a *quencher,* in this case designated as Q. The quencher must have a thermodynamically accessible excited state, one whose energy is lower than that of S*. A donor molecule must possess sufficiently long lifetime to be an efficient sensitizer. The reaction of energy transfer can be illustrated as follows:

$$S* + Q \longrightarrow S + Q*$$
$$S* \longrightarrow S_0 + h\nu$$

where * designates an excited state. In the process of energy transfer, S* returns (or relaxes) to the ground state S. Energy transfer is further categorized as involving singlet (paired electron spins) or triplet (unpaired electron spins) states. Symmetry rules, as explained above, require a singlet S* to produce a singlet Q* and a triplet S^{3+} to produce a triplet Q^{3+}.

The quenching reaction of the excited state is expressed in a equation by Stern-Volmer. The reaction shown below is based on a quenching reaction that is accompanied by a release of heat:

$$S* + Q \longrightarrow Q* + S_0 \quad (k_2)$$

$$S* \longrightarrow S_0 + heat \quad (k_3)$$

The equation is written as follows:

$$\Phi_0 / \Phi_Q = \{k_1 + k_2[Q] + k_3\} / (k_1 + k_3)$$

In experimental studies of energy transfer it is convenient to express the experimental results in an other form of the Stern-Volmer equation, as follows,

$$\Phi_0 / \Phi_Q = 1 + k_q \tau [Q]$$

where Φ_0 is the quantum yield for a particular process in the absence of a quenching molecule,

Φ_Q is the quantum yield of the quenched process

k_q is the bimolecular rate constant for the quenching process

τ is the lifetime of the state in the absence of a quenching molecules. It is equal to $1 / (k_1 + k_3)$, and [Q] is the concentration of the quenching molecules.

Two processes were proposed to explain the mechanism of energy transfer. In the first one, energy transfers result from the interactions of the dipole fields of the excited donors and ground state acceptor molecules (long-range: Forster (dipole-dipole))[5, 9]. This is referred to as the *resonance transfer mechanism.* Such transfer is rapid when the extinction coefficients for absorption to the donor and acceptor excited states involved in the process are large (10^4 to 10^5 at the maximum). When the dipolar interactions are large, resonance transfers are possible over distances of 50-100 Å. Close proximities of donors and acceptors, however, are required for weakly absorbing molecules. In the second mechanism [9] (short-range: Dexter (exchange)), the excited donor and acceptor are in very close proximity to each other, (up to ≈ 15 Å) such that their electronic clouds overlap slightly. In the region of the overlap, the location of the excited electron is indistinguishable. It may be at any one instant on either the donor or acceptor molecule. Should the pair separate when the excited electron is on the acceptor molecule, energy transfer has been achieved by the mechanism of electron transfer.

Both absorption and emission processes may be intramolecular, localized in a single molecule. On the other hand, they can also involve whole crystals that may act as absorbers and emitters. Such energy transfers can manifest themselves in different ways that include sensitized fluorescence or phosphorescence, concentration depolarization of fluorescence, photo-conduction, and formation of triplet acceptor molecules.

Intermolecular energy transfer can be electronic and vibrational, and can take place in solid, liquid and gaseous phases. In addition, the sensitized excitation of Q by S* has to take place within the time that the molecule S remains in the excited state. In summary, theoretical and empirical considerations suggest two modes of transfer, described above:

    1. Only when the two molecules are in very close proximity to each

other and their centers are separated by the sum of their molecular radii will transfer take place.

2. When the two molecules are at distances that exceed their collision diameters, resonance transfer, or long range electronic excitation takes place though Coulombic interactions.

The transfers that take place by mechanism 1 are limited by diffusion of molecules in solution and should be affected by the viscosity of the medium. Transfers by mechanism 2, on the other hand, should be much less sensitive to the viscosity of the medium It was shown by Foster[5] that the rate constant of resonance-energy transfer (mechanism 1), as a function of distance, is:

$$\textbf{Rate Constant } (S^* \rightarrow Q^*) = 1(R_0/R)^6/\tau \text{ S}$$

where τ S is the actual mean lifetime of S^*, R is the separation between the centers of S^* and Q, and R, is the critical separation of donor molecules and the acceptor molecule. The efficiency of energy transfer is expressed by Turro, Dalton, and Weiss[13] as follows:

$$\Phi_{et} = k_{et}[S^*][Q] / \{k_{et}[S^*][Q] + k_d[S^*]\}$$

The transfer by long range excitation or mechanism 2 can be in the form a *singlet-singlet* transfer, a *triplet-singlet* transfer, and a *triplet-triplet* transfer. Due to the fact that the lifetime of triplet state of molecule is longer than the singlet one, it is more probable to be the one to participate in energy transfer. Molecules that undergo intersystem crossing with high efficiency, like benzophenone, are efficient triplet sensitizers. Such molecules must possess high energy in the triplet state and a life time of at least 10^{-4} seconds.

The two types of intermolecular energy transfers can be expressed as follows:

Forster (dipole-dipole) long-range: $k_{SQ}(R) = k_S^0(R_0^{SQ}/R)^6$
Dexter (exchange) short-range: $k_{SQ}(R) = k_{SQ}^0 \exp(-\alpha R)$

The nomenclature that was developed in connection with energy and charge transfer processes is as follows, an *eximer* is a transient dimer formed by the combination of an excited (usually aromatic) molecule and a second similar (usually unexcited) molecule. Such a dimer bonds only in the excited state and promptly dissociates in losing its excitation energy. The term *exiplex* was explained by Birks[11] to describe a complex between two molecules, one a donor and the other one, an acceptor, which subsequently dissociate in a deactivation process. One of the components of the *exiplex*, either the donor or the acceptor, is in excited state while the counterpart, acceptor or donor are is in the ground state. An *eximer* is then just a special case in which the two constituent molecules are identical. While numerous charge-transfer complexes can form between certain molecules in the ground state, a number of compounds can form only charge-transfer complexes when either the donor or the acceptor is in an excited state . Formation of eximers was observed in a number of aromatic polymers, such as polystyrene, poly(vinyl naphthalenes), poly(vinyl toluene) and others.[12]

An *exterplex* is composed of three molecules and often takes an important role in pbotophysical and photochemical processes. Polymers with pendant aromatic chromophores and dimeric compounds often show efficient *exterplex* formation due to high local chromophore concentration in their structure. It was observed that *exiplex* emission spectra from a chromophore is usually broad, structureless, and red shifted to the corresponding monomer fluorescence. The extend of such a shift is a function of the distance between the two components of the complex. It is also strongly affected by the polarity of the media. Martic et al., [14] obtained emission spectra of the exiplexes of anthracene and N,N, dimethyl-*p*-toluidine in toluene and in polystyrene. While the maximum band of the emission spectra in toluene at 30 °C is at 616 nm, in polystyrene it is shifted to 400 nm. The exiplex emission spectra in a copolymer of styrene with 4-N,N-diaminostyrene is at 480 nm. The maxima of the emission spectra are temperature dependent. The maxima shifts in toluene solution to shorter wave length and in polystyrene it is the opposite, it shifts to longer wave length with an increase in temperature. The maxima approaches common value at the glass transition temperature of polystyrene. Similar results were reported by Farid *et al.*, [15] who studied formation of exiplexes of 4-(1-pyrenyl)butyrate in different solvents and in polymers.

Chemical and physical changes take place in molecules when they absorb energy and reach and excited state. This is particularly true of carbonyl compounds. There is a change in the dipole moments of the molecules. This is due to the fact that dipole moments depend upon the distribution of the electrons. In carbonyl compounds this change is particularly large. Also the geometry of the molecule changes from the ground to the excited states. In addition, the chemical properties of the molecules change. Thus phenol, for instance, is a weak acid, but in the excited state it is a strong acid. This can be attributed to the $\pi \longrightarrow \pi^*$ transition where one of the pair of π electrons is promoted to an anti bonding orbital.

By the same reason, the acid strength of benzoic acid is less than in the excited state because the charge in this case is transferred to carbonyl group. The excited states of both phenol and benzoic acid can be illustrated follows [12]:

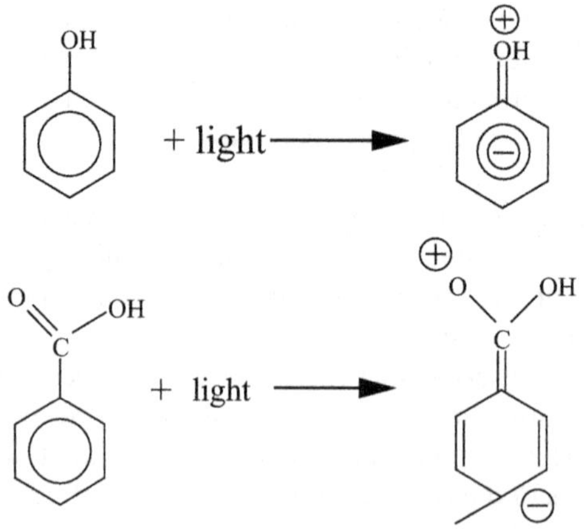

1.5. Electron Transfer Process

Simple migration of energy is a thermodynamically neutral process. It allows the excitation energy deposited at a site in a solid or in a concentrated solution to move to another position by transferring the excitation energy in the absence of an intermediate quencher. Electron transfer, however, is a process by which an electron is passed from an electron-rich donor to an electron-deficient acceptor. [12] This reaction is substantially accelerated when the donor or acceptor is excited. Electron transfer from an excited-state donor molecule D* to a ground-state acceptor A generates a radical cation D$^+$• and a radical anion A$^-$•. The resulting radical ion pair exists as a charge-separated pair of ions:

$$D* + A \longrightarrow D^+• + A^-•$$

The oxidized and reduced species are usually highly energetic, storing a substantial fraction of the energy absorbed from the photon. The charge separation that occurs in such a photoinduced electron transfer provides a way to convert the excitation energy of the excited molecule to a chemical potential in the form of a radical ion pair.

Electron migration can also be a movement of an electron either to a neutral electron donor from an oxidized one (D + D• $\longrightarrow$ D• + D) or from a reduced acceptor to a neutral one (A• + A $\longrightarrow$ A + A•). [12] These thermoneutral processes, that are called *hole* and *electron migrations*, respectively, permit further spatial charge separation between an excited donor, D* and a reduced acceptor, A•. This separation is beyond one that is initially produced in an ion pair by photoinduced electron transfer. After the absorption of light by A to A$^-$ sensitizer, the energy migrations or the energy transfer moves the excited state site where the excitation energy is converted to a radical ion pair by photo induced electron transfer. A kinetic competition then takes place between the rates of several possible next steps. These steps can be chemical reaction of the radical ions, or they can be further charge migrations by *sequential electron or hole transfers*, or actually nonproductive charge recombination, called *back-electron transfer*. The back-electron transfer regenerates the ground states of both the donor and the acceptor.

1.6. The Charge Transfer Processes in Polymeric Materials

Charge transfer in polymers is either electronic (transfer of electrons or of positive charges alone) or it is ionic (transfer of protons or larger charged species). Electronic conduction can be also of two types. One type is conduction due to diffusion of electrons that are not localized on any particular molecule (this is usually found in liquids or in gases). The other type can be by conduction due to positive or negative charges that are localized on any particular part of the molecules. Such charges can be exchanges between like polymeric molecules (or between segments of single polymeric molecules). This can occur without any net energy loss (*resonant charge transfer*). It was shown experimentally that the electrical conductivity in many polymeric materials, subjected to short irradiation pulses, consists initially of a "prompt" component. That means that very rapid transfer of a considerable amount of charge takes place over a comparatively short distance (≈ 100 Å). The movement of the charge is then terminated as a result of trapping in "shallow" traps. [9,12,16] This is followed by a "delayed" component that is very temperature dependent and probably indicates a thermally activated charge-hopping process between the shallow traps. This continues until terminated (after ≈ 1 µ) by trapping in deep traps or by recombination. [9,12,16]

There is a major difference between *eximers* of polymers and those of small molecules. The difference is that at least in some polymers a large part of the excitation of the excimer site appears to be a result of singlet energy migration. [12] Also, in polymeric materials with a number of identical chromophores, either in the backbone or as pendant groups, when photons are absorbed, the excited states cannot be considered as localized. In simple cases of rigid lattices the excitations are distributed over the entire volume of the material as a wave-like linear combination of local excitations. [7, 9,10,13] They are referred to as *tight-binding excitations*. [9,10,13] As one might expect, excimer formations in polymers depend upon the properties of the chromophores and upon their location on the polymeric chain. [9] In addition, polymer tacticity, conformation, and distance between chromophores can greatly affect the formation of eximers. Also, it is possible to distinguish between two different types of energy transfers in polymeric materials In the first one the transfer of excitation can take place either from or to large molecules from small ones. Thus, for instance, a polymer transfer of the excitation energy can be localized from a chromophore on one polymeric chain to another. An example of a transfer to a small molecule is an energy transfer from a polymer, like polystyrene to a scintillator molecule, like 1,4-bis[2-(5-phenyloxazolyl)]benzene shown below[14]:

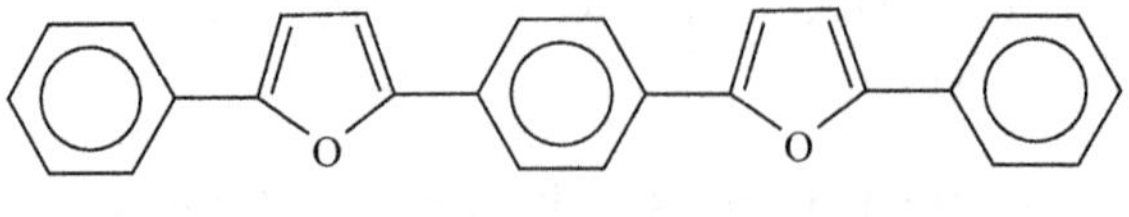

More than that, transfer can also take place from one group of atoms, or from a chromophore, located on a polymeric chain in one section of the molecule, intramolecularly, to another one located at another section of the same polymer. Thus, in copolymers from monomers with two different chromophores groups, the energy absorbed by one group of chromophores can be transferred to the chromophores from the other group. This can take place by either Foster or exchange mechanism. The possibility of energy transfer from one chromophore to an adjacent different chromophore in polymeric chains depends to a large extent upon the lifetime of the excitation and its alternative modes of deactivation. For this reason the most readily observed form of energy migration is one that occurs through the mechanism of the triplet. [7,9,12,16]

Intermolecular energy transfer from one polymeric material to another while the molecules are in solution or in the melt can also take place. [17] This was demonstrated on an intramolecular *excimer* and *exiplex* formation in solutions of polyesters containing naphthalene or carbazole moieties in their chemical structures [17]:

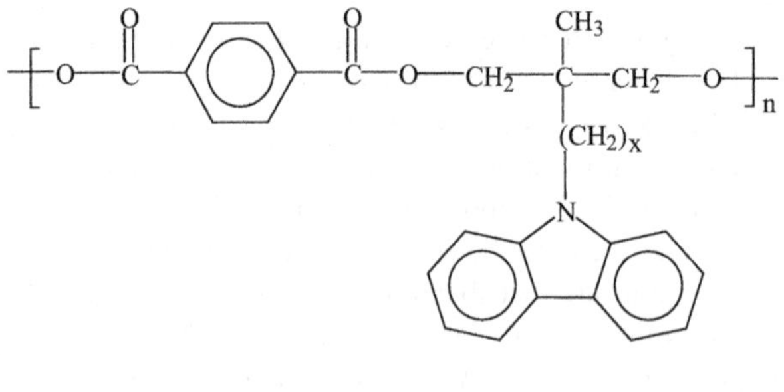

In general, the migrations of energy in polymers is somewhat more complex, because chain folding and conformations are additional factors that enter into the picture. The separation between interacting units can be affected by the composition of the polymer, the geometry of the polymeric chains, and the flexibility of the backbones.[18]

There are two limiting cases for the effects of polymer folding on energy-transfer efficiency. Folding of a polymer before excitation into a conformation in which the sensitizers are held within a hydrophobic pocket improves the efficiency of energy migration when a large number of intramolecular hops or through bond interactions intervene between the sensitizer and the ultimate trap. [12] If the polymers are flexible, however, they can also bend after photoexcitation to bring otherwise distant chromophores close enough so that energy can hop from one to the other, skipping intervening units and thereby considerably shortening the effective migration distance along an individual polymer chain. [12] For flexible polymers in solvents that promote folding, this motion can take place even faster than excited-state decay.[18]

Intramolecular singlet energy migration can also proceed via electronic coupling through the bonds that form the polymer backbone. In a random walk, the excitation energy migrates without directional control, moving back and forth along a chain or across space. Through-space interactions between pendant chromophores are also common in polymers with large numbers of absorbing units.[18] One should also include movement of excitation across folds or loops that can form in polymeric chains. Such folds can be the result of packing into crystalline domains or simply from temporary collisions.

In principle, the excitation can be localized for some finite time (however small) on a particular chromophore before it is transferred to another one in the chain. Guillet defines *intramolecular energy migration* as any process that involves more than one exchange of excitation energy between spetroscopically identical chromophores attached by covalent bonds to a polymeric chain. [12] He further terms "energy transfer" as a single step migration between two chromophores, while one that involves several or more chromophores as "energy migration".[12]

The, polymers with multiple sensitizers offer several routes for energy migration This can be illustrated as follows [18] :

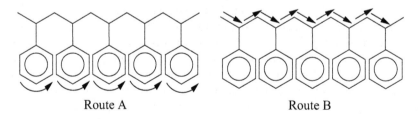

Route A   Route B

A very common arrangement is for the photosensitive groups to be aligned outside of a spiral arrangement of the polymeric chain in close enough proximity to each other for energy transfer. Also, as mentioned earlier, folding of a polymer before excitation into such a conformation that the sensitizers are held within a hydrophobic pocket improves the efficiency of energy migration with a large number of intramolecular hops. Efficiency of energy migration is also helped through-bond interactions that intervene between the sensitizer and the ultimate trap. [18] Also, as mentioned before, flexible polymer frameworks can bend the polymeric chains in such a manner as to bring otherwise distant chromophores close enough together so that after excitation the energy can hop from one to another. In such a case the energy migration can skip intervening units and thereby considerably shorten the effective migration distance along a single polymer chain. As stated above, for flexible polymers in solvents that promote folding, this motion can be even faster than excited-state decay. [18]

Intermolecular energy migration can also occur between two different polymeric molecules. Thus, for instance, Turro and coworkers investigated inter- and intramolecular energy transfer in poly(styrene sulfonate). They found that excimer formation between adjacent phenyl groups is a dominant reaction both along a single chain and between two different chains. [13] At low densities of excited states, singlet energy transfer between a sensitizer and its nearest quencher (perhaps on another chain) dominates, whereas at high excited state densities, energy migration takes place through the series of donors. [18]

Webber reports that he uses the following equation (that he calls crude but useful) to obtain rough estimates of the energy migration diffusion rate along the polymer backbone [18] :

$$k_q = 4\pi N_0 (D_Q + \Lambda_g)PR / 1000$$

where D_Q is the normal diffusion constant of the quencher and k_g is the energy migration diffusion rate along the polymer.

In some aromatic vinyl polymers excimer emission can occur after an initial excitation of an aromatic chromophore. This is followed by intramolecular singlet energy migration, either along the polymer chain, or intermolecularly along the chromophores. Here too, it can be to different chains in a polymer that is in bulk and the chains are in close proximity to each other. The process generally continues until the excitation is trapped at some chain conformation that is suitable for excimer formation. Such a chain conformation is referred to as *eximer-forming site.* If the polymer is in solution and viscosity is low, interconversion of chain conformations proceeds fairly rapidly. In such cases the lifetimes of any particular conformation is limited by the collision processes as well as by the magnitude of the rotational barriers with respect to thermal energy. [12] In the solid state, however, the rotational freedom of the polymeric chain is considerably reduced. Large scale conformational changes are unlikely. There still is the possibility, however, that adjacent chromophores will be in a marginal eximer forming site. [13]

The carbazole molecule is a good example of a chromophore that readily forms excimers. This makes polymers that bear carbazole moieties useful as photoconducting materials . It is described in Chapter 5.

1.7. The Antenna Effect in Polymers

It was originally observed by Schneider and Springer [19] that efficient fluorescence occurs from small amounts of acenaphthalene that is copolymerized with styrene. Fox and coworkers [18] observed the same effect in a copolymer of styrene with small amount of vinyl naphthalene. The emission of naphthalene fluorescence is much higher than from solution of a mixture of the two homopolymers. It was suggested by both groups that this phenomenon is due to energy migration between styrene sequences to the naphthalene moieties. Guillet and coworkers carried out quantitative studies of this phenomenon with various polymers that contained naphthalene or phenanthrene as the donors and anthracene as the trap. [12] This effect is similar to one observed in ordered chlorophyll regions of green plant chloroplasts (antenna chlorophyll pigments). It was, therefore, named the *antenna effect*.

Work by Guillet demonstrated that the effect is not entirely due to energy migration among the chromophores that form the antenna, but rather a combination of migration and direct Forster energy transfer to the trap. [12] Guillet concluded that energy migration and transfer in such systems are primarily due to long-range Forster transfer by dipole-dipole mechanism (discussed earlier). In the absence of any trap in the polymer the energy will migrate along the backbone of the polymer chain until it is deactivated by some other processes. In the presence of a singlet energy trap, the lifetime of the excitation will be reduced and the length of energy migration will be reduced. The difference between this form of energy transfer and one observed in solid aromatic polymers is that the photon energy is collected within a single polymer molecule and all energy transfer is intramolecular. The antenna effect permits collection of the photon energy from the entire region of space (the hydrodynamic volume of the polymer) and transmitting it to the traps located on the polymer chain. The efficiency is relatively independent of concentration and can be very efficient even in dilute solutions. [12]

References

1. J.G. Calvert and J.N. Pitts, *Photochemistry*, Wiley, New York, 1967
2. F.K. Richtmeyer and E.H. Kennard, *Introduction to Modern Physics*, McGraw-Hill, New York, 1947
3. N.J. Turo, *Molecular Photochemistry*, W.J. Benjamin/Cummings, New York
5. Cowan and Drisko, *Elements of Organic Photochemistry*, Plenum, NewYork, 1978
4. Rohatgi-Mukherjee, *Fundamentals of Photochemistry*, Wiley, New York, 1978

5. T. Foster, *Disc. Faraday Sec.,* **1959**, *27*, 1; *Radiation Research, Supplement,* **1960,** *2,* 326 (1960); S.E. Webber, *Chem. Rev.*, **1990,** *90*, 1460-1482

6. W.J. le Noble, *Highlights of Organic Chemistry*, Dekker, New York, 1974

7. M. Orchin and H.H. Jaffe, *The Importance of Antibonding Orbitals*, Houton Miffin Co., Boston, 1967

8. D. Gegiou, K.A. Muszkat, and E. Fischer, *J. Am. Chem. Soc.*, **1968**, *90*, 12

9. R. Srinivasan, *J. Am. Chem. Soc.*, **1964**, *86*, 3318 ; K.E. Wilzbach and L. Kaplan, J. Am Chem. Soc., **1964**, *86*, 3157; D. Schulte-Frohlinde, *Ann.*, **1958**, *615*, 114

10. A.S. Davydov, *Theory of Molecular Excitations,* McGraw Hill, New York, 1962; R.S.H. Liu and G.S. Hammond, *J. Am. Chem. Sec.*, **1964**, *86*, 1892

11. J.B. Birks, *Photophysics of Aromatic Molecules,* Wiley-Interscience, London, 1970

12. J. Guillet, *Polymer Photophysics and Photochemistry,* Cambridge University Press, Cambridge , 1985

13. N.J. Turro, J.C. Dalton, and D.S. Weiss in *Oranic Photochemistry*, O.L. Chapman, ed., Dekker, New York, 1969

14. P.A. Martic, R.C. Daly, J.L.R. Williams, and S.Y. Farid, *J. Polymer Sci., Polym. Lett. Ed.*, **1977**, *15*, 295

15. S. Farid, P.A. Martec, D.R.Thompson, D.P. Specht, S.E. Hartman, and J.L.R. Williams, *Pure. Appl. Chem.*, **1979**, *51*, 241

16. T.J. Meyer, "Photoinduced Electron and Energy Transfer in Soluble Polymers," *Coord. Chem. Rev.*, **1991**, *111*, 47; W.G. Herkstroeter, Chapter 1, *Creation and Detection of the Excited State,* A.A. Lamola, ed., Dekker, New York, 1971

17. S. Tazuke and Y. Matsuyama, *Macromolecules*, **1975**, *8*, 20; *ibid.*, **1977**, *10*, 215

18. M.A. Fox, W.E. Jones Jr., D.M. Watkins, Chem. and Eng. News, 1993, (March 15), 18

19. F. Schneider and J. Springer, *Makromol. Chem.* **1971**, *146*, 181

Chapter 2

Photosensitizers and Photoinitiators

2.1. Phortosensitizers

As explained in Chapter 1, photosensitizers are molecules that absorb the energy of light and act as donors by transferring this energy to acceptor molecules. The molecules that receive the energy may in turn undergo various reactions, such as polymerizations, isomerizations, couplings and others. Many different molecules can act as photosensitizers, but the most useful ones are various aromatic compounds. In Table 2.1. are listed some common photosensitizes that appeared in various publications in the literature. The process of photosensitization and energy transfer involves formation of charge transfer complexes. A good photosensitizer, therefore, is not only a molecule that readily absorbs light energy, but also one that readily transfers it to another molecule. Some compounds are capable of forming such transfer complexes in the ground state, but many more form *exiplexes* in the excited state. Others can form complexes between a compound in the ground state and another one in the excited state. Such complexes are called *excimers* or excited *dimers*. The difference between the excited state of a dimer and an exiplex is that the dimers possess binding energy in the ground state, while exiplexes lack any binding energy in the ground state. This is described in Chapter 1. The emission spectra from two molecules that are capable of forming *exciplexes* depend upon the distances between the two molecules. An equation for the excited state wave function of a one-to-one exciplex that forms from a donor molecule D and an acceptor molecule A was written by Guillet as follows[1]:

$$\psi_E = \alpha\psi_1(D^{+\cdot}A^-) + \beta\psi_2(D^-A^+) + \gamma\psi_3(D^* A) + \delta\psi_4(D A^*)$$

The first two terms on the right side of the above equation correspond to charge resonance states and the last two to the excitation resonance states. Thus a photosensitizer can act in two ways, by energy transfer and by electron transfer. To be exact, one may feel that a true photosensitizer is one that acts by energy transfer alone. This, however, is not always the case. Also, in the event of electron transfer, the process can lead to photoinduced decomposition via electron transfer. [2]

The rate of absorption of light by a sensitizer that corresponds to excitation from the ground state to the excited singlet can be expressed as [1]:

$$I_{abs} = d[S_0]/dt - d[S_1]/dt$$

23

Table 2.1. Some Often Used Photosensitizers

Photosensitizer	Chemical Structure	Φ
Benzophenone	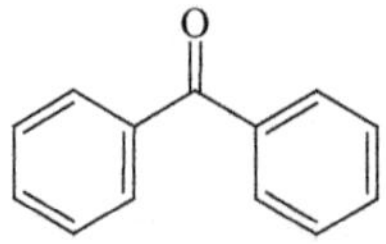	1.0
Acetophenone	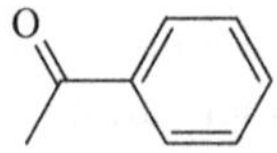	0.99
Triphenylene		0.95
Fluorenone	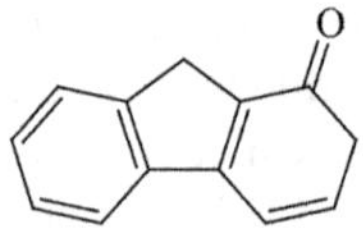	0.93
Anthraquinone	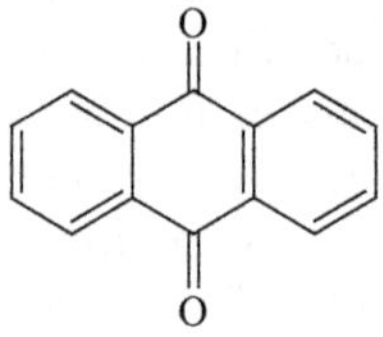	0.88
Triphenylamine	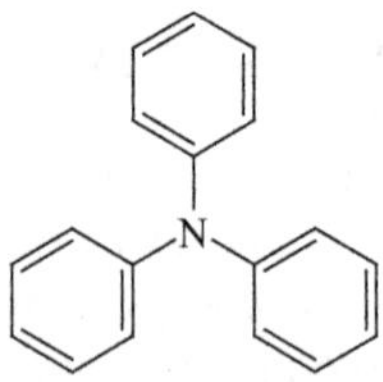	0.88
Phenanthrene	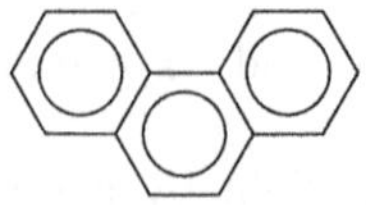	0.76

Table 2.1. **(Continued)**

Photosensitizer	Chemical Structure	Φ
Benzil	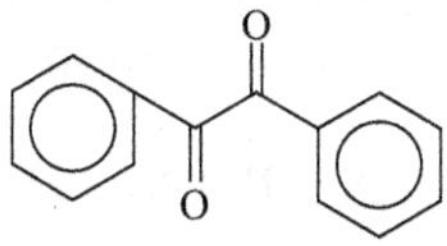	0.87;0.92
Pyrene	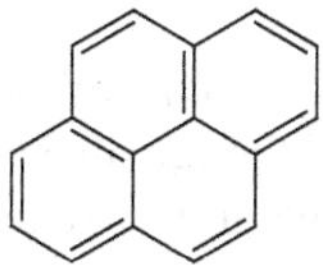	0.40
Naphthalene	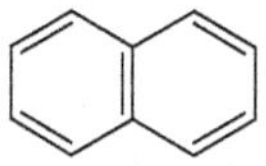	0.40
Durene	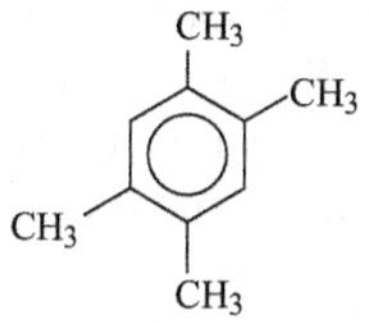	
Anthracene	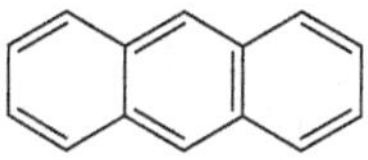	

from various literature sources. Φ represents the quantum yield of triplets

The measurement of fluorescence and phosophorescence spectra of photosensitizers is very important in providing information about the energy of the excited states. It also allows identification of the phenomena.

The process of energy transfer requires that the excited donor diffuse to the proximity of an acceptor within the time period of its excited lifetime. This is subject to the viscosity of the medium and the efficiently of the collision process and the range r in which the collisions can occur. The observed rate constant for energy transfer k_{ET} is governed by the molecular rate constant k_{diff} for diffusion controlled reaction. This is defined by the Debye equation:

$$k_{diff} = 8RT/3000\ \eta$$

$$k_{ET} = \alpha\ k_{diff}$$

where α is the probability of energy transfer. R is the universal gas constant, T is the temperature in kelvins, η is the viscosity of the medium in poise. The Schmoluchowski [1] equation defines the diffusion constant in terms of the diffusion coefficient of the sensitizer and the acceptor:

$$k_{diff} = 4\pi/1000\ (R_s + R_a)(D_s + D_a)\ N_a\ /2)\ \{1 + [\ R_s + R_a\ /(\tau_0\ (D_s + D_a)/\ 2)^{0.5}]\}$$

where D_s and D_a are the diffusion coefficients of the sensitizer and the acceptor R_s and R_a are the molecular radii of the sensitizer and the acceptor, N_a is the Avogadro number and τ_0 is the lifetime of the excited state of the sensitizer.

2.2. Photoinitiators

The photoinitiators are compounds, usually organic, that, upon absorbing light energy, form polymerization initiating species. Such species can be free radical, ionic, or both. Some molecules can function as both, photosensitizers and photoinitiators. An example of such a compound is benzophenone. It can absorb light energy and transfer it to another molecule and it can also cleave to form initiating species:

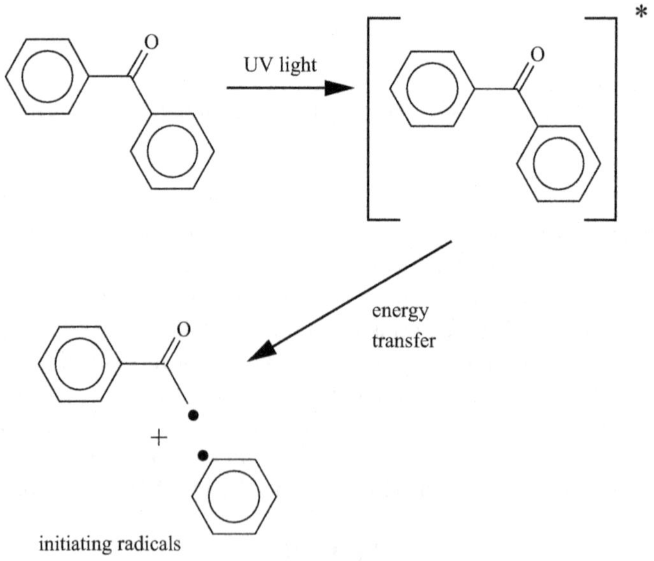

An example of an ionic initiator is an onium salt:

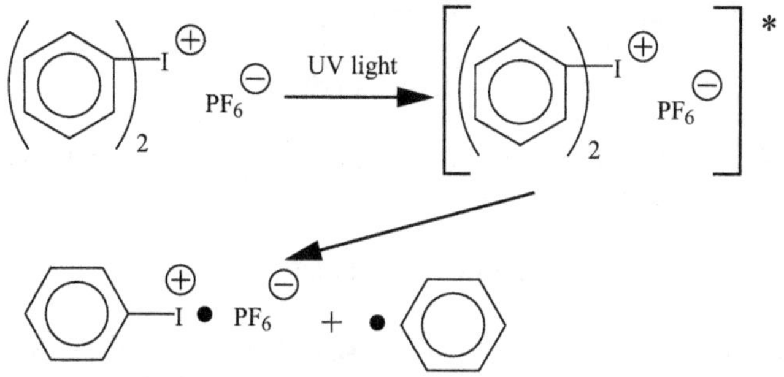

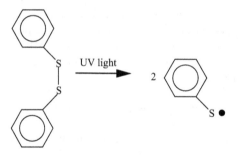

initiating ions

2.2.1. Free-Radical Photoinitiators

Many compounds can fit the definition of free-radial photoinitiators. They can be peroxides, disulfides, azo compounds, ketones, aldehydes and other. One example is a diphenyldisulfide that has been utilized in photocurable systems based on styrene-unsaturated polyester compositions:

Many other examples can be sited. Following are some basic kinetic considerations of the reactions of photoinitiators and photoinitiating processes. Based on Beer's Law, the fraction of absorbed light by a solution (or a light curable composition) can be expressed as follows:

$$I_{abs} = I_0 (1 - e^{-2.3 \, \varepsilon lc})$$

Where I_0 is the intensity of the incident light in photons per square centimeter per second at a given wave length; ε is the molar extinction coefficient per mole per centimeter; l is the length of the optical path; and c is the concentration of the initiator. The above defines the fraction of light that is absorbed by one cubic centimeter the light reactive material per second.

Following light penetration, the quantum yield of initiation by many free-radical photoinitiations can be shown in terms of the quantum yield of formation of charge transfer complexes (exiplexes)[3] :

$$T_1 \rightarrow \textbf{Exiplex } (\Phi_{CT})$$

This is followed by formation of free-radicals, in a reaction that involves proton abstractions (see section 2.2.3)[3] :

$$\textbf{Exiplex} \longrightarrow \textbf{R} \bullet (\Phi_H)$$

the quantum yield for radical formation, Φ_R becomes Φ_{CT}, Φ_H. The cleavage of the initiator in the triplet state and formation of initiating radicals can be expresses as follows [3]:

$$\Phi_R = \Phi_c = k_C / (k_C + k_q [M])$$

and the kinetic expression for the formation of initiating species in a system that is composed of a ketone and an amine (see section 2.2.3), can be shown as follows, [3]

$$\Phi_R = \Phi_{CT}, \Phi_{H\bullet} = k_C[AH] / k_d[AH] + k_d + k_q [M])$$

where k_c is the rate constant for quenching by amines, $[AH]$ is the concentration of the amine, k_d is the rate constant for decomposition of the photoinitiator, and k_q is the rate constant for monomer quenching with $[M]$ representing the concentration of the monomer. Turro[3] gives the quenching rate constant for diffusion controlled reactions as,

$$k_q = 8RT/3,000 \; \eta \; \textbf{liters mole}^{-1}\textbf{sec}^{-1}$$

where R is the gas constant, T is the temperature, and η is the solvent viscosity in poise.

2.2.1.1. Aromatic Ketone Photoinitiators

The diphenyldisulfide, shown above, although useful in some applications, is not efficient enough to be used as an initiator in many commercial applications. Aromatic ketone compounds, however, are a common choice. The ketone carbonyl orbitals that are important in photochemistry are shown in Figure 2.1. The picture is similar to the one shown for formaldehyde in Chapter 1. In the ground state, the π orbital is localized on the carbonyl oxygen. In the excited state, however, the π^* orbital is delocalized over the entire carbonyl function. This means that in an n $\longrightarrow \pi^*$ transition the electron is further away from the oxygen and the molecule behaves more like a diradical:

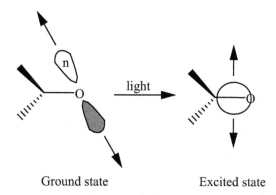

Ground state   Excited state

Figure 2.1. Illustration of the ketone carbonyl orbitals

The photoinitiators with carbonyl groups can be either small molecules or they can be polymers, because it was demonstrated that ketones on polymeric compounds undergo the same photoreactions that do small molecules [1] Commercially, the most important photoreactions of ketones are the Norrish type I and II processes and the Fries rearrangement. Many of them will undergo Norrish I type α-cleavage in the triplet state when irradiated with light of the appropriate wave length. In some cases, however, β-cleavage may takes place instead. Both types of cleavages are also possible in some cases. This is discussed further in this section. Tables 2.2.and 2.3. give some examples of typical aromatic ketone initiators that are available commercially.

The benzoin derivatives, shown in Table 2.2., are an important family of photoinitiators. The unsubstituted benzoin molecule itself is believed to fragment from the triplet state. On the other hand, the benzoin ethers are thought by some to fragment from singlet excited states.[4,5] Others, however, disagree and believe that they also fragment from the excited triplet state.[4] The benzoin ether derivatives vary considerably in their photo initiating efficiency. [4,6] The nature of the monomer and the environment exert an effect as well. Also the efficiency decreases when radical recombination is favored by the cage effect due to an increase in viscosity of the medium. This is also true of many other photoinitiators, like, for instance, benzyl monoxide and its derivatives. As a whole, however, many benzoin ethers are efficient photoinitiators for acrylate and methacrylate esters, again, depending upon chemical structure, concentration, and the intensity of the light.[7, 8] Generally, they undergo a high rate of photo cleavage and compete effectively with bimolecular quenching by oxygen and by monomers. These ethers are often utilized in particle board finishing compositions.

As state above, the cage effect detracts from initiation efficiency. It was shown by Pappas, however, that both radicals produced by photo cleavage of bezoin ethers, the benzoyl radical and the diether radical are comparably efficient initiators for the polymerization of acrylates and methacylates.[9] This is contradicted by work of Turro and coworkers, [10] who used time-resolved infrared spectroscopy to study reactivity of some photoinitiators. A series of substituted benzoyl radicals were generated by laser flash photolysis of α-hydroxy ketones, α-amino ketones, and acyl and bis(acyl)-phosphine oxides. The absorbtion rate constants for their reaction with *n*-butyl acrylate, thiophenol, bromotrichloromethane and oxygen were measured in acetonitrile solution. They concluded that the rate constants of benzoyl radical addition to *n*-butyl acrylate range from 1.3×10^5 to 5.5×10^5 M^{-1} s^{-1} and are about 2 orders of magnitude lower than for the *n*-butyl acrylate addition to the counter radicals that are produced by (α-cleavage of the investigated ketones).[10]

Other groups of efficient photoinitiators are benzil ketals and acetophenone based compounds. Table 2.3. lists some of these materials. In the benzoyl radicals the unpaired electrons are not delocalized over the whole aromatic ring. As a result, they are active in initiating polymerizations. The benzyl ether radicals, on the other hand, are more stable. They will initiate some polymerizations, but can also contribute to chain termination reactions. The free-radicals shown above in Tables 2.2. and 2.3. that form through primary bond cleavage, however, are not necessarily the initiating species. Instead, secondary products that form as a result of further decompositions may actually do the initiating. Thus, for instance, the high efficiency of benzyl dialkyl ketal shown

Table 2.2. Some Examples of Benzoin Derivatives Used in Photoinitiation Reactions[a]

Benzoin Ethers	Primary Photodecomposition Products	λ_{max}
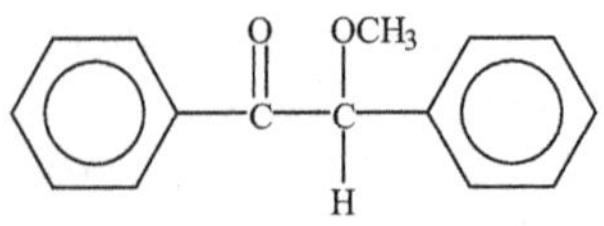	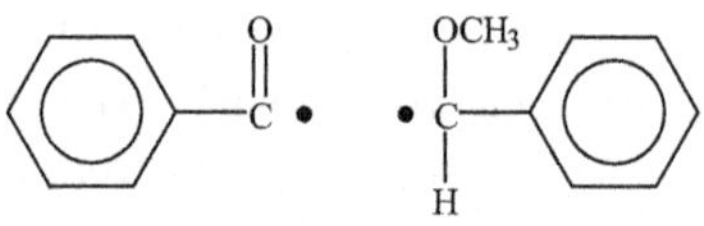	323
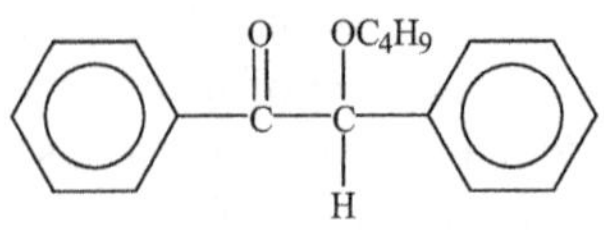	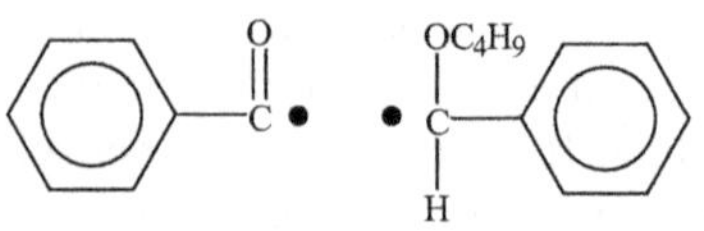	
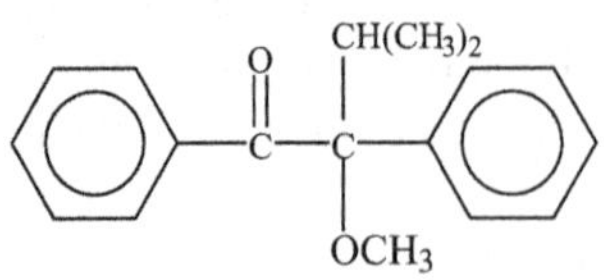	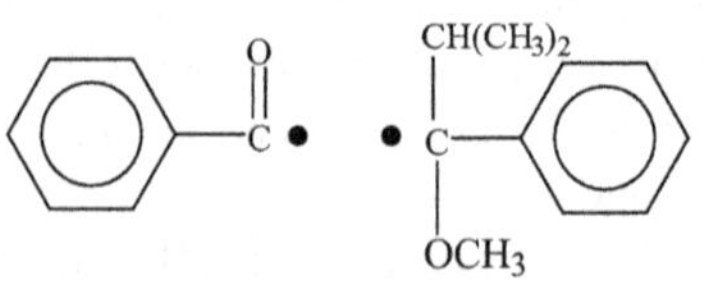	

Table 2.3. Some Benzil Ketals and Acetophenone Based Photoinitiators [a]

Initiators	Primary Decomposition Products	λ_{max} (nm)
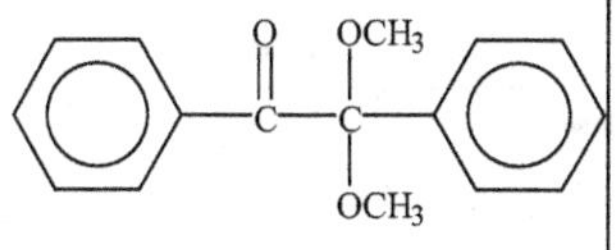	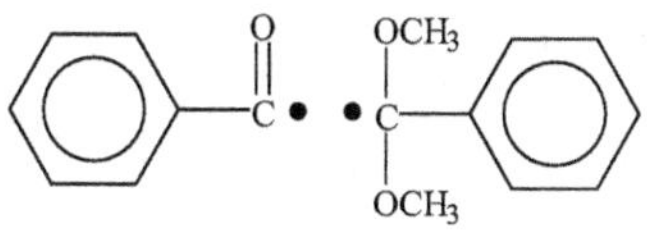	335
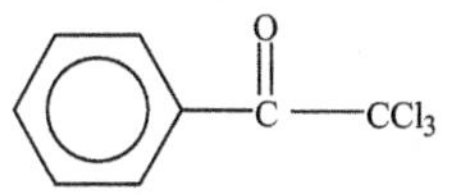	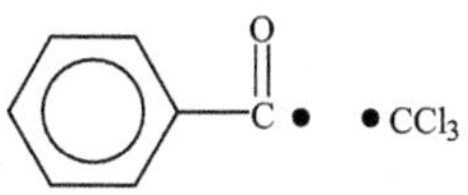	
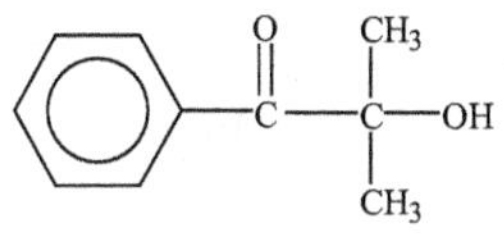	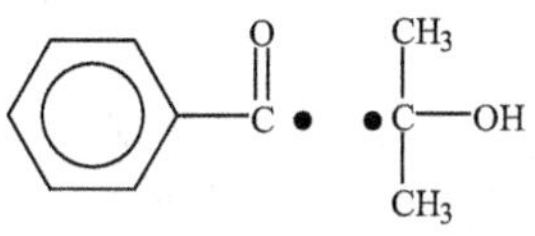	320
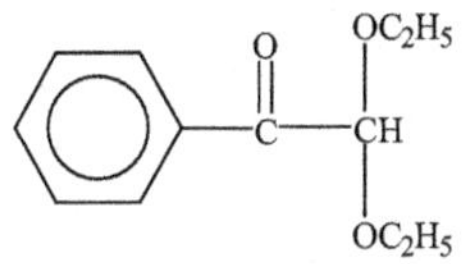	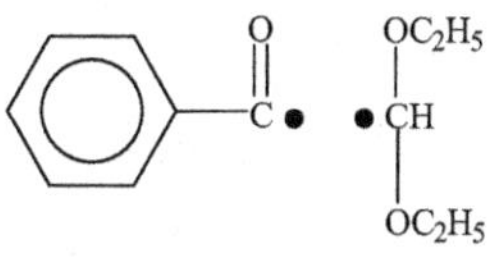	323

[a] from various sources in the literature and from sales brochures. λ_{max} is in nm.

above is attributed to formation of a very active methyl radical:

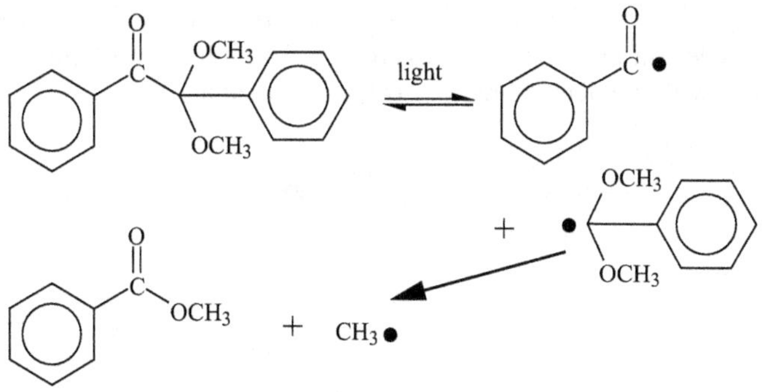

The benzoyl radical can actually undergo further decomposition reactions and the initial cleavage can be followed by many side reactions [11]:

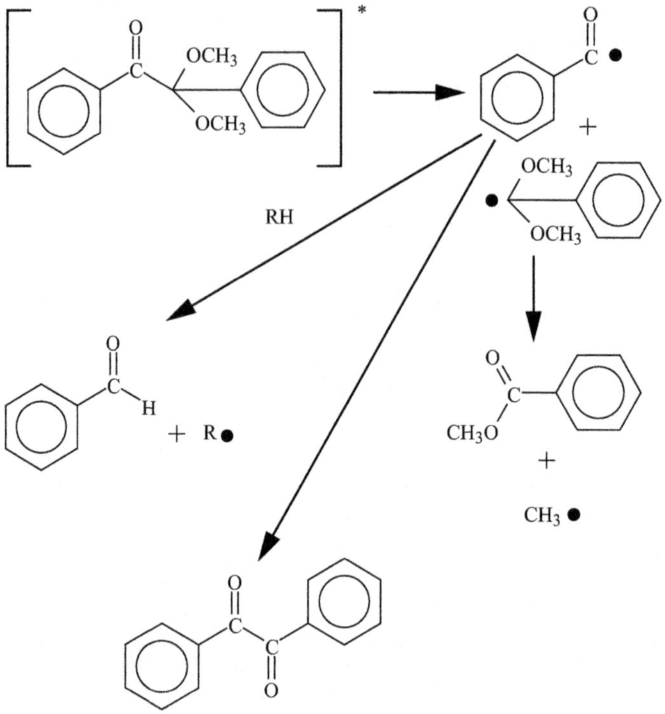

The benzaldehyde, as shown above, is a product of a side reaction. Benzaldehyde can also form as a byproduct from photodecomposition of other compounds with the benzoyl radical as the intermediate [11]:

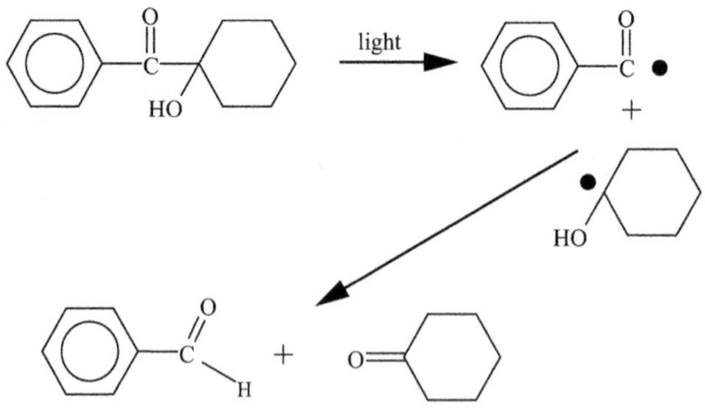

Also, side reactions of the products form photodecomposition of benzil ketals that can detract from their initiation efficiency are recombination reactions of radicals [11,12,13]:

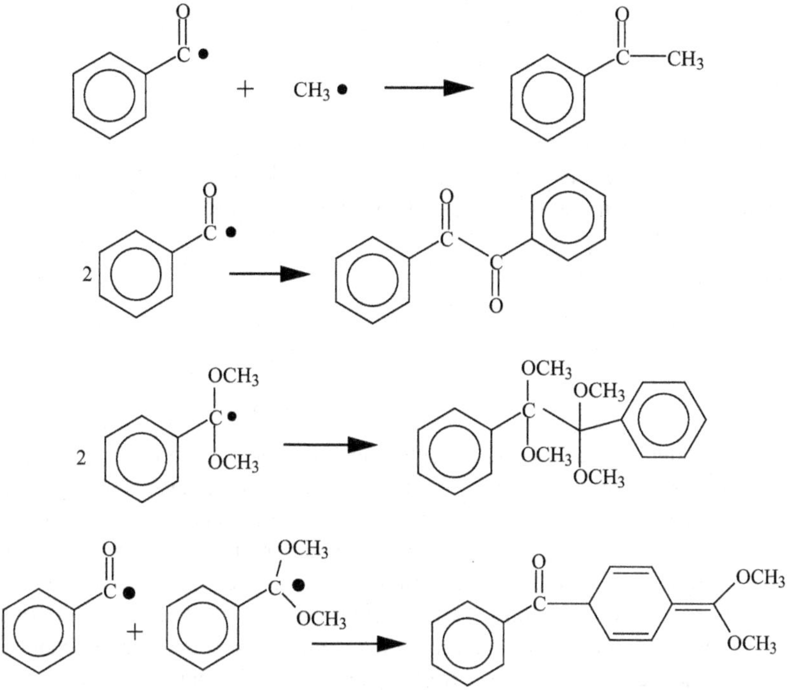

Norrish II type hydrogen abstraction and Norrish I photo cleavage were shown to take place in the triplet state of dialkoxy acetophenones[7]:

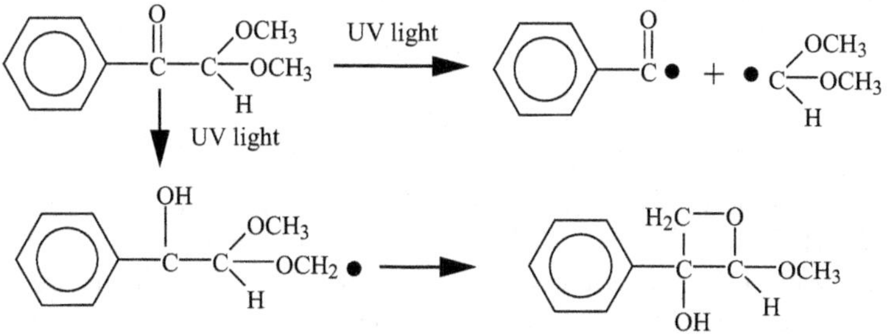

In addition to the above reaction, the benzoyl radical was also shown to undergo recombination with ether radicals at the *ortho* and *para* positions.[14] The products are semiquinoids:

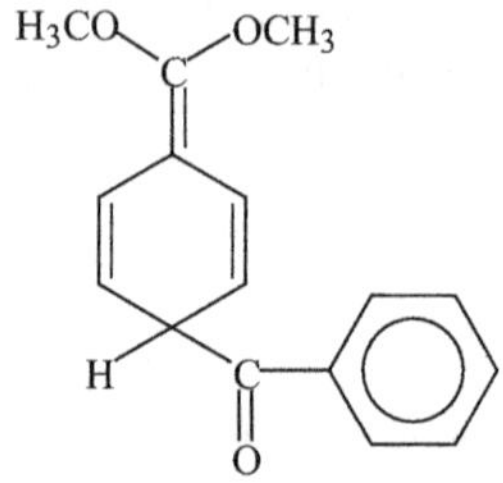

The para isomer shown above is known to contribute to yellowing of coating during cure. On the other hand, β-cleavage is the predominant mode of decomposition of compounds, like, α-halogen acetophenones [8,15]

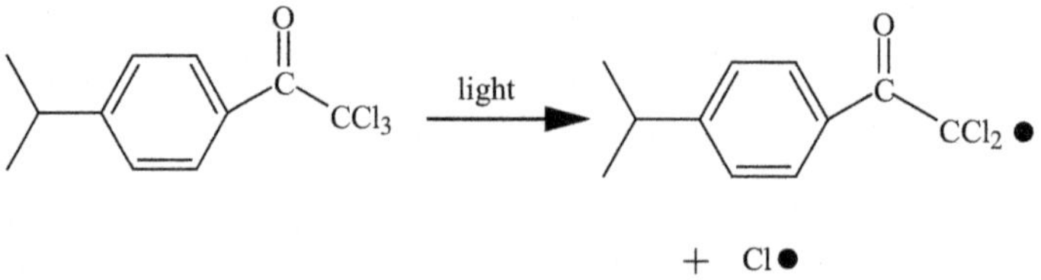

$$+ \quad Cl\bullet$$

As stated earlier, the photo initiating efficiencies of aromatic ketones, and this is true of all other photoinitiators, are affected by their chemical structures. The nature of the transition state strongly influences the process of α-cleavage. It is also affected by substituents, whether electron withdrawing or releasing, because that affect stabilization. For instance, investigation of the photochemistry of five substituted benzoin ethers and their activities in initiating polymerizations of acrylates were reported recently.[16] When the photolysis of such compounds leads to very short triplet states, rapid cleavages and formations of very efficient initiating radicals result.[16] That, however, is not true of all benzoin ethers, even if they exhibit strong near ultraviolet-visible light absorptions. [16] An interesting correlation of rate constants for photo cleavage with substituents on acetophenone [8]:

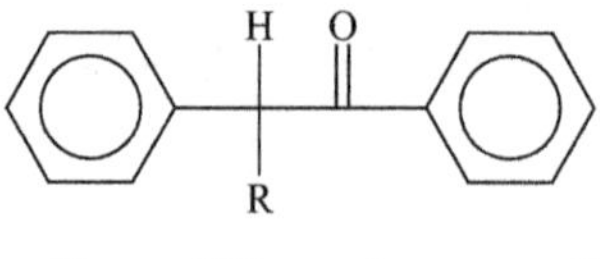

shows the accelerating effect of R on the rate of cleavage as follows: OCH_3 >OH > $OCOCH_3$ > CH_3 > C_6H_5 > H. Thus, as stated earlier, in general, the process of α-cleavage in aromatic ketones is strongly affected by the nature of the transition state and by the stabilization of the partial charges by electron releasing of withdrawing groups.

Benzophenone and its derivatives are also affected by substituents. Thus, 4'-(4'-methylphenylthio)-benzophenone, exhibits higher photo activity than its parent compound.[17] The incorporation of a thioether group results in strong $n \rightarrow \pi^*/\pi \rightarrow \pi^*$ mixing and a change of the properties in the excited state.[17] Substitution with long alkyl groups also increases activity, but, on the other hand, nitro substitution in the 4 position,

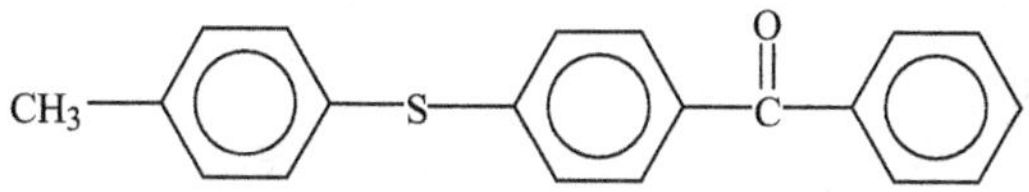

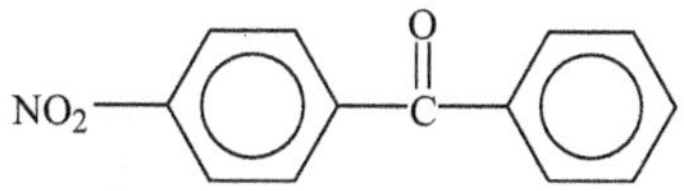

decreases the activity.[17]

Another interesting example is a bifunctional photoinitiator that is based on ketosulfone-benzophenone. It is reported to be 20% to 40% more active than dimethoxy phenylacetophenone[18]:

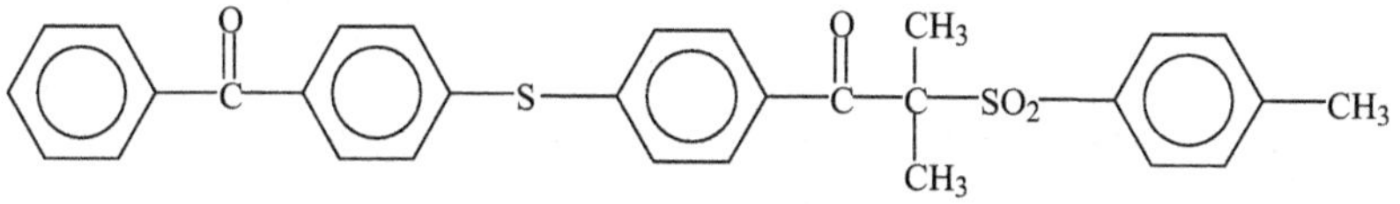

The triplet state of this compound undergoes β-cleavage reaction at the ketosulfone moiety. Upon addition of an amine this material generates a ketyl radical. The quantum yield of intersystem crossing is 0.7 in acetonitrile and the triplet energy is reported to be around 275 kJ mol^{-1}.[18]

Turro et al.[19] studied the effect of substituents upon the efficiency of 2-hydroxy-2-methyl-l-phenyl propanone in initiation of polymerization of acrylic monomers. The parent compound decomposes in the following manner:

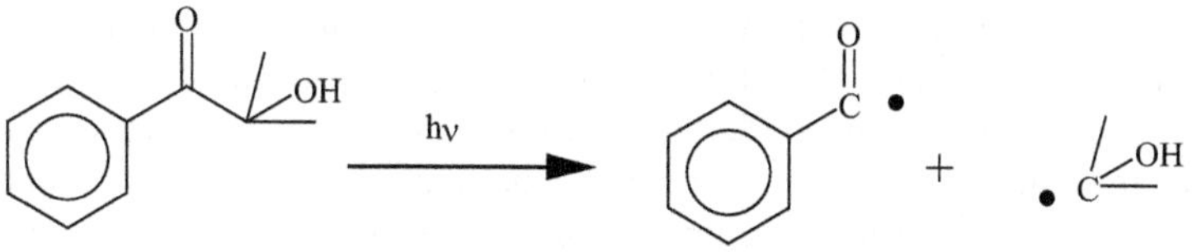

The comparison was carried out on the following derivatives[11]:

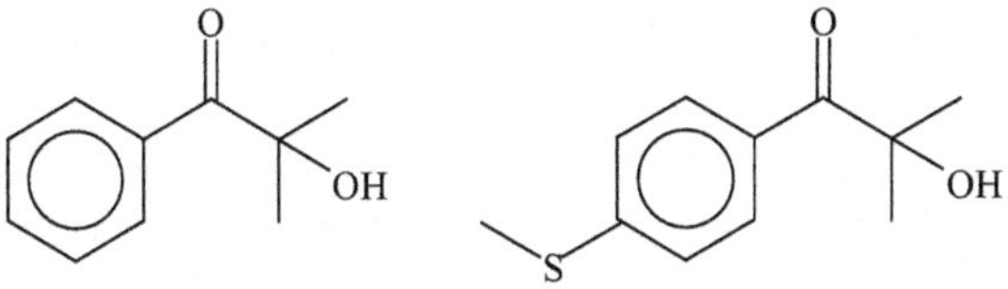

A   B

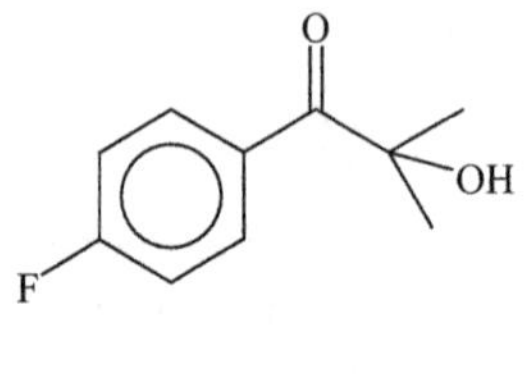

C 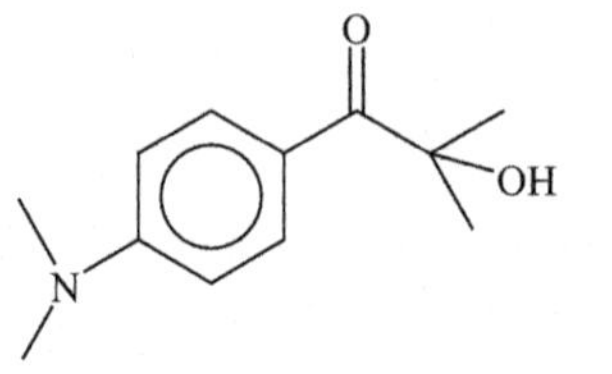   D

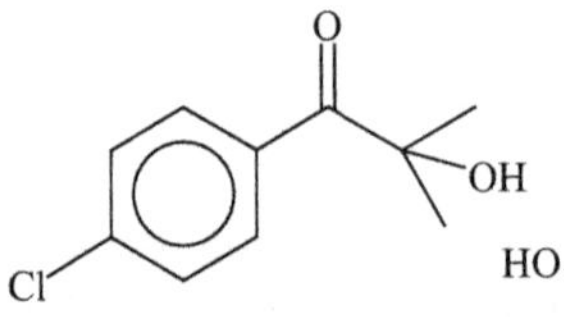

E 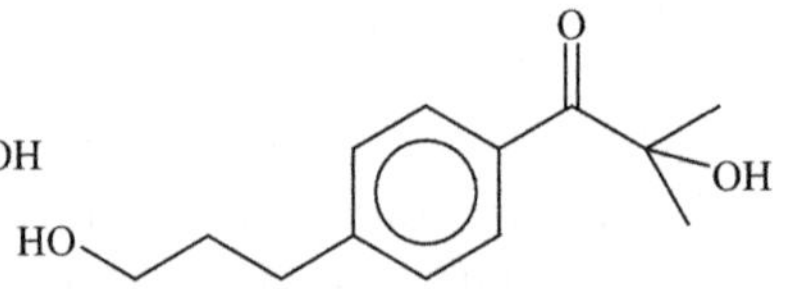   F

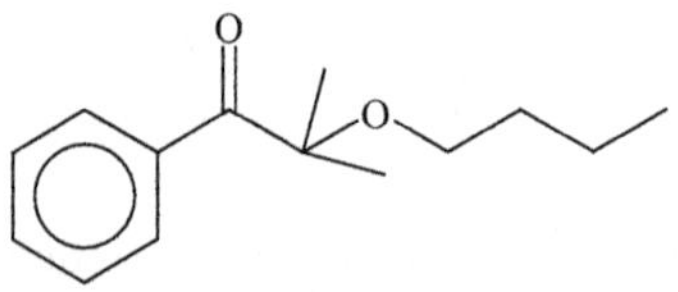

G 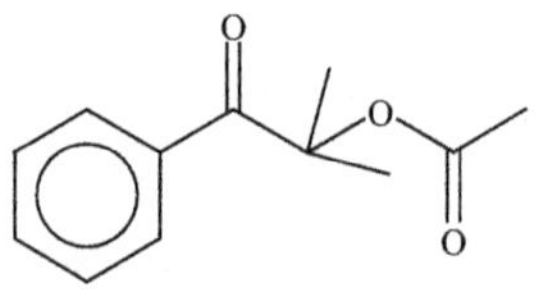   H

Evidence was presented[19] that substituents at the benzoyl moiety can change the photochemistry and the photophysics of these compounds significantly, because they change the nature of the lowest triplet state from an $n \rightarrow \pi^*$ to a $\pi \rightarrow \pi^*$. Upon irradiation the hydroxy ketones undergo fast and efficient $n \rightarrow \pi^*$ transitions in the lowest triplet state ($k_\alpha > 10^9$ s^{-1}) and efficient α-cleavage ($\Phi_\alpha \sim 0.4$—0.6). The alkyl ether derivative, shown above, also exhibit fast α-cleavage upon irradiation,[19] Slow α-cleavage was observed, however, for the ester derivative.[19] The conclusion is therefore, that all but one of the above undergo rapid intersystem crossings from the S_1 to T_1 states and exhibit relatively pure $n \rightarrow \pi^*$ character of the T_1 state. An exception, however, is the compound that has the hydroxy function replaced by an acetate one (compound H). The esterification slows the rate of cleavage by a factor of nearly 1000 [19].

It was also shown[20] that the initiating performance of compounds, like 1-chloro-thioxanthene-9-one, is enhanced by the presence of an acyloxy group in the four position. This was attributed to the greater triplet n→π* activity because of the electron withdrawing effect of the 4-acetoxy group. This high photoactivity was also observed in the presence of oxygen and under visible light. That, however, is believed to be due to the 1-chloro group that produces active chlorine radicals.[20]

Cyclic benzyls will initiate photo polymerizations of acrylates The process of photodecomposition and formation of initiating radicals was reported by Pappas to be as follows [21]:

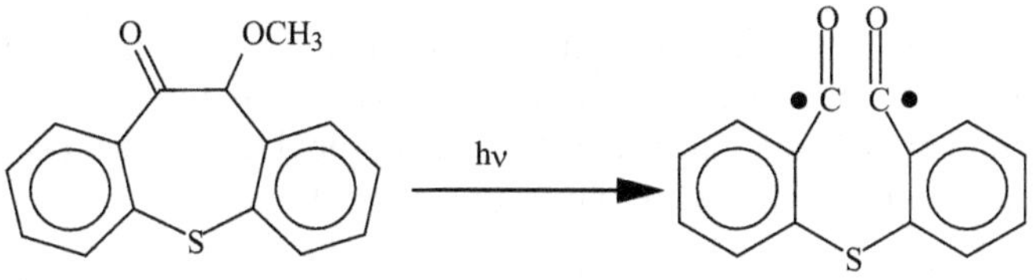

Other cyclic benzils that are shown below behave similarly [18]:

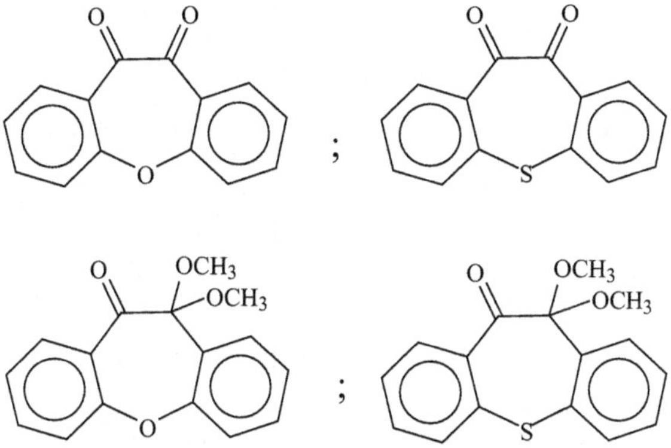

Acyloxime esters are also efficient photoinitiators. They were investigated by Delzenne.[22] He demonstrated that they undergo the following cleavage reaction:

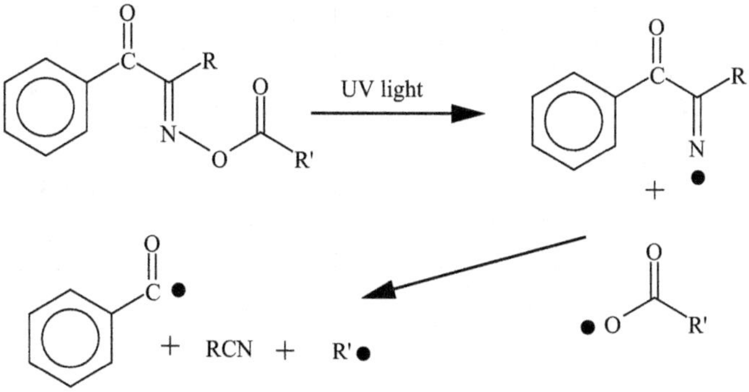

The monomer can interact with the triplet state of the initiator by either quenching it or by completing the step of initiation. Quenching of the excited state is detrimental as it can cause deactivation of the excited state and prevent formation of monomer radical, RM● Stern-Volmer plots of the reciprocal values of the lifetimes of the triplet states or radicals as a function of monomer concentrations allow us to determine the quenching rate constants, k_q .[7] The structures of both, the aromatic ketone and the monomer influence the rate of quenching.

Many pigmented coating formulations make use of α-amino ketone photoinitiators.[23-25] The effectiveness of these materials is further enhanced by addition of photosensitizers, such as methyl or isopropyl thioxanthene-9-one.[23-25] Some typical α-amino-ketones are shown in Table 2.4.

Table 2.4. Typical α-Amino Ketone Photoinitiators [a]

Photoinitiators	Primary Decomposition Products
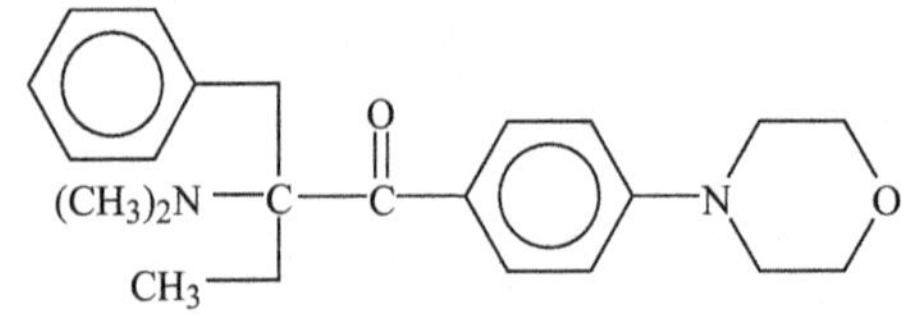	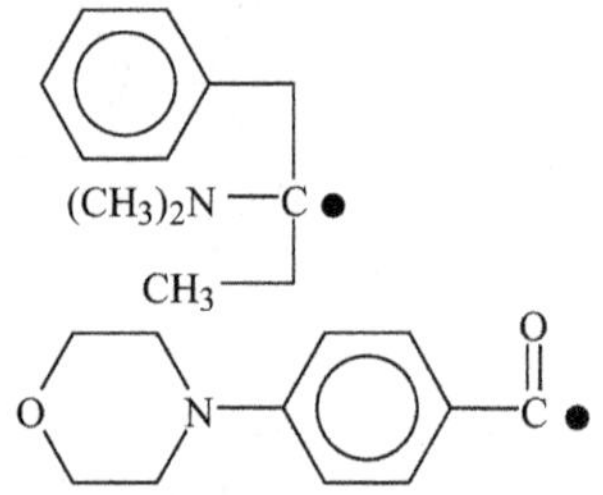
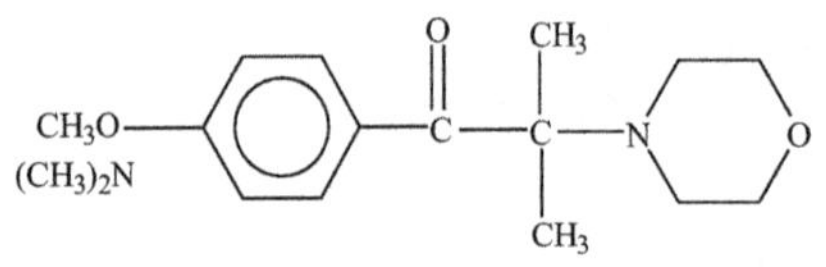	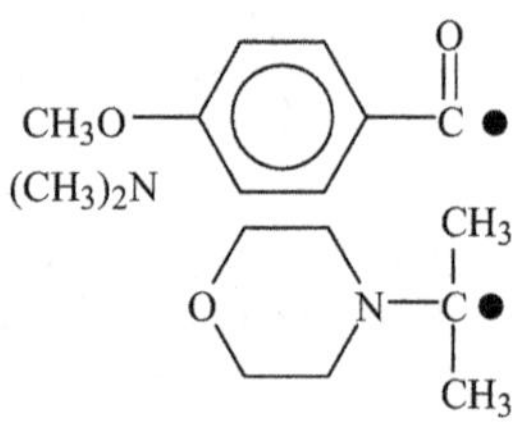
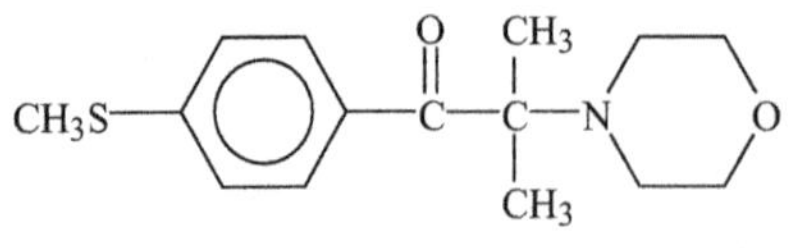	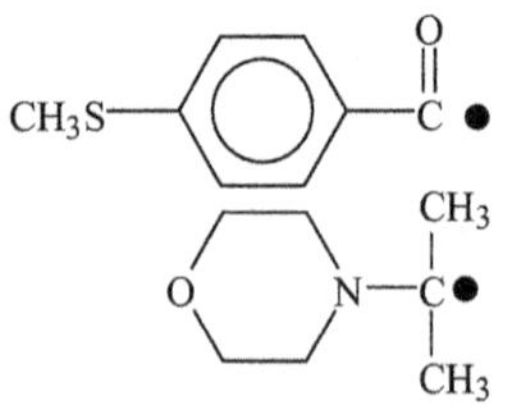

Table 2.4. (continued)

Photoinitiators	Primary Decomposition Products
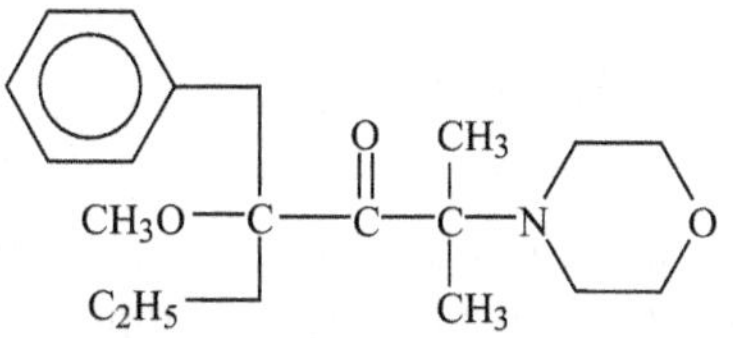	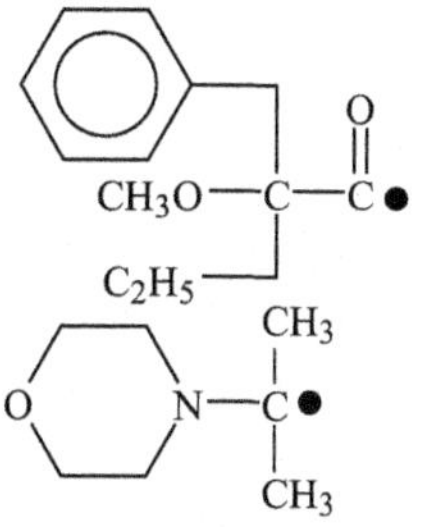

[a] from various sources in the literature

Although α-cleavage is the main photodecomposition product of morpholino ketones, some β-cleavage might also be taking place.[26,.27] Also, side reactions that can result in loss of initiating radicals can be illustrated as follows:

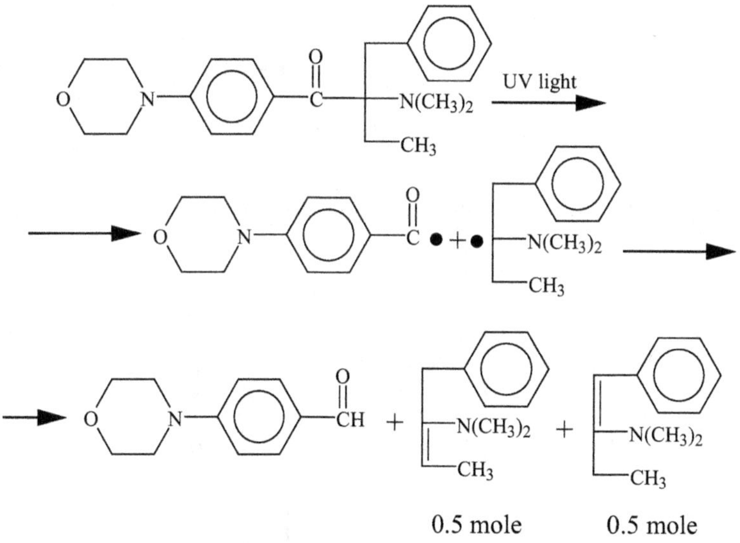

An interesting photoinitiator was reported by, Liska and Seidl who found that a diynone based compound shown below yields surprisingly high reactivity and initiation in polar monomers. Compared to benzophenone-amine system (described in the next section), an increase, up to two times in reactivity is claimed for this compound in the absence of an amine.[28] Addition of an amine was found to actually decrease reactivity. It is suggested that the high hydrogen abstraction properties of this material might be the cause of this high reactivity [28] :

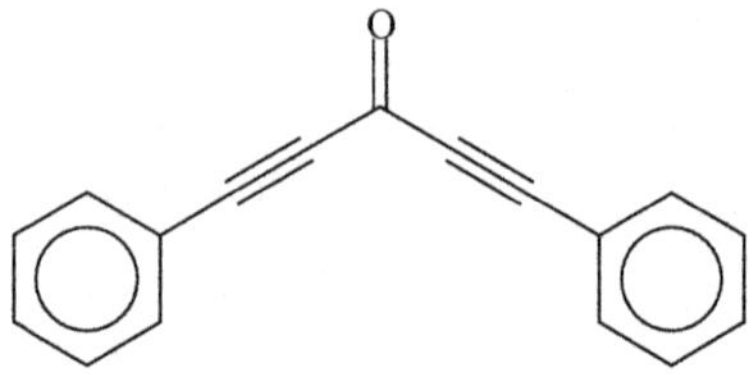

Phosphorus containing photoinitiators are useful in pigmented as well as in water based systems. Compounds like trimethylbenzoyl diphenyl phosphine oxide and its derivatives generally undergo α-cleavage.

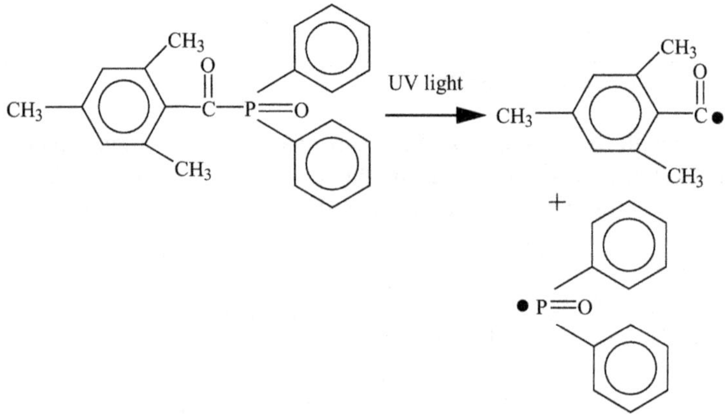

The acylphosphonates, like the other photoinitiators, can also undergo side reactions:

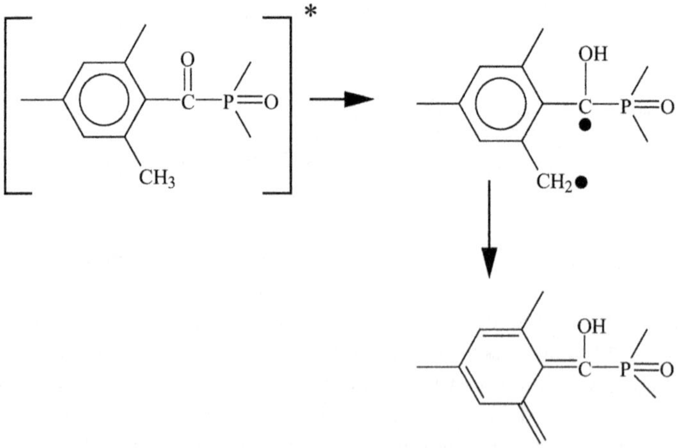

Methyl substituent on the benzoyl group in the ortho position leads to competitive enolization in compounds like 2,4,6-trimethylbenzoylphosphonic acid. [29,30] This, however, is not observed in compounds like 2,4,6-2-trimethylbenzoyl phosphine oxide:

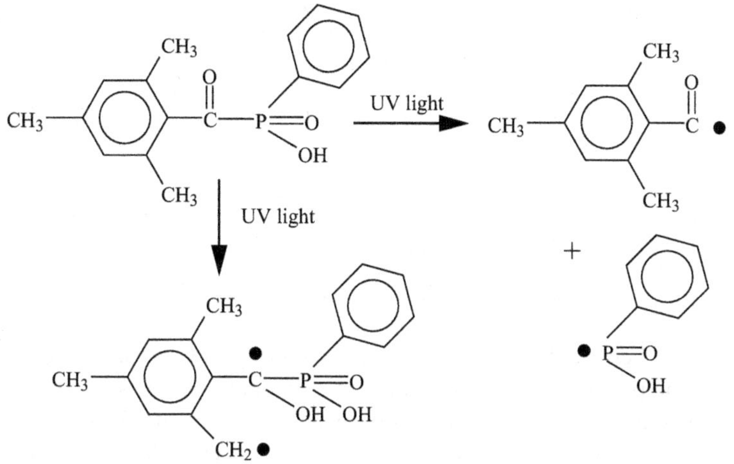

There are some comments in the literature that phosphorus containing photoinitiators are not as effective in imitating free-radical polymerizations as are other aromatic ketones.[31] That is contradicted, however, by others. Thus Decker rates bisacylphosphine oxide, [32] shown below, as being on par with morpholinoketones in photoinitiations of the polymerizations of urethane acrylates (see Chapter 3) The compound probably cleaves into two initiating radicals upon exposure to light[32] :

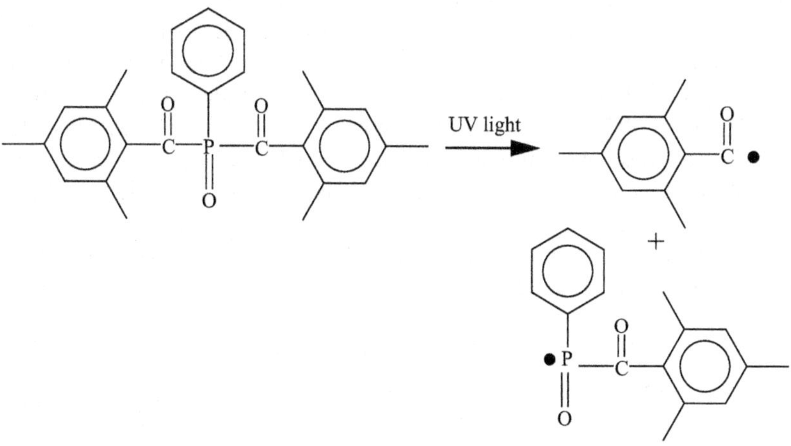

Similar photocleavage might perhaps be occurring with the following phosphine oxide:

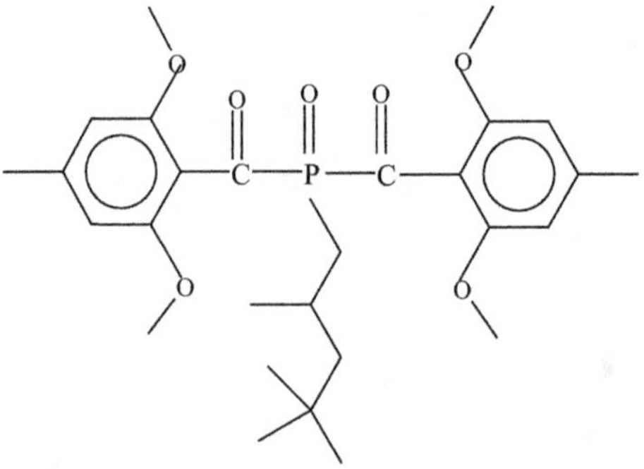

Decker et al.,[32] claim that when the initiation efficiency of acyl phosphine oxides was compared to typical radical photoinitiators (aromatic ketones shown above), it was found that acyl phosphine oxides are most efficient ones with respect to both the polymerization rate and the extent of cure, mainly because of their fast photolysis. Frontal polymerizations, they found, proceed readily, allowing thick specimen to cure through by simple exposure to sunlight. Because they absorb in the region of 350-400 nm, acyl phosphine oxides are claimed to be particularly well suited to initiate polymerizations of both pigmented systems and protective coatings used in exterior applications that contain UV-absorber-type light stabilizers. Superior performance was also reported for these photoinitiators in the photocrosslinking of functionalized polymers, such as acrylated polyisoprene and polybutadiene with pendant vinyl groups. Beyond that, addition of small amounts of a trifunctional thiol was shown to drastically speed up the crosslinking polymerization and insolubilization. That was reportedly achieved within a 0.3 seconds of UV-exposure, even for initiator contents as low as 0.1 weight per cent.[32]

  A patent was issued to BASF [33] that describes two component photoinitiators containing mono- or a diacylphosphine oxide, which is described as R1R2P(O)C(O)R3 . The other component is mono or polysubsituted benzophenone. These photoinitiators are low in volatility. It is claimed that they are not inhibited by air and are particularly suitable for ultraviolet light curing of coatings.

  Liska et al., [34] studied pyridine ketones initiators using laser flash photolysis. The materials were illustrated as follows:

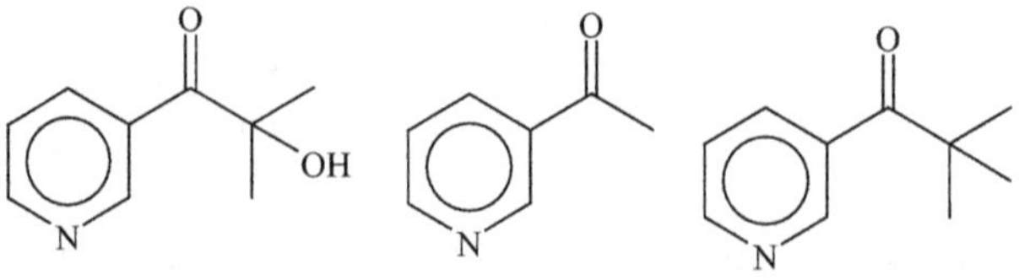

The first compound, 2-hydroxy-2-methyl-1-pyridin-3-yl-propan-1-one was previously found to be water soluble and very reactive. [35] The activity of this compound was found to be comparable to its phenyl analog, 2-hydroxy-2-methyl-1phenyl-propane-1-one. All compounds show absorption maxima at about 410 to 430 nm.

2.2.1.2. Initiators Based on Maleimide and Similar Compounds

Combinations of donor/acceptor systems, comprised of at least one multifunctional monomer, are actually capable of sustaining rapid free-radical polymerization without external photoinitiators. Donor monomers can be vinyl ethers, N-vinylformamides, and N-vinylalkylamides. The acceptor monomers are maleic anhydride, N-arylmaleimides, N-alkylmaleimides, dialkyl maleates, and dialkyl fumarates. N-alkylmaleimides can participate in excited state hydrogen abstraction from diacrylates. [36] The reaction proceeds either in the presence or in the absence of oxygen. [36]

The use of N-substituted maleimides as photoinitiators for radical polymerization. has gained attention as an alternative material to ketone based photoinitiators. [37] Sensitization of maleimides to the triplet state markedly increases polymerization efficiencies. The maleimide amine system produces two radicals, one centered on the amine and the other on the maleimide. Both radicals are capable of initiating polymerization [36, 37] Also, ketone based sensitizers with varying triplet energies were reported to initiate the polymerization. of hexamethylene diacrylate more efficiently in the presence of a maleimide/amine coinitiator system. In certain cases, extremely rapid rates were reported. [37] Additional studies showed that charge transfer complexes of maleimides and vinyl ethers initiate polymerizations efficiently and rapidly in the absence of any other photoiniiators. [38]

The process by which substituted maleimides initiate polymerization of acrylic esters in the presence of a sensitizer/hydrogen atom donor was studied by Ngyuen et al.. [38] The initiations were shown to be through an energy transfer mechanism that is followed by an electron transfer/proton transfer. The high rates of initiation and polymerization are affected greatly by the substituents due to the change in the stability of the radical that is produced and steric hindrance at the radical site. Also, the high rates are attributed to efficient energy transfer from the triplet sensitizer to the maleimides and efficient electron transfer from the tertiary amines to the excited triplet state maleimides. [38]

Hoyle *et al.*,[36] concluded that   N-aryl and N-alkyl substituted maleimides, when excited by UV light, can initiate free-radical polymerization by either hydrogen abstraction (when alcohols or ethers are used as coinitiators) or by electron/proton transfer (when tertiary amines are used as coinitiators). When an ether or an alcohol is used, as the coinitiator, the mechanism is believed to occur by the excitation of the maleimide to the excited singlet state followed by a somewhat inefficient intersystem crossing to the excited triplet state. The triplet species then abstract labile hydrogen atoms from the ethers or from the alcohols. On the other hand, when the hydrogen atom donor is a tertiary amine such as N-methyl-N.N-diethanolamine, the substituted maleimide is reduced via an electron transfer, followed by a proton transfer, that is very similar to that of the reaction of isopropylthioxanthone/amine systems (discussed in the next section).[37] Additional investigations [38] confirmed other observations, that triplet sensitization of N-substituted maleimides dramatically increases the observed rate of initiation. The triplet sensitizer transfers energy followed by an electron transfer/proton transfer:

$$
\text{(maleimide)} \; + \; N\!-\!CH_2CH_2OH \;\; \xrightarrow{\text{UV light}}
$$

$$
\left[\text{(maleimide)} \; + \; N\!-\!CH_2CH_2OH \right]^{*} \longrightarrow
$$

$$
\left[\text{(maleimide radical)} \; + \; \overset{\oplus}{N}\!-\!CH_2CH_2OH \right]^{*} \longrightarrow \text{(product)}
$$

$$
+ \;\; N\!<\!\!\begin{array}{l} CH_3 \\ CH_2CH_2OH \\ \dot{C}HCH_2OH \end{array}
$$

Jonsson et al.,[38] studied a difunctional maleimide with two different electron donors, a vinyl ether and phenyl dioxolane. They observed that structural modifications increase the electron density in the vinylic C=C bond of the donor monomer and promote higher rates of co-polymerization with maleimides. They also compared a vinyl ether with exomethylenic dioxolane and showed increased "reactivity" for the dioxolane monomer. As a result, the mechanism of reaction of maleimide with a donor monomer, a vinyl ether was illustrated by them as follows[38]:

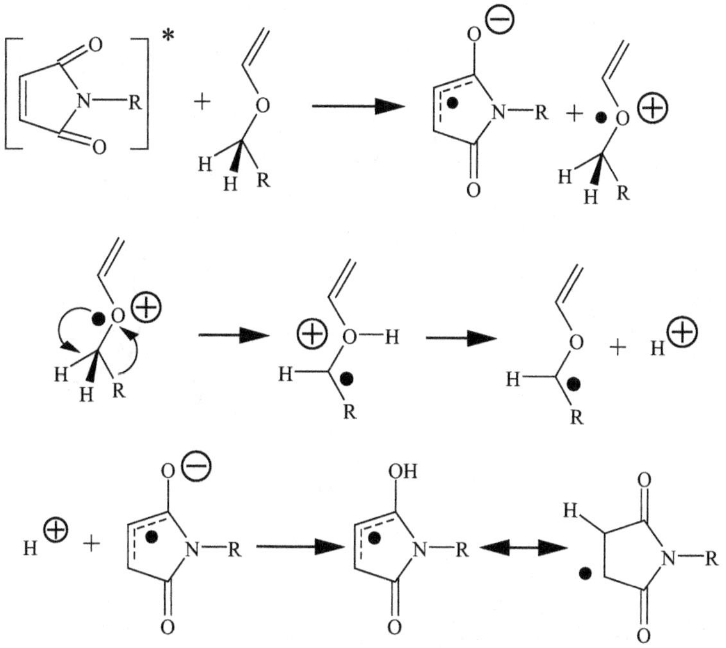

Jonsson et al.,[36] extended the study to the synthesis of polymers that function as hydrogels for controlled release studies.[36] Hydroxypentyl maleimide and N-vinyl pyrrolidone were the acceptor and donor respectively. Glucose, 1,1 diethoxyethane, and isopropyl alcohol were the hydrogen donors. In this reaction, glucose was shown to be the most efficient hydrogen donor. Also, the triplet state of maleimide is quenched by vinyl pyrrolidone trough an electron transfer process and presumably formation of 2+2 cycloadducts of maleimide and vinyl pyrrolidone. When the concentration of N-vinyl pyrrolidone is high enough to quench the singlet state of maleimide, rapid proton transfer takes place and results in formation of initiating radicals.[36] Furthermore, it was shown[26] that N-aromatic maleimides can be segregated into two groups. Into the first one belong those that can adopt a planar conformation and into the second one, those that can not adopt such a conformation.. Planar N-aromatic maleimides have a relatively low excited-state triplet yield They show a significant shift of the primary maleimide absorption band in the UV spectra with changes in solvent polarity, and do not initiate free radial polymerization upon direct UV excitation. Twisted N-aromatic maleimides, one the other hand, have higher relative triplet yields, show negligible shift of the primary maleimid UV absorption band, with changes in solvent polarity, and initiate free radical polymerization upon direct excitation. Addition of benzophenone, a sensitizer dramatically increases the initiation efficiency of both planar and twisted N-aromatic maleimides.[26]

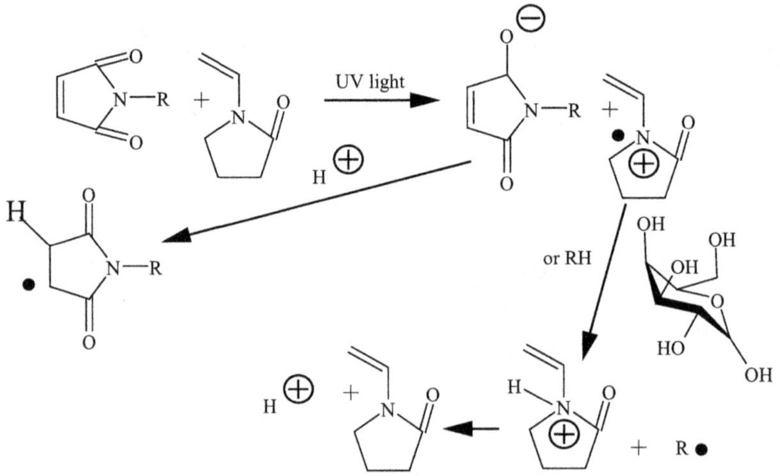

In addition, when used to photoinitiate the polymerization of acrylic monomers, low concentrations of N-substituted maleimide coupled with isopropylthioxanthone and a tertiary amine system result in markedly increased rates of polymerization, faster than the traditional isopropyl-thioxanthone/amine system alone (see the next section).[36] Maleimides can also be used in combinations with a diarylketones and amines to initiate polymerization.[36]

The maleimides are not the only imides that can accelerate the photoinitiation process. For instance, it was reported earlier that N-phenylphthalimides function in the same capacity as the N-substituted maleimides.[36] Phthalimide was also found to be capable of enhancing the photo initiated polymerizations of acrylic monomers.[42] When sensitizers like isopropylthioxanthone, 4-benzoylbiphenyl, or benzophenone are used in addition to N-phenylphthalimides, rapid rates of polymerizations of the acrylates are attained. A tertiar amine must be present as a hydrogen source. A suitable hydrogen source can be N-methyl-N,N, diethanolamine. Electron withdrawing substituents on the phenylphthalimide accelerate the reaction. Thus considerable enhancement was observed for N-(3,4-dicyanophenyl)phthalimide. [42] Also, when isopropylthioxanthone is used with substituted N-phenylphthalimide photo initiators, rapid rates of acrylate polymerization are attained in the presence of a tertiary amine that also acts as a hydrogen source. Use of N- phenylphthalimide with electron withdrawing substituents on the N-phenyl ring in the presence of a combination of isopropylthioxanthone and N-methyl- N,N-diethanolamine results in an increase in the maximum rate of polymerization of 1,6-hexanediol diacrylate. This increase is by a factor of two over the coinitiator and N-methyl-N,N-diethanolamine alone. [42]

Other charge transfer complexes are described in the literature. Thus, recently, it was reported that cyclic N-vinyl amides were also observed to enhance the relative rate of polymerization of acrylates in nitrogen and even more so in air.[39] The mechanism of the reaction is not yet explained, but the following cyclic vinyl amides were reported as being useful enhancers:

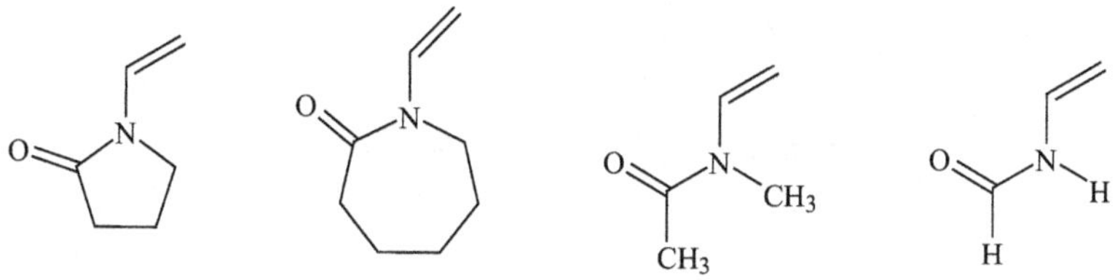

2.2.1.3. Two Component Photoinitiators

Many aromatic ketones are very efficient initiators of photo polymerizations as a result of photoreductions in the triplet state[40-46] Upon irradiation of the ketone, one of the n nonbonding electrons of the oxygen atom undergoes a $n \rightarrow \pi^*$ transition. This was discussed in Chapter 1 and, as it is explained in the beginning of this Chapter, makes the oxygen in the excited state electron deficient and gives it the ability to react similarly to an electrophilic alkoxy radical. This can be illustrated as follows:

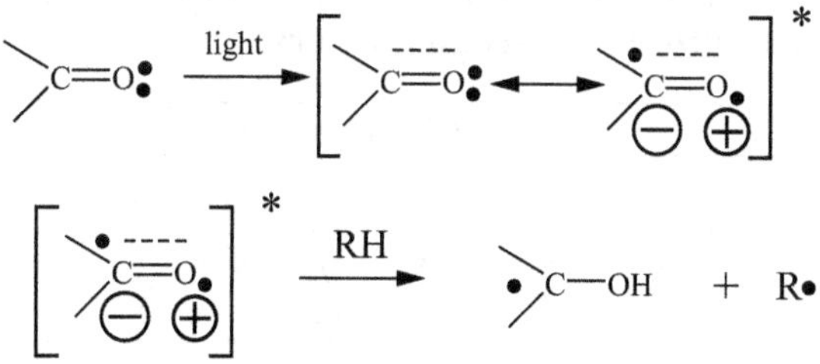

The RH hydrogen donor shown in the above reaction can be any source of labile hydrogen. It can be a solvent, an alcohol, or even a trace of moisture. The use of an alcohol, like isopropyl alcohol is particularly favored as a coinitiator in such reactions. An example used here is photo reduction of benzophenone by isopropyl alcohol. This reaction can be illustrated as follows,

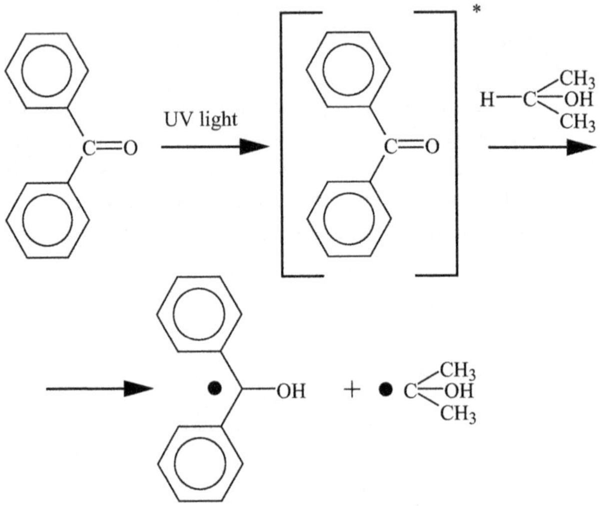

In the above reaction, it is believed that the isopropyl alcohol radical does the initiation of the free radical polymerization while the ketyl radical that forms is so resonance stabilized that it mainly dimerizes to benzpinacol or participates in chain terminations.[45]

Many other aromatic ketones, besides benzophenone, can be photo-reduced in the triplet state by electron donors. The transfer can take place in the lowest–lying triplet $n{\rightarrow}\pi^*$ or in the $\pi{\rightarrow}\pi^*$ excited states. Examples of such ketones are shown in Table 2.5.

Table 2.5. Examples of Ketones that Are Photoreduced in the Triplet State by Hydrogen or Electron Donors[a]

Ketone	Chemical Structure
Benzophenones	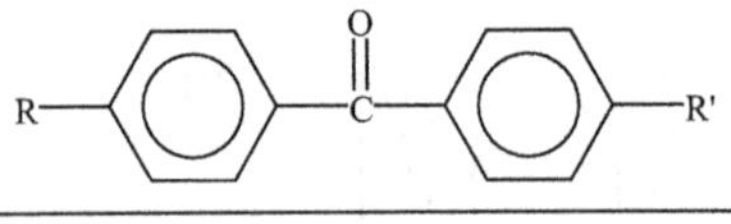
Benzyls	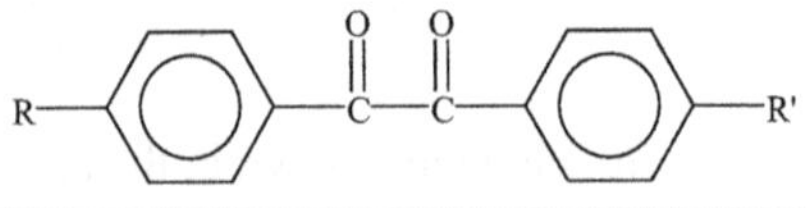
Camphorquinones	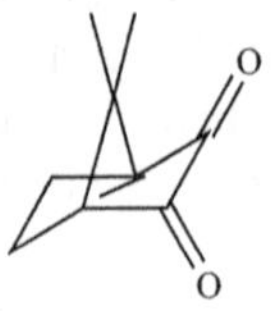

2. Photosensitizers and Photoinitiators

Table 2.5. (Continued)

Ketone	Chemical Structure
Thioxanthones	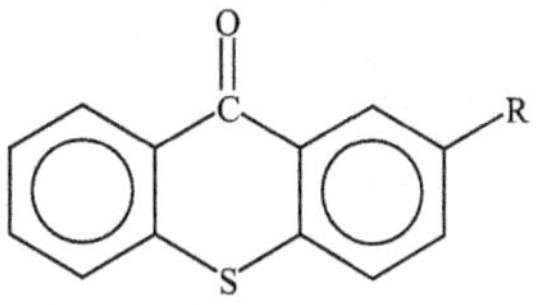
Ketocoumarins	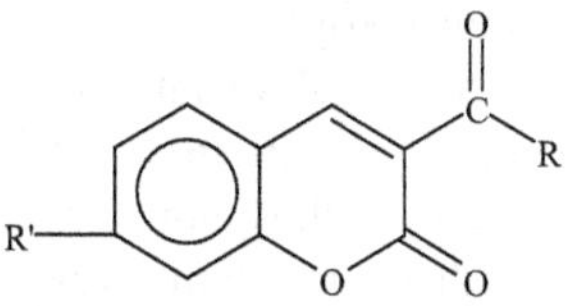

^a from various sources in the literature

Recently, Turro *et al.*,[46] reported preparation of a new photoinitiator for free-radical polymerization, 2-mercaptothioxanthone, :

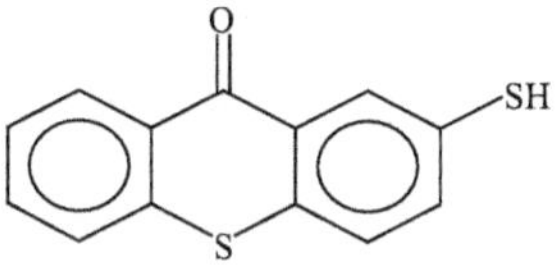

This compound is an efficient photoinitiator in the presence of a coinitiator, like N-methyldiethanolamine. [46] The thioxanthone derivative shown above is a hydrogen abstraction type photoinitiator. The postulated mechanism is [46]

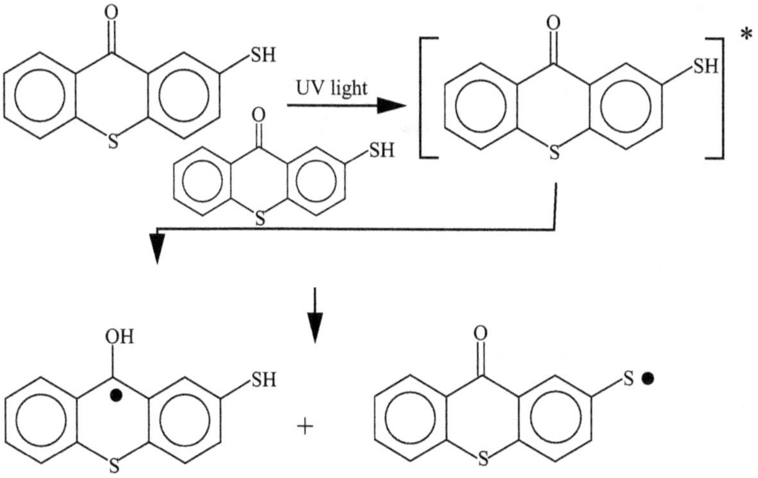

Yagci, Turro and coworkers also reported [47] a mechanistic study of photoinitiation of free radical polymerization with thioxanthone thioacetic acid as a one-component, Type II, photoinitiator.

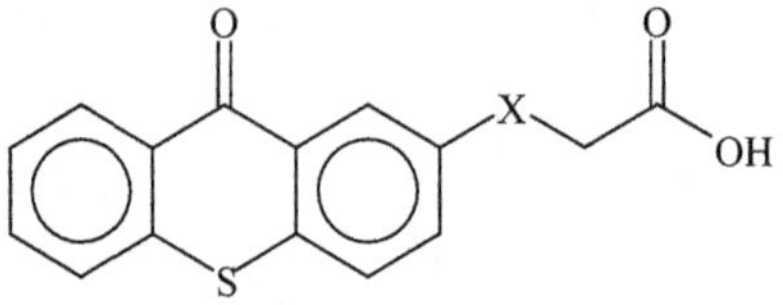

X = O,S

The initiator undergoes efficient intersystem crossing into the triplet state and the lowest triplet state possesses a $\pi \longrightarrow \pi^*$ configuration. In contrast to the unsubstituted thioxanthone, the thioacetic acid derivative shows an unusually short triplet lifetime (65 ns) indicating an intramolecular reaction. From fluorescence, phosphorescence, and laser flash photolysis studies, in conjunction with photopolymerization experiments, the authors concluded that the molecules in the triplet states undergo intramolecular electron transfer. This is followed by hydrogen abstraction and decarboxylation, producing alkyl radicals. These radicals are active initiator radicals in photo induced polymerization.

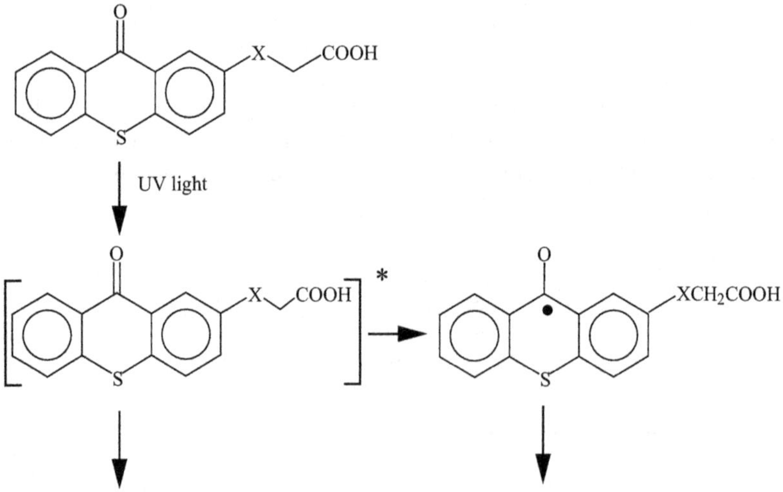

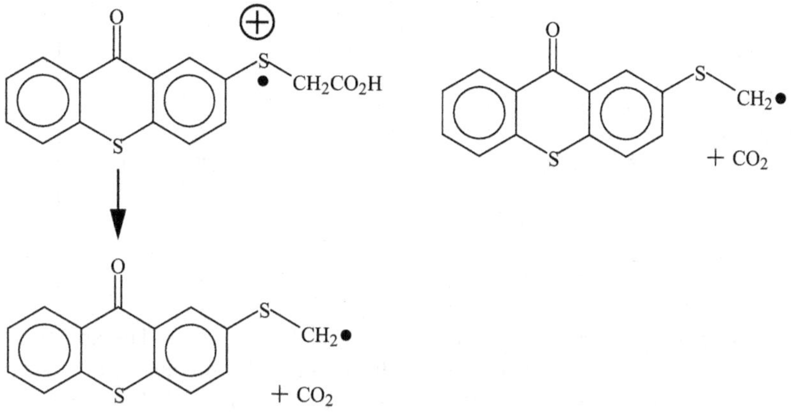

It was also concluded that at low concentrations of this initiator (below 5×10^{-3} M) the intramolecular reaction, shown above, is the dominant path. At concentrations above 5×10^{-3} M, however, the respective intermolecular reactions may be operative. [47] In effect, the above can be looked upon as a two component photoinitiator, where both components are combined into one molecule.

The interactions between aromatic ketones in the triplet state and hydrogen donors are not as efficient as reactions between such molecules in the triplet state with electron donors. As a result, the triplet state ketone reactions with hydrogen donors are used commercially mainly in initiating photocrosslinking reactions. On the other hand, reactions between triplet state ketones and amines find much wider applications in light curable coatings and other formulations. Furthermore, higher conjugated aromatic carbonyl compounds, such as p-phenylbenzophenone or fluorenone, exhibit lowest $\pi \rightarrow \pi^*$ excited triplet states and, actually, do not abstract labile hydrogens from compounds like ethers or alcohols. They are, however, readily photo reduced by electron donors,[41] like, ground state amines or sulfur compounds. The exact mechanism of the reaction is still not fully elucidated. It is believed, however, that electron transfer takes place after the excited triplet carbonyl compounds form *exiplexes* with the donors. The collapses of the complexes result in formations of pairs of radical anions [41] :

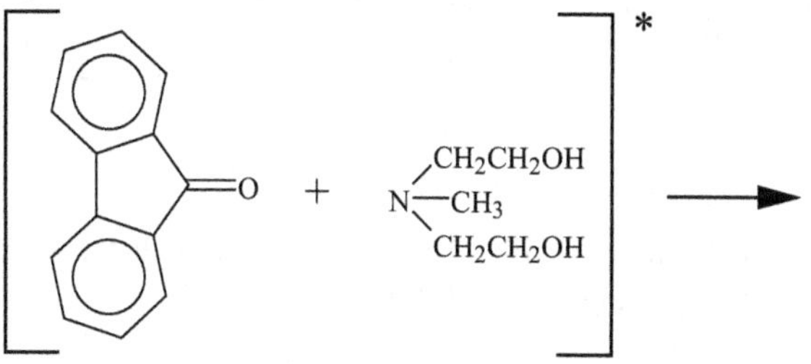

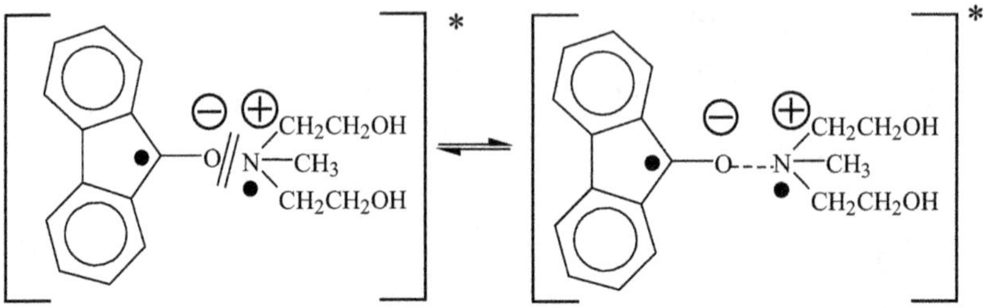

solvated ion pair   contact ion pair

It is believed that both pairs are in equilibrium with each other. The above ion pairs can lead to formation of different radical pairs. The solvated ion pair can lead to back electron transfer and /or free ion formation.[48] The hydrogen bonding character of the other components in the reaction mixture, such as monomers, solvent, or prepolymers can affect the efficiency of the proton transfer. The substituents on the nitrogen of the amine can affect the proton transfer as well. The overall reaction can also be illustrated as follows:

charege transfer complex

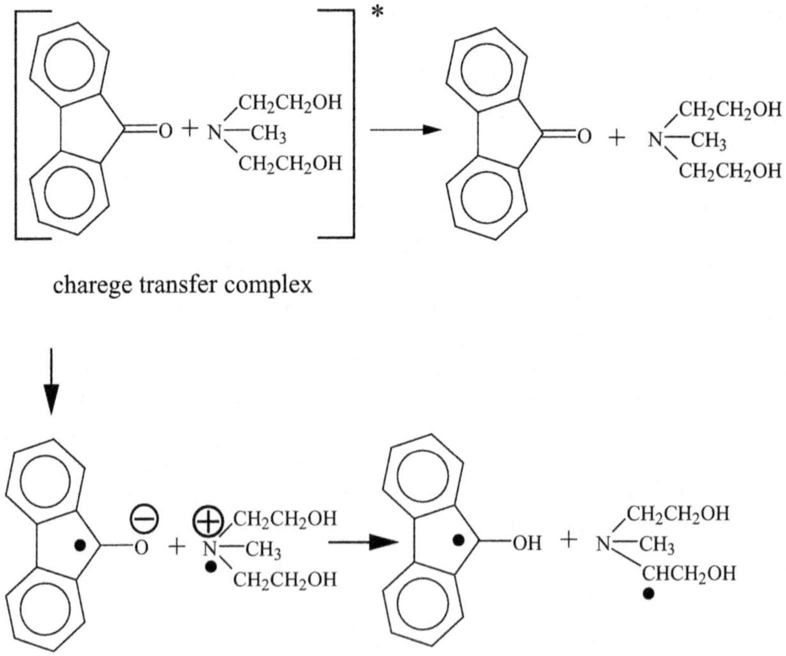

Various side reactions are possible. One such reaction that results in loss of initiating radicals can be illustrated as follows:

It is commonly believed that only the dihydroxyethylmethyl amine radicals shown above are the ones that initiate chain growth.[49] This, however, has not been fully established either. In general, as stated earlier, ketone-amine photoreactions are very efficient initiators for both photocrosslinking and for photocuring reactions, regardless of whether they reach the excited state through $n{\rightarrow}\pi^*$ or through $\pi{\rightarrow}\pi^*$ transitions. This efficiency of initiation, however, is also subject to intermediate reactions, such the reactivity of the α-amino alkyl radical and other side reactions, as shown above.[50] Quenching by amines of the excited triplet should be taken into consideration. When efficient cleavable photoinitiators are used, however, amine quenching does not usually interfere with the photo scission process.[8]

Several studies were carried on the relationship between the chemical structures of ketones and of amines and the rates and efficiencies of photoreductions.[51,52] It was observed, for instance, that Norrish I reaction of aromatic ketones is affected by electron donating and electron withdrawing substituents of the aromatic portions of the ketones. That, of course, is a result of stabilization or destabilization of the partial charges in the transition state by the substituents.[53] The rates of electron transfer reactions in photo reduction of various ketones, like benzophenone or thioxanthones and their derivatives by some amines can be found in the literature. Some examples are listed in Table 2.5.

Various derivatives of thioxanthone are widely used in photoreductive type free radical initiations in curing many commercial coatings and inks.[21] These compounds exhibit extended absorption in the near ultraviolet-visible region of the electromagnetic spectrum. That makes them also effective for curing pigmented films. The efficiency of the thioxanthone chromophores however, depends upon many factors. These include the nature and type of the substitution, the environment (the monomers and prepolymers present) and last but not least on the amine co synergist used. Radicals or ions derived from the amine play a major role in the mode of action and in the overall efficiency of the reaction.[21] Several studies were carried out to determine the influence of substituents on the efficiency of initiation by photoreactions of thioxanthones with amines. In one investigation, the excited state characteristics of thirteen derivatives of 1-chloro-4-oxy/acyloxy thioxanthone were studied [54] All the compounds that were tested were found to exhibit high photoinitiation activity, with the exception of two, 2-methyl-4-*n*-propoxy and 4-hydroxy derivatives.

chloro,2~methyl-4-*n*-propoxythioxanthone   1-chloro,4-hydroxythioxantheneone

The results are consistent with the character of their closely located mixed triplet states of $\pi{\rightarrow}\pi^*$ and $n{\rightarrow}\pi^*$ transitions. In photo reductive solvents such as methyl alcohol and 2-propanol a longer lived ketyl radical forms.[54] In the presence of a tertiary amine, however, no ketyl radical was observed. [54] Triplet lifetimes increase with solvent polarity, confirming the presence of mixed $\pi{\rightarrow}\pi^*$ and $n{\rightarrow}\pi^*$ states where vibronic coupling influences the rate of intersystem crossing to the ground S_0 state. Bimolecular triplet quenching rate constants indicate that all the thioxanthone derivatives studied, except for 2-methyl-4-n-propoxy and 4-hydroxy interact strongly with a range of tertiary amines. On the other hand, triplet quenching constants for 2~methyl-4-*n*-propoxy and 4-hydroxy derivatives indicate weaker interaction (an order of magnitude[36]) with the amine and this is consistent with their lower photoinitiation activities.[36] Low triplet quenching rates were also observed in the presence of a monomer (methyl methacrylate).[54]

    Similarly to the above, in another earlier study, activities of six 1-chloro-4-oxy substituted thioxanthones were compared for their initiating efficiencies.[55] It was found that in the case of the 4-propoxy derivative, when the 1-chloro group was replaced with a 1-phenylthio substituent, a high rate of intersystem crossing to the triplet state resulted. The 1-chloro-4-hydroxy and 1-phenylthio- 4-propoxy derivatives, however, were found to be less efficient photoinitiators than 4-oxy derivatives. This is consistent with an observed marked enhancement in the photo reduction quantum yields, especially in the presence of an amine cosynergist[55] It was also noted that high photo conversion during polymerization occurred in the presence of oxygen for all the 1-chloro derivatives, except for the 4-hydroxy one. That difference was attributed to photodehalogenation, which is enhanced by the presence of a tertiary amine co-synergist.[55] Alkoxy substitution in the 4-position was found to enhance this mechanism. On the other hand, replacement of the chloro group in position 1 by a phenylthio moiety significantly reduces photoinitiation activity. It was also concluded that photopolymerization is more effective with polychromatic visible irradiation than with UV light.[55] Also, when thioxanthone initiators are used in the presence of sensitizers, the nature of the solvent plays a major role in controlling the energy transfer process.[55]

It is interesting that comparable conclusions were also drawn from a different, earlier study of the spectroscopic data from seven oil soluble, substituted thioxanthone compounds in various solvents.[56] Generally, all seven compounds exhibited low fluorescene and high phosphorescence quantum yields, but the ratio is solvent dependent. That is consistent with the high photoreactivity of the molecules operating in the lowest excited triplet state which is $n \rightarrow \pi^*$ in character. Activation of the thioxanthone molecule in the 3 and 4 positions with a methyl group enhances initiator activity, whereas substitution in the 1 position deactivates it through intramolecular hydrogen abstraction.[56]

Several 2-substituted derivatives of anthraquinones were also compared in photoactivity, subject to the nature of the light source, the amine co-synergist and the type of the monomer used.[57] All were found to be effective triplet state initiators. The anthraquinones, however, with electron withdrawing substituents were more active. This suggests that electron transfer is also an important part of the process of initiation.[57] In addition, when halogen substituents are present, upon irradiation they give rise to halogen radicals and further enhance the polymerization rate.[57]

The affect of variations in chemical structure of amines on the photoinitiating ability of 4-*n*-propoxythio-xanthone was also investigated in photopolymerizations of n-butyl methacrylate and a commercial triacrylate resin in isopropyl alcohol solution.[58] The conclusion from that study is that the activity is highly dependent on the ionization potential of the particular amine, the formation of a triplet exciplex and an electron transfer process. Analyses of chloroform extracts of the cured resin confirmed that the alkylamino radical is the initiating radical.[58]

Valderas et al.,[59] also studied the photopolymerization of methyl methacrylate initiated by 2-chlorothioxanthone in the presence of various amines of different structures. Here too, the photoinitiation efficiency of these systems was found to be highly dependent on the structure of the amine. The polymerization rate increases with the amine concentration and reaches a constant value at an amine concentration range of 10—30 mm. At these amine concentrations, aliphatic hydroxyalkyl amines are more efficient photoinitiators than the corresponding trialkyl-substituted compounds. Dimethylanilines with electron acceptor substituents in the 4-position give higher polymerization rates than electron donor substituted anilines. Their data also show that the singlet and triplet excited states of thioxanthones are efficiently deactivated by the amines. Rate constants correlate well with the oxidation potentials of the amines.[59] The effects of the chemical structure of the amine on the polymerization rates of 2-

hydroxyethyl methacrylate was also investigated in still another study .[60] In this work amines of different structures were used as the coinitiators with riboflavin:

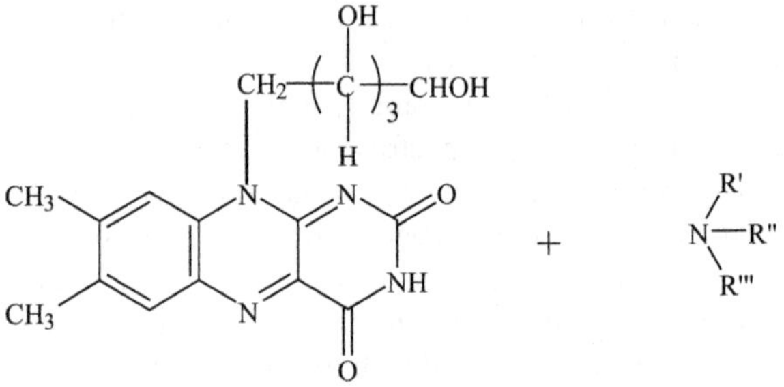

riboflavin

The results show a marked dependence of the formation of amine radicals upon the reactivity toward the monomer double bonds and depends upon the structures of the amines.[60] Table 2.6. gives some rate constants for electron transfer from amines to some aromatic ketones.

Table 2.6. Rate Constants for Electron Transfer from Amines to Some Aromatic Ketones

Ketone	Amine	$10^{-9}k_e(M^{-1}\,s^{-1})$	Ref.
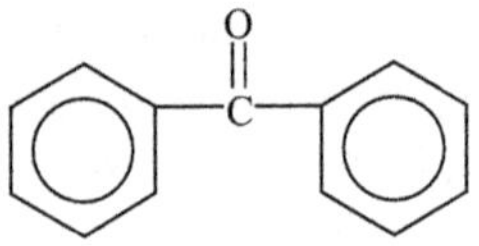	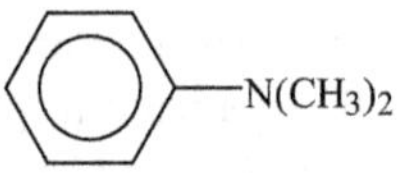	2.7	41-43
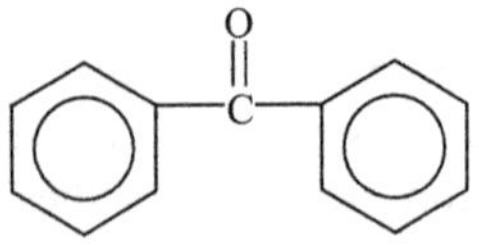	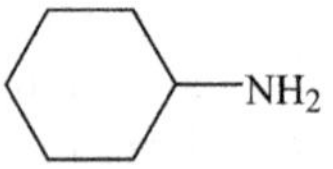	0.32	41-43
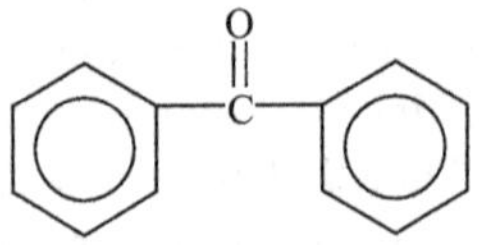	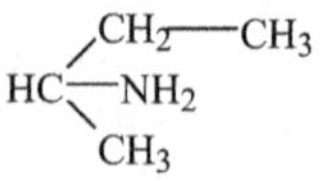	0.24	41-43

Table 2.6. (Continued)

Ketone	Amine	$10^{-9}k_e(M^1\,s^{-1})$	Ref.
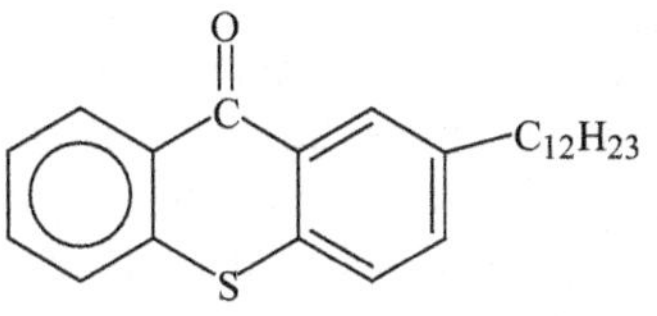	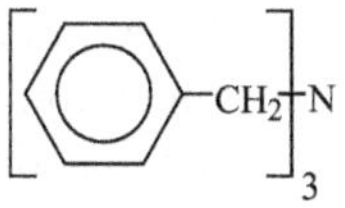	0.04	44
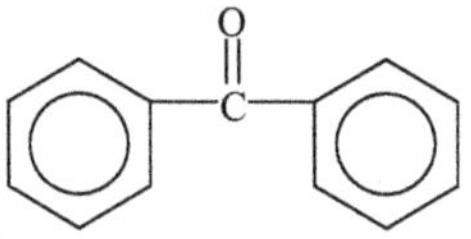	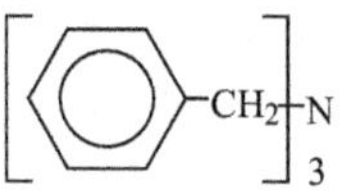	1.1	41-43
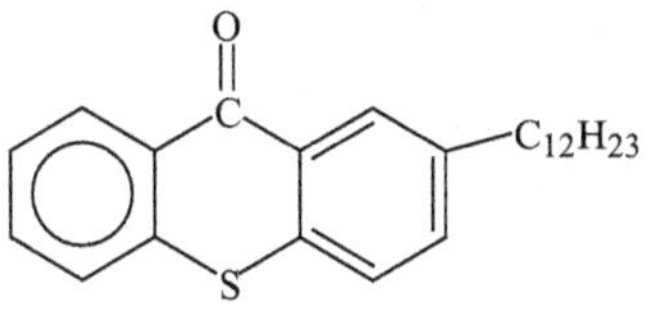	$CH_3N(C_2H_5)H$	2.7; 1.3	44
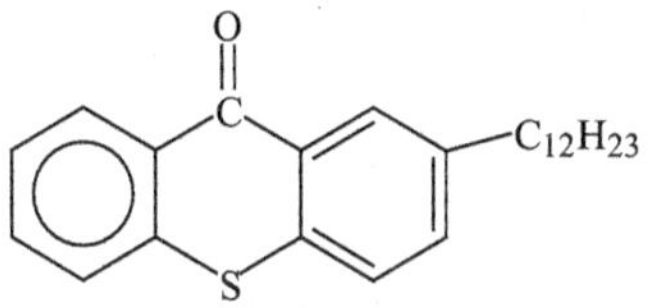	$(C_2H_5)_3N$	2.5; 3.0	44

[a] in toluene.

In addition, a systems for radical polymerization was reported that consists of a radical-generating reagent, 2-[(p-diethylaminostyryl)benzoyl)]-4,5-benzothiazole combined with 3,3',4,4'-tetrakis(tert-butylperoxy-carbonyl)-benzophenone in toluene.[61]

Bradley and Davidson[62] pointed out that for any particular aromatic ketone, the efficiency of photoinitiation is determined by the structure of the ketone and by the molecular geometry of the α-aminoalkyl radical produced. On the other hand, Paczkowski et al,[63] suggested that the initial electron-transfer reaction from aromatic amines to the excited state of xanthene dyes, used in their study, is responsible for the variation of the photoinitiation efficiency.

Aromatic amines were also examined as possible oxygen scavengers during the process of photocuring.[64] How efficient the aromatic amines are in scavenging oxygen, however, is not clear. In general, the reaction of oxygen with the amines appears to be complex.

A study was also made that of the excited states of three photoinitiators derived from sulfonyl acetophenones.[65] It was observed that interactions with hydrogen donors, monomers, and amines lead to bimolecular quenching. It was also observed that in this composition the triplet state of one of the ketones has charge transfer character which confers peculiar interactions with electron donors, e.g., amines. [65]

Camphoquinone is often used in dental filling composites in combination with an amine coinitiator, like ethyl 4-dimethylaminobenzoate. Liska and coworkers [35] reported that they found significant improvement in reactivity of camphoquinones, actually by a factor of two, when the amine is covalently bonded to the acetyl derivative of the quinone in the 10 position. The camphoquinones can be illustrated as follows, with n equal to one or two:

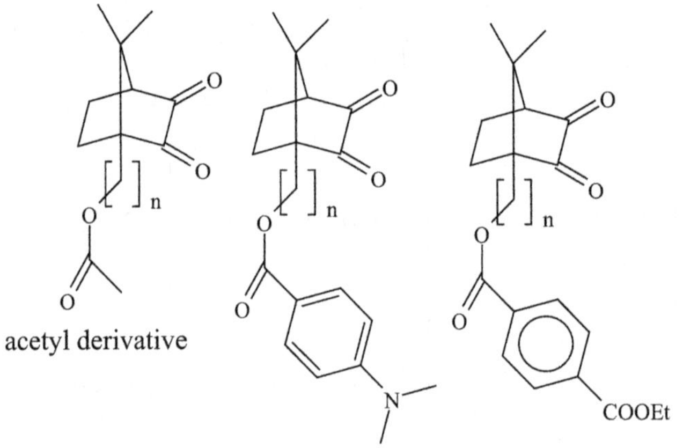

acetyl derivative

covalently bonded camphoquinones

Extension of the spacer between the acetyl moiety and the camphoquinone (n = 2 rather than 1) reduced the activity of the quinone. Liska and coworkers [35] suggest that several factors might be responsible for the higher activity. They feel that one possible explanation is that when the amine is bonded hydrogen abstraction between the ester and the camphoquinone moiety is hindered:

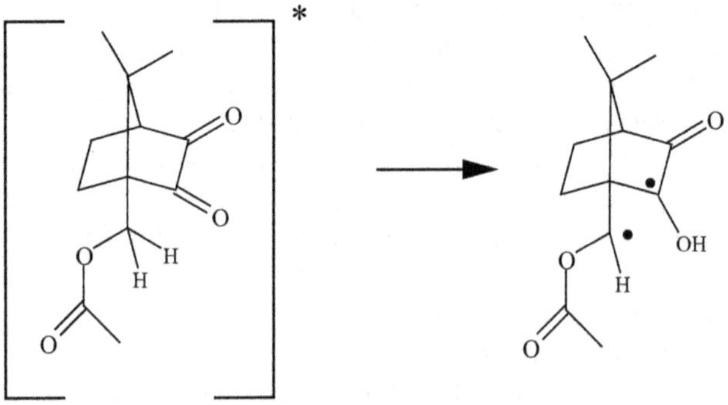

Another possible explanation, suggested by them, [35] is stabilization of the transition state after electron transfer from the amine :

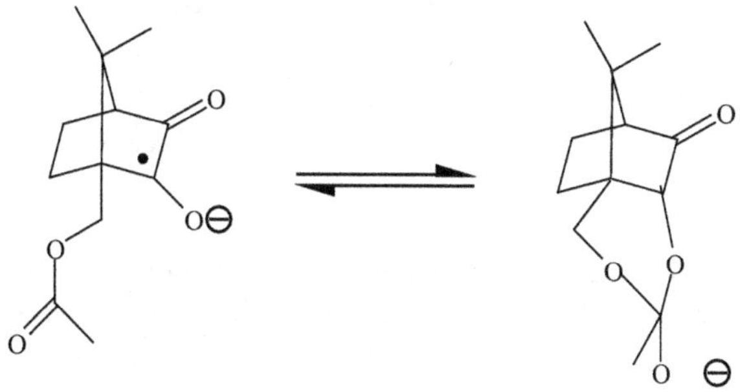

    Liska also investigated several benzophenone- and thioxanthone-based photosensitizers that were covalently bonded to hydroxyalkylphenone and amino-alkylphenone based photoinitiators. [66] This was done to enhance the rate of the excitation-transfer effect due to the close vicinity of the photosensitizer to the photoinitiator. Selective excitation of the photosensitizer chromophore revealed that the energy transfer significantly increases in covalently bonded initiators by comparison to their physical mixtures. This effect, however, is most pronounced in hydroxyalkylphenones that are sensitized by suitable benzophenone derivatives, especially at low photoinitiator content. [66]

    Among other two component photo initiating systems, combinations of photoinitiators with peroxy compounds, like with pyrilliuim peresters,[67, 68] or peroxides with Michler's ketone [69] were reported to be effective. The same is true of combinations of aromatic ketones with dyes. [70] Further discussions of the use of dyes in combination with ketones can be found in sections 2.2.4 and 2.2.7.

    Various derivatives of thioxanthone are often used as photosensitizers with substituted morpholino ketones as energy acceptors.[23] Actually, the two materials by themselves are efficient initiators. It was shown that when combinations of thioxanthone derivatives with morpholino ketones are irradiated with light, two types of reactions can take place.[71] If the triplet-triplet energy transfer occurs when the energy level of the thioxanthone is higher than that of the morpholine ketone, cleavage of the morpholino ketone is the result:

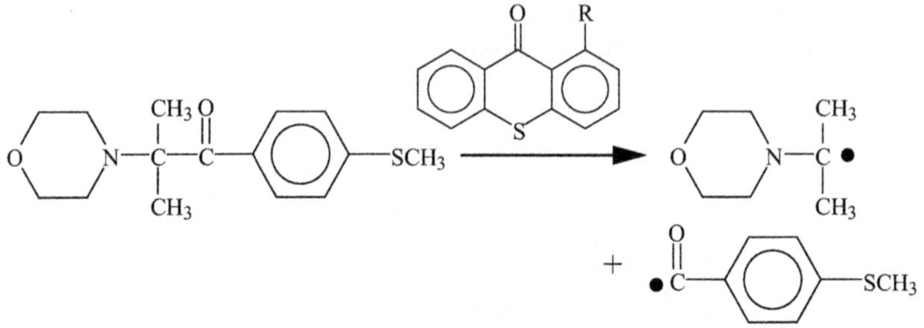

On the other hand, reactions of thioxanthones with morpholino ketones can also results in electron transfers without cleavages. This is shown below [8,25]:

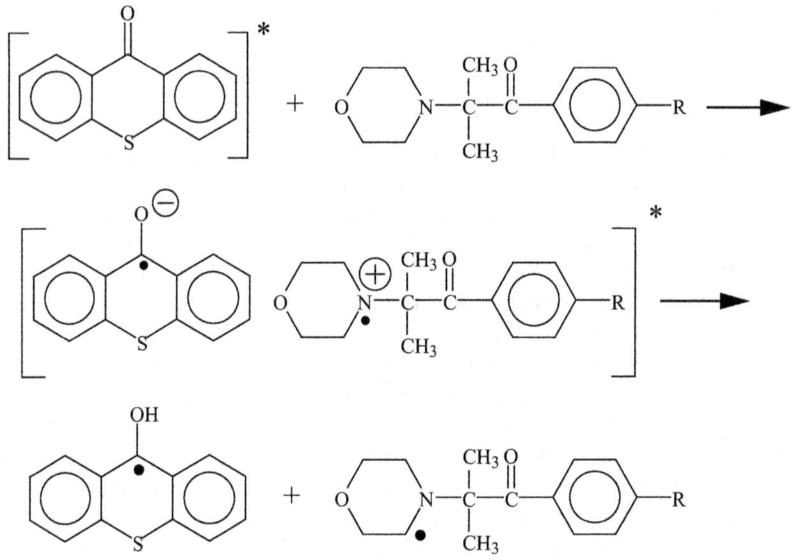

where R is a methyl or a hydrogen. Both reactions can take place simultaneously and the balance between the two depends on the polarity of the medium. [71] This can lead to a different formation of free radicals: [72,73]:

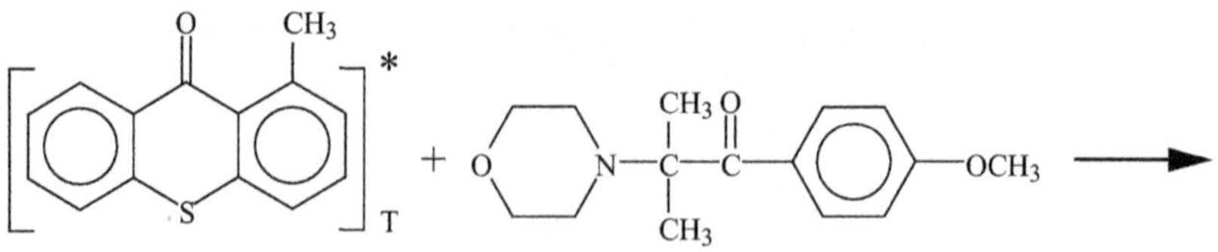

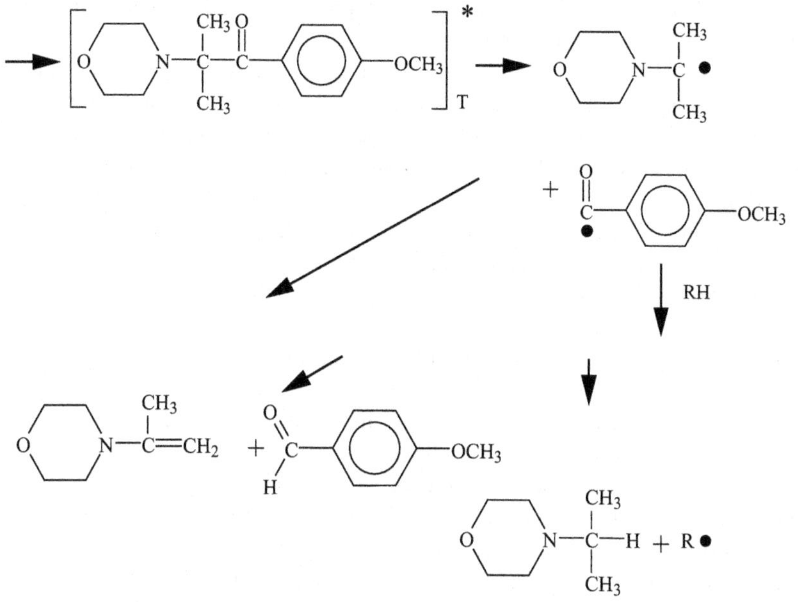

The balance between the two competitive processes depends strongly upon the polarity of the medium. [8] The same is true of other photoinitiators.

Another example is a two component system that was used to cure isobornyl acrylate, [72]

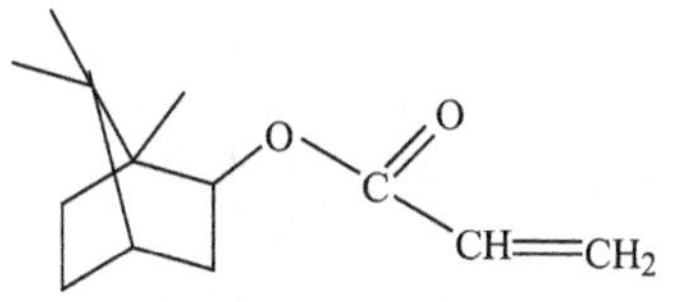

isobornyl acrylate

It consists of thioxanthone or 2-isopropylthioxanthone in combination with (2,4,6-trimethylbenzoyl)-diphenyl-phosphine oxide or bis(2,4,6-trimethyl-benzoyl)-phenylphosphine oxide. The mechanism of sensitization was reported to involve triplet-triplet energy transfer from the thioxanthones to the phosphine oxides.[52] That is followed by formation of radicals through α–cleavage of the photoinitiators. Direct photolysis of the phosphine oxides results in an absorptive, chemically induced, dynamic electron polarization due to the triplet mechanism of polarization of the substituted benzoyl. The phosphorus-centered radicals produced by α–cleavage of the photoinitiators are the same radicals that are produced either by direct or by sensitized photolysis.[72] But the sensitization increases the efficiency of the process.

It is interesting to note that a recent patent[75] describes a light curable coating composition that includes cationic photoinitiator and/or free radical photoinitiator, in combination with a charge transfer complex. The charge transfer complex is described as an singlet electron withdrawing reactant component.

For the UV light photo initiated polymerization of acrylonitrile, with 3-amino-9-ethyl-carbazole as a sensitizer,[76] the following kinetic equation for the polymerization was developed [76]:

$$R_p \propto [\text{aminoethylcarbazole}]^{0.17} [\text{acrylonitrile}]^{0.60}.$$

Above equation is based on an analysis that indicates charge transfer and exciplex formation of acrylonitrile with carbazole forms as the intermediate. [76] Gao et al.,[76] carried out kinetic studies of photopolymerization of methyl methacrylate by using piperazine sulfur dioxide charge-transfer complex as a photoinitiator. The polymerization rate (Rp) is dependent on the molar ratio of piperazine to sulfur dioxide, and the complex with a composition of piperazine to sulfur dioxide in a molar ration of 1:2 is the most effective. By using the complex as the photoinitiator, the polymerization kinetics was expressed as,

$$Rp = kp[I]^{0.34} [MMA]^{1.06}$$

The apparent activation energy (E_a) value was found to be 23.7 kJ/mol.

The photoreducing behavior of p-nitroaniline was also studied in the presence of N.N-dimethylaniline .[77] The stoichiometry of the photoreduction reaction shows that several amino radicals derived from dimethylaniline are generated by each photoreduced nitroaniline molecule. The rate of polymerization of lauryl acrylate was found to be proportional to the square root of both the incident light and the concentration of the co-initiator, dimethylaniline. The polymerization efficiency of this system was claimed to be higher than that obtained with conventional aromatic ketone photoinitiators.[77] Similar results were obtained [78] with another bimolecular photoinitiator systems, consisting of $p\text{-}O_2NC_6H_4NH_2$ and its derivatives and tertiary amines acting as reducing agents. In addition, it was found that the 2,6-dihalogen derivative of p-nitroaniline is more photoactive in initiating polymerizations than a number of aromatic ketones.[78]

Rodriguez and coworkers[71] compared the behavior of other nitroaromatic amines, like 4-nitro-1-naphthylamine, N-acetyl-4-nitro-1-naphthyl-amine, and N,N-dimethyl-4-nitroaniline with 4-nitroaniline. The rate and quantum yield of the polymerizations as well as the residual unsaturation in the products were found to be strongly dependent on the nature of the photoinitiator. Also, the authors found that formulations based on 4-nitro-1-naphthylamine are superior to those based on *p*-nitroaniline, and provide a conversion of nearly 85%.[71]

Among novel photoinitiators is one that is based on a ketosulfone-benzophenone structure that was reported by Fouassier et al.,[18] At a given amount of compound, this photoinitiator is claimed to be 20% to 40%. more active than such ketones as dimethoxy phenylacetophenone for the photocuring of multiacrylates in bulk and in open air. The triplet state of this compound undergoes a β cleavage reaction at the ketosulfone moiety and generates ketyl radicals upon addition of an amine. The quantum yield of intersystem crossing was reported as 0.7 in acetonitrile. The triplet energy level is located around 275 kJ mol[-1].

2.2.1.4. Multicomponent Photoinitiating Systems

Multicomponent photoinitiation systems have emerged as an improvement over the two-component electron transfer initiating systems. Dyes are often used as light absorbing moieties. They may be added to expand and match the sensitivity of the initiating combination to the available light source. Multicomponent systems can be very flexible because a wide variety of materials may be used. They can also be very fast. The selection of the dye determines the useable light wavelength. Many multicomponent systems are also effective in visible light. This is discussed in section 2.2.6 Mixtures of photoinitiators appear to result at times in complex sequences of reactions. Details of such mechanisms are still not fully understood. Three types of mechanisms were suggested for the electron transfer processes that can take place when multiple components, including dyes, are present and reactions occur between an electron donor in the ground state and the acceptor.[79] These are: (1) donor-acceptor pairs with electrostatic interaction in the ground state; (2) donor-acceptor pairs without electrostatic interaction; and (3) donor-acceptor pairs, neutral in the ground state, but charged after electron transfer.[79] The electron donors that are commonly used include aromatic ketones, xanthenes, thioxanthenes, xanthones, thioxanthones, coumarins, ketocoumarins, thiazines, merocyanines, and many others. An example is a mixture of Michler's ketone with benzophenone. The interaction results in formation of a triplet exciplex that dissociates into two different free radicals [80] :

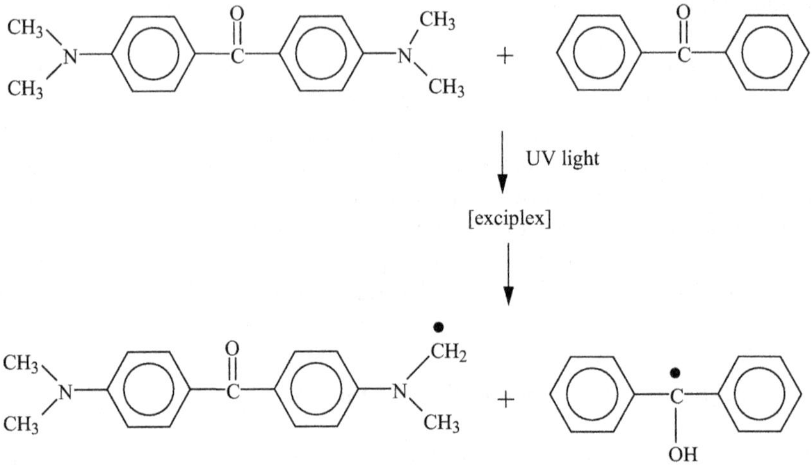

Another examples is a combination of methylene blue and perylene that was shown to be effective in photoinduced polymerizations of acrylic esters.[81] Also, dyes, like toluidine blue and thiazine were reported to be effective in the presence of different alcoholamines.[82,83]

Still another example is an initiating system composed of 7-diethylamino-3-(2'-N-methyl-benzimidazolyl)-coumarin and diphenyliodonium hexafluorophosphate. This composition initiates the polymerization of methyl methacrylate in visible light. After the dye absorbs the light energy, quick electron transfer takes place from the dye to the iodonium salt to produce free radicals.[84] The light induced reaction is claimed to occur mainly through the excited singlet state of the coumarin and results in low sensitive to O_2. The fluorescence of the coumarin compound was reported to be quenched efficiently by the iodonium salt.[84] The reaction was observed to be in accord with the Stern-Volmer equation. The influence of the concentration of coumarin on the polymerization rate of methyl methacrylate led to the conclusion that the free radicals from coumarin act mainly as chain terminators.[84]

The influence of oxygen on the rate of consumption of methylene blue dye in the presence of an iodonium salt and an amine, a typical three component system, was studied.[85] Oxygen quenches the triplet state of the dye, leading to retardation of the reaction.[85] This is followed by rapid exponential decay of the methylene blue fluorescence after the oxygen is depleted. On the basis of the impact of the amine and iodonium salt concentration on the fluorescence intensity and the duration of the retardation period, a mechanism was proposed that includes an oxygen-scavenging pathway, in which the tertiary amine radicals formed in the primary photochemical process consume the oxygen via a cyclic reaction mechanism.[85] The iodonium salt is an electron acceptor, acting to re-oxidize the neutral dye radical back to its original state and allowing it to reenter the primary photochemical process.[85]

Ivanov and Khavina [86] reported that their data suggests a common character of synergism in photoinitiating systems that are composed of an aromatic ketone, a halomethyl aromatic compound, and an aliphatic or an aromatic amine. They developed a kinetic model that takes into consideration the influence of the amine on the quantum yield of the primary radical pairs and the enhancement of the radical escape from the cage by the action of the halomethyl compound.[86]

Combinations of several coumarin or ketocoumarin/additives with bisimidazole derivatives and mercaptobenzoxazole, or titanocene, or oxime esters were found to be capable of efficient initiating of free radical polymerizations. Interactions of excited states of a coumarin or ketocoumarin with mercaptobenzoxazole, and titanocene were also investigated. This led to the conclusion that the coumarin forms radicals through an electron transfer reaction, while the ketocoumarin undergoes an energy transfer reaction with bisimidazole and a hydrogen abstraction reaction with the benzoxazole derivative.[87] Some examples of three component initiating compositions found in the literature are presented in Table 2.7.

Spectroscopic investigation of a three-component initiator composition was carried out. [88] This system consists of (1) methylene blue/N-methyl-diethanolamine/diphenyliodonium chloride, (2) eosin/methyl-diethanol-amine/diphenyliodonium salt, and (3) and alcohol solution of eosin/methyl-diethanolamine/diphenyliodonium salt. The kinetic studies revealed that the photoinitiations with the aid of eosin dyes or with methylene blue are quite similar. The fastest polymerization rate is obtained when all three components are present, the next fastest with the dye/amine pair, and the slowest with the dye/ iodonium salt pair. In the case of methylene blue/N—methyl-diethanolamine/diphenyliodonium salt system, it was concluded that the primary photochemical reaction involves electron transfer from the amine to the dye. [88] Also it was suggested that the iodonium salt reacts with the resulting dye-based radical (which is active only for termination) to generate the original dye and simultaneously produce a phenyl radical that is active in initiation. Moreover, oxygen quenches the triplet state of the dye leading to retardation of the reaction. Padan et al.,[88] propose a sequence of steps in which the dye is rapidly oxidized by diphenyliodonium salt in a reaction that is not highly efficient at producing initiating radicals. The dye may, however, also be reduced by the methyldiethanolamine amine producing active initiating amine radicals. Another active radical may be produced if the dye radical is further reduced to its leuco form by a third molecule, the amine. In addition, the reduction of the bleached dye radical (formed via reaction with the iodonium salt) by the amine has a two fold effect. First it regenerates the original eosin dye, and second, it produces an

Table 2.7. Some Examples of Three-Component-Initiating Compositions

Component #1	Component #2	Electron donor	Ref.
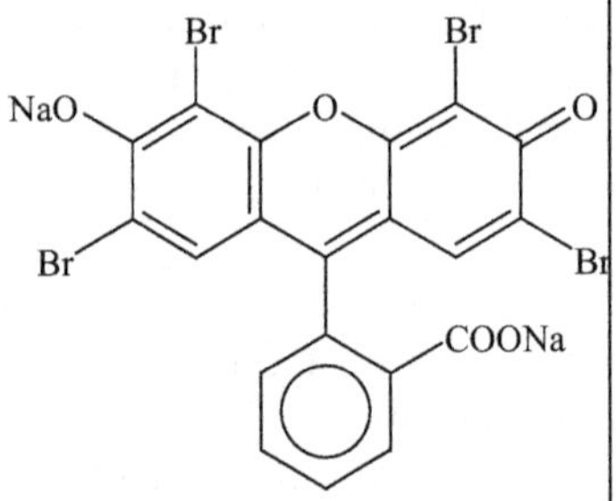	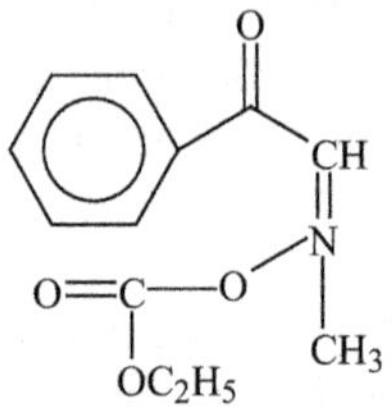	amine	223
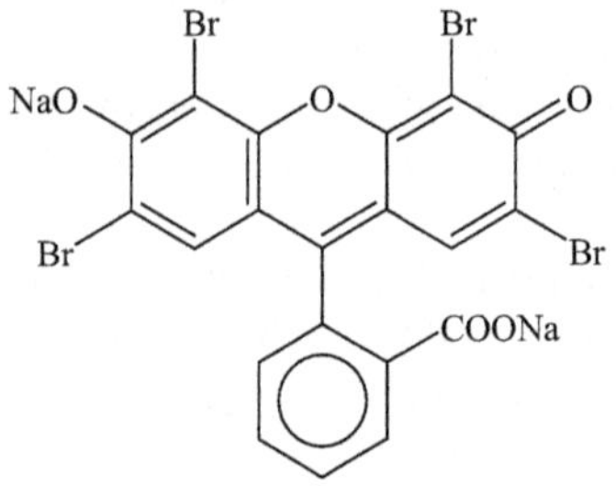	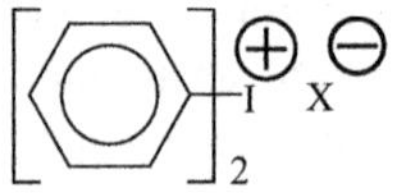	amine	226
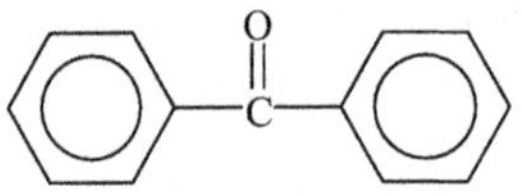	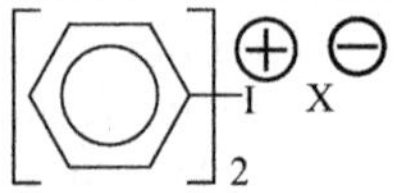	amine	61
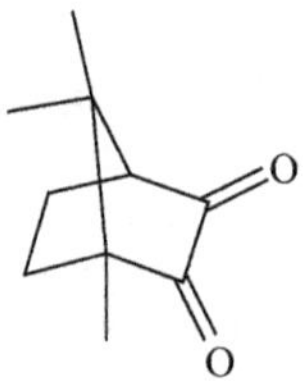	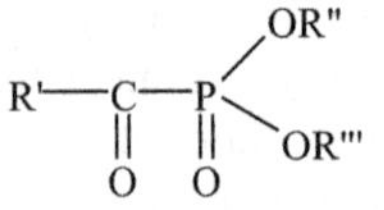	amine	17
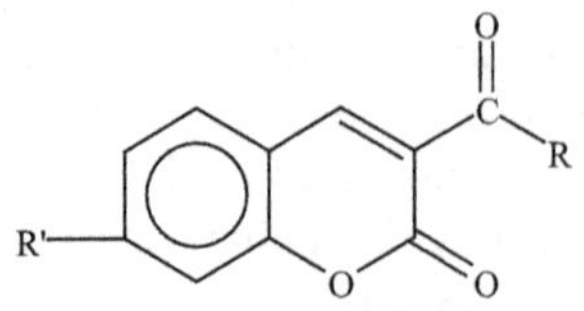	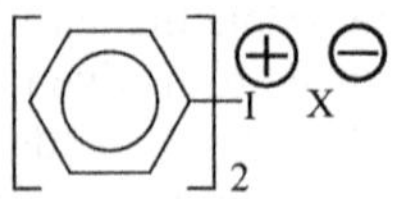	amine	86

Table 2.7. (Continued)

Component #1	Component #2	Electron donor	Ref.
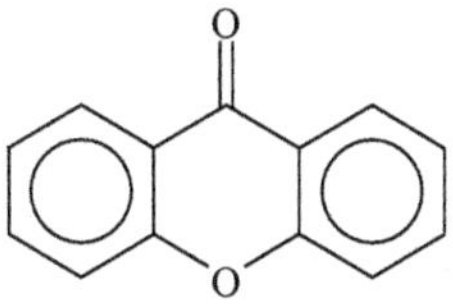	substituted triazine	amine	64
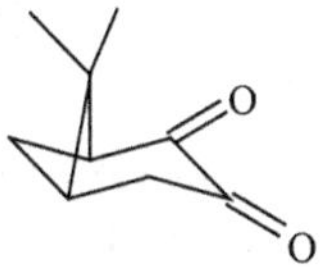	acyl phosphate	amine	24
merocyanine	substituted triazine	amine	61
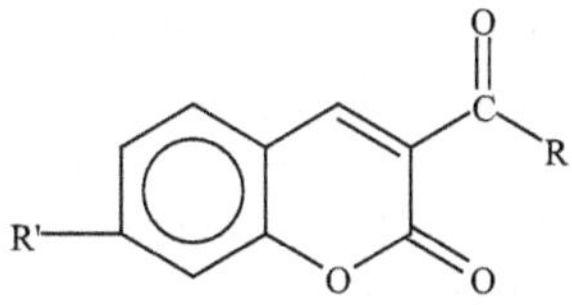	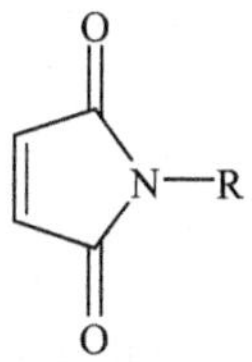	amine	272

active amine radical in place of the less active dye radical. Dual cell UV-visible absorption spectroscopy yielded no evidence of formation of a ground state complex between the eosin dyes and the iodonium salt.[88]

A composition of an electron transfer free radical photo initiating system was reported recently.[89] This composition uses a light absorber, a dye, and an electron donor; The donor is a sulfur-containing aromatic carboxylic acid. The structure of the donor is such that upon photocleaving, the leaving group forms free radicals. It is claimed that the experimental results show a transformation of the sulfur-containing aromatic carboxylic acid into an ammonium salt. This is accompanied by a substantial increase in photoinitiation ability [89] :

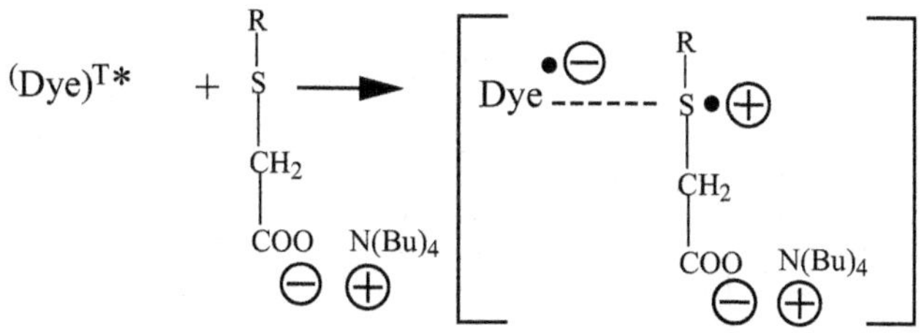

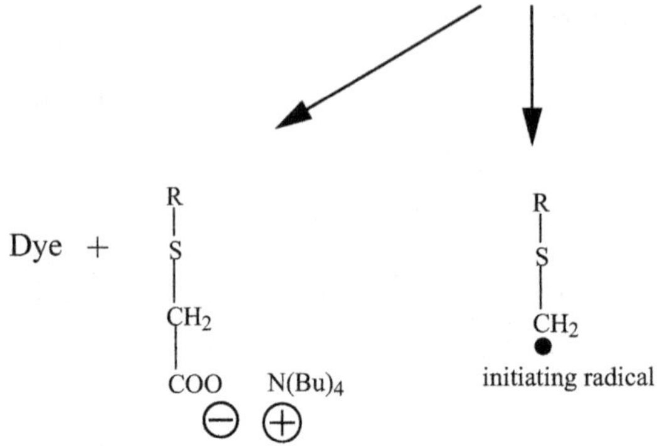

$$\text{Dye} \ + \ \begin{array}{c} R \\ | \\ S \\ | \\ CH_2 \\ | \\ COO \quad N(Bu)_4 \\ \ominus \quad \oplus \end{array} \qquad \begin{array}{c} R \\ | \\ S \\ | \\ CH_2 \\ \bullet \\ \text{initiating radical} \end{array}$$

The photoreduction of the above dye, 5,7-diiodo-3-pentoxy-9-fluorenone, in the presence of (phenylthio)-acetic acid and its tetrabutylammonium salt occurs via a photoinduced electron transfer process. On the basis of the known photochemistry of sulfur-containing aromatic carboxylic acids, it is postulated [89] that the existence of the carboxyl group in an ionic form allows a rapid decarboxylation, yielding a neutral very reactive α-alkylthio-type radical (R-S-CH$_2$•). [89]

When photoinitiators consisting of mixtures of benzophenone, or 4-benzoylbiphenyl, or isopropylthioxanthone with a tertiary amine are combined with an electron deficient anhydride, rapid photoinitiations of polymerizations of acrylate esters result.[90] Thus, additions of less than 0.1 wt. percent 2,3-dimethylmaleic anhydride to 1,6-hexanediol diacrylate containing any of the above diarylketones and N-methyldiethanolamine result in an increase in the polymerization rate maximum by a factor of as much as three that is attained for the same reaction without the anhydride. Laser flash photolysis results show that benzophenone, 4-benzoylbiphenyl, and isopropylthioxanthone triplets are readily quenched by dimethylmaleic anhydride. [90]

Fouassier et al.,[91] reported a four component photoinitiating system that consists of a photosensitizer, Rose Bengal, ferrocenium salt, an amine and a hydroperoxide, such as cumene hydroperoxide:

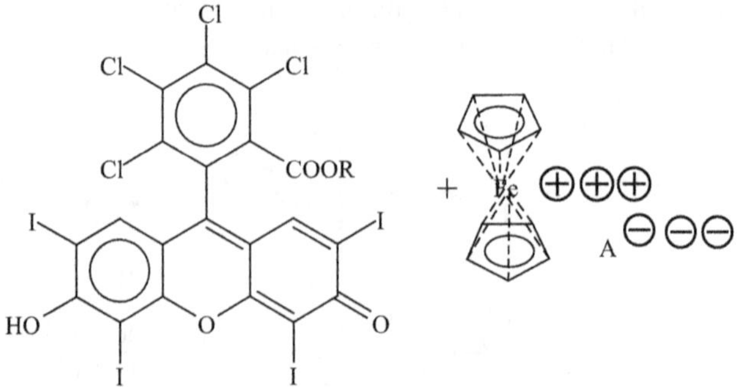

+ amine + cumene hydroperoxide

This initiating system is reported as being capable of initiating photopolymerizations of thick pigmented coatings, useable as paints in wood furniture industry.

2.2.1.5. Photoinitiators for Water Borne Systems

The appeal of UV curable coatings and inks to many is in the fact that they can be formulated into solventless systems. Nevertheless, water borne compositions are needed and have been developed due to many requirements. Such systems find application in pigmented water based paints, in textile printing, in coatings for glass reinforced fibers, for sunlight curing of waterborne latex paints, in light curable inks for jet printing, and in high speed photopolymers used for laser imaging. Many water soluble and hydrophilic photoinitiators were developed by simply incorporating water solubilizing groups into the chemical structures of known hydrophobic initiators. Thus, for instance, various salts of benzophenone [71, 72], benzil [71, 72], thioxanthone,[71, 74, 15] and α-hydroxyalkylphenone were formed.[71]

The effects of water, used either as a solvent or as a dispersing agent, on the photocuring reactions was studied earlier by Encinas and coworkers.[75] They demonstrated that water solvated species, when irradiated with light of 300 nm, can produce significant amounts of polymer from monomers like vinyl acetate.

A number of water soluble thioxanthone derivatives were prepared specially for water borne systems. They can be illustrated as follows[76]:

where R = O(CH₂)₃SO₃⁻ Na⁺
O(CH₂)₃N(CH₂)₃SO₃Me
O(CH₂)₃N(CH₂)₃OSO₃Me

These compositions were reported to react in the lowest excited singlet state. When amine coinitiators are present, they abstract an electron via a singlet, and to some extent a triplet exiplex to produce the radical anion. In the absence of the amines, however, hydrogen abstraction is the dominant initiation step and necessitates the presence of efficient hydrogen donors.[76]

Another group of seven water soluble initiators is based on methyl substituted salts, 3-(9–oxo-9H–thioxanthene–2-yloxy)-N,N,N-trimethylpropane ammonium. These compounds are both water and isopropyl alcohol soluble.[77]

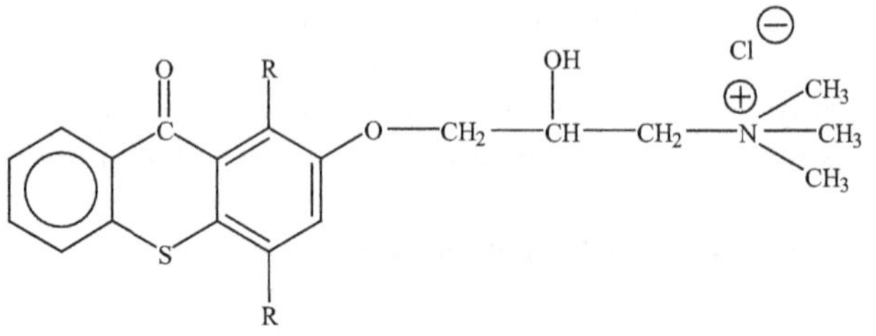

where, R = H, CH$_3$. It was shown that in all cases methyl substitution enhances the photopolymerization of 2-hydroxyethyl methacrylate in water. The 4-substitution is the most effective.[79] The longest wavelength absorption maxima and extinction coeffients of these materials are similar to oil and water soluble types.[77, 55] The fluorescence and phosphorescence properties, on the other hand differ significantly. Both are strongly quenched by substitution of a methyl group in the 1-position α to the carbonyl. There is evidence of intramolecular hydrogen atom abstraction as being the primary initiating process. Also, transient absorption is only via the lowest excited singlet state.[74]

Still another group of water soluble initiators were prepared by Wang and coworkers, [78] who converted the primary hydroxy group in a commercial photoinitiator, Diarocur 2959 (2- hydroxy-1-[4-(2-hydroxyethoxy)phenyl]-2-methylpropan-1-one) to various amine and sugar moieties:

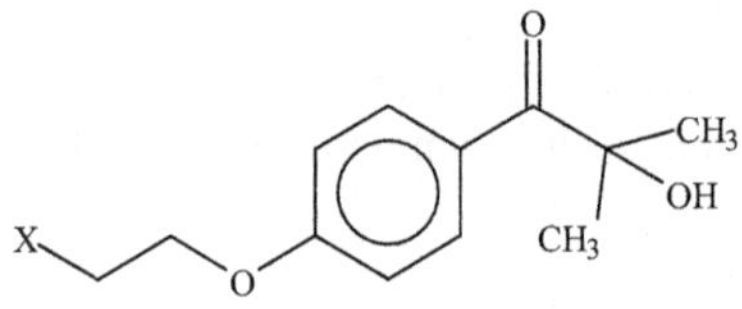

where X = OH. Nine hydrophilic compounds were formed. The effectiveness of these materials in curing epoxy acrylate resin was found to depend on the substituents and on the counter ions of the salts.[78]

Liska reported preparation of water-soluble photoinitiators that contain carbohydrate residues as well as copolymerizable derivatives of the carbohydrate residues.[79] These materials consist of alkylphenones, benzophenones and thioxanthones. To these compounds were attached carbohydrates like glucose, cellobiose, and 1-amino-1-deoxy-D-glucitol. In addition selected initiators were reacted with methacryloyl chloride to form polymerizable photoinitiators. The glucose modified photoinitiators were claimed to yield the best results with respect to compatibility with resins, reactivity and gel content. [79]

Some additional water soluble photoinitiators that were reported in the literature are shown in Table 2.8.

Table 2.8. Some Examples of Water Soluble Photoinitiators

Photoinitiator	R	Ref.
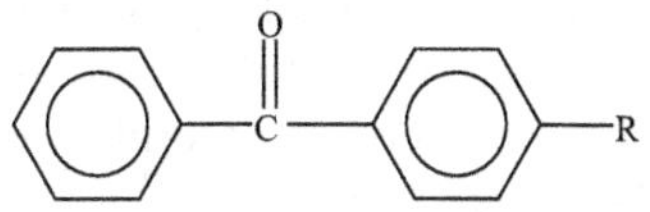	$CH_2SO_3^{\ominus} Na^{\oplus}$ $CH_2N^{\oplus}(CH_3)_3Cl^{\ominus}$	80
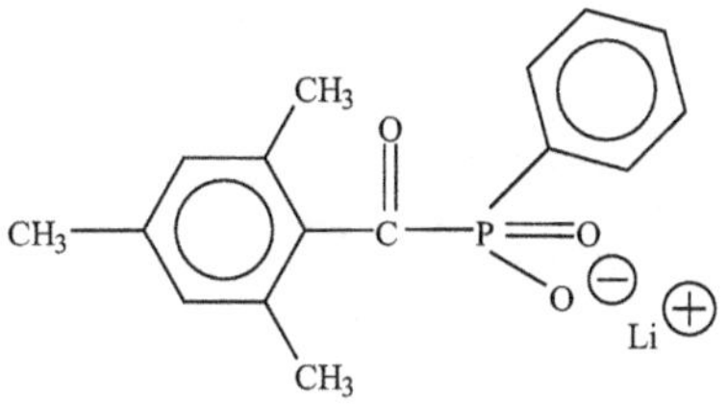		30
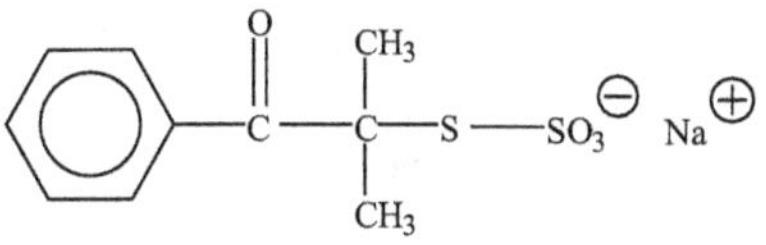		81
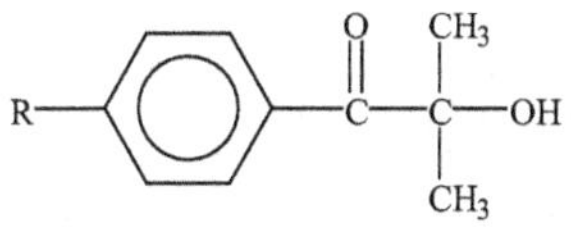	OCH_2CH_2OH $OCH_2COO^{\ominus} Na^{\oplus}$	82
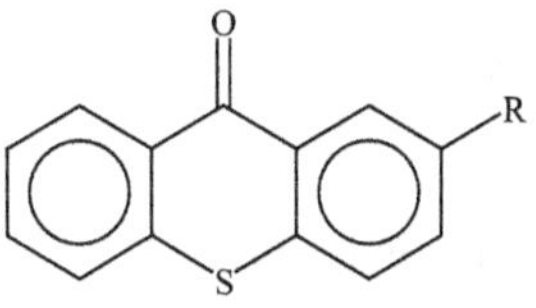	$CH_2SO_3^{\ominus} Na^{\oplus}$ $CH_2N^{\oplus}(CH_3)_3Cl^{\ominus}$	83
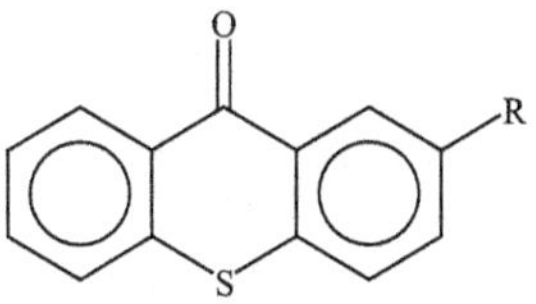	$OCHC_2OOH$ $O(CH_2)_3N^{\oplus}(CH_3)_3SO_3^{\ominus}CH_3$	84

2.2.1.6. Oligomeric and Polymeric Photoinitiators

Various considerations led to design of polymerizable photoinitiators, as well as to oligomeric ones. In the case of surface coatings, separation and subsequent migration of the remaining photoinitiators, or their unattached fragments from their decomposition to the surface can be a problem. This can be particularly severe when the migration of these materials proceeds to and is deposited at the interface, between the substrate and the film and subsequently interferes with adhesion. Volatility and extractability in the case of films used for food packaging materials is also a serious consideration.

A common approach to the preparation of polymeric photoinitiators is to attach photoinitiator molecules, like benzophenone, or benzoin ether, or thioxanthones to some polymerizable groups For instance, acryloyl chloride was reacted with hydroxythioxanthone and then copolymerized with styrene[84]:

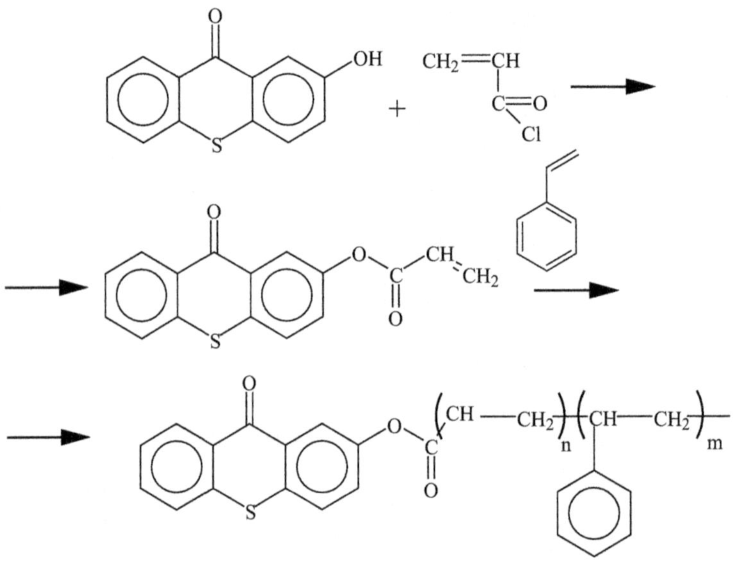

Hydroxythioxanthone was also reacted with poly(chloromethyl styrene)[85]:

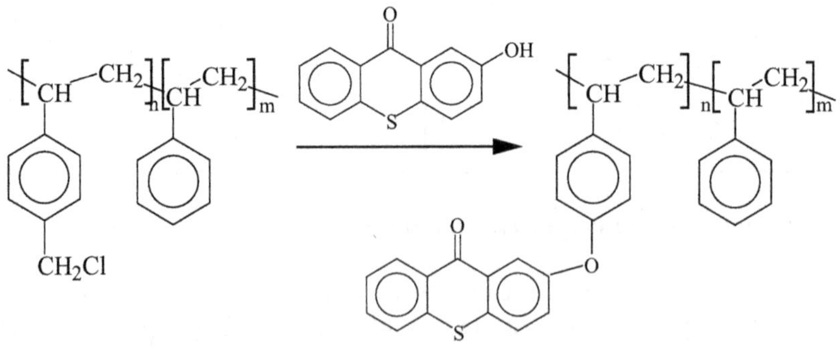

In a number of instances these polymer attached photoinitiators were reported as being somewhat less effective than the original compounds.[86] That, however, is not always the case. An example is the work by Ahn and coworkers[87] who prepared polymers and copolymers from monomers prepared by the following reaction?

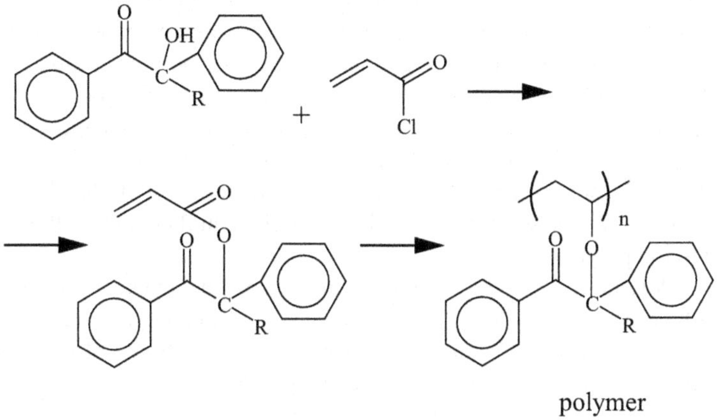

polymer

The polymerized material was claimed to exhibit greater efficiency in initiating photopolymerizations than does methyl benzoin ether.[87] Also, laser photolysis of polymers with side chains consisting of thioxanthene-9-one and morpholinokeone were shown to possess excitation energy for transfer from the thioxanthone to the morpholine groups as being two orders of magnitude greater than that of low molecular weight mixtures of these two componenets.[54]

On the other hand, Martinez-Utrilla and coworkers[88] studied soluble polymer bound photosensitizers and concluded that the sensitization efficiency diminishes when the material it is bound to a polymer.[88] The same conclusion was reached by Tsuchida and coworkers[89] who studied electron transfer in a heteroexcimer system of N-ethylcarbazole, a donor, and poly(vinyl methyl terephthalate), an acceptor. Behavior of the ion-radicals that form was compared with model compounds. The experiments showed that these ion-radicals do not transfer readily from the polymer chains to acceptors. By comparison, anion-radicals formed on the model monomeric or dimer systems transferred easily.

A patent issued to Coates Brothers in England describes oligomeric initiators that contain thioxanthone molecules[40]:

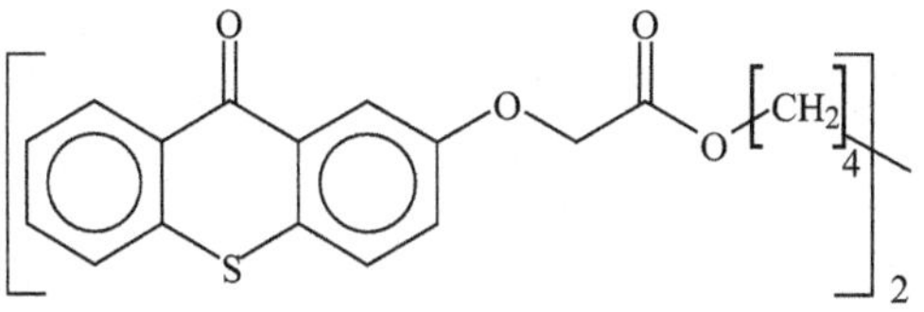

This and similar compounds are claimed to yield fast cures and low migration in ink formulations.

Also, Corrales *et al.,*[136] reported a study of photopolymerization of methyl methacrylate using initiators based on thioxanthone chromophores and low molecular weight tertiary amines bound to polymeric chains as co-initiators. The efficiency of the catalyst systems was compared to that of the corresponding low molecular weight analogs. Higher polymerization efficiency was obtained with system consisting of thioxanthone bound to polymeric chains and free amines. The catalyst efficiency was found to be independent of the amount of amine units in the polymeric chains. In addition, the polymeric chains were found [91] not to affect the emission characteristics of the thioxanthone chromophore. The ketyl radical yield was slightly higher for the thioxanthone bound to the polymer. [91]

Polymeric structures containing benzoin ether,[92] benzophenone tetracarboxylic dianhydride,[93] polysilanes,[94] camphoquinone,[95] aromatic aliphatic azo compound,[96] and others were reported.[97] Some of the interesting copolymerizable photoinitiators are listed in Table 3.9.

Table 2.9. Some Examples of Copolymerizable Photoinitiators

Photoinitiator	Ref.
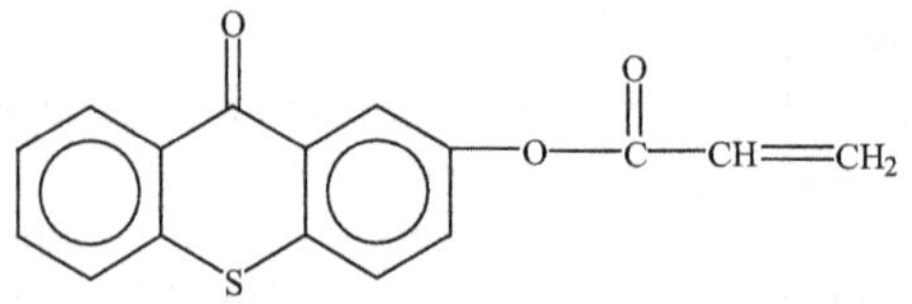	98, 99
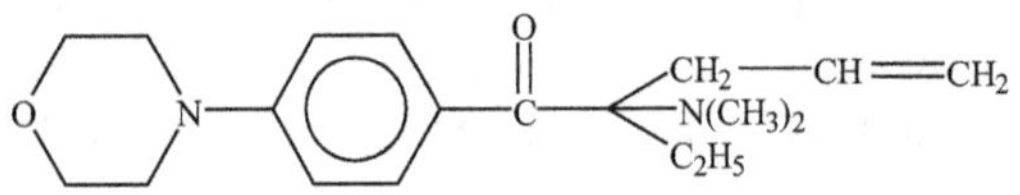	97
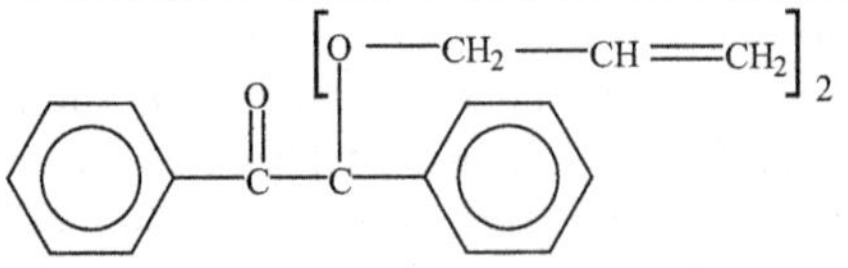	100
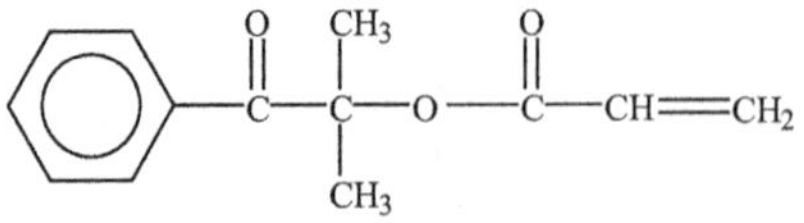	99
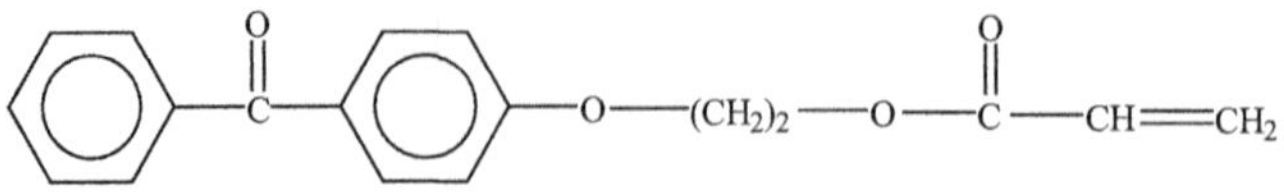	97

Table 2.9. (Continued)

Photoinitiator	Ref.
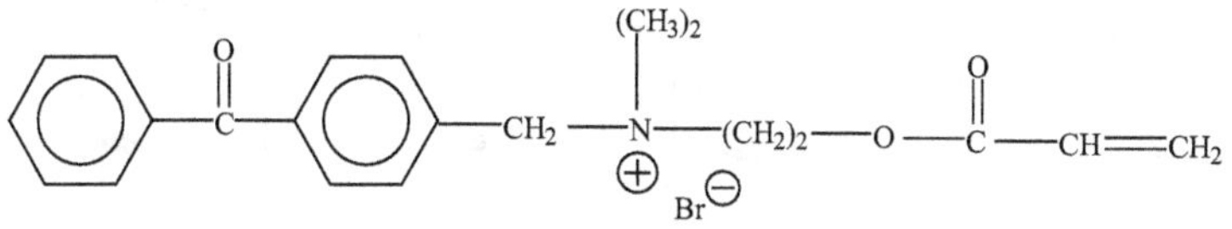	100
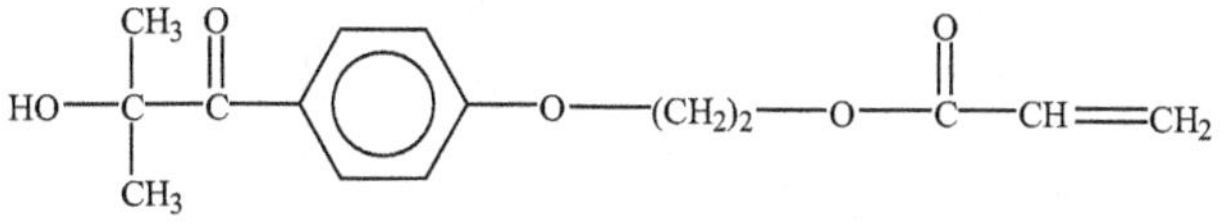	97
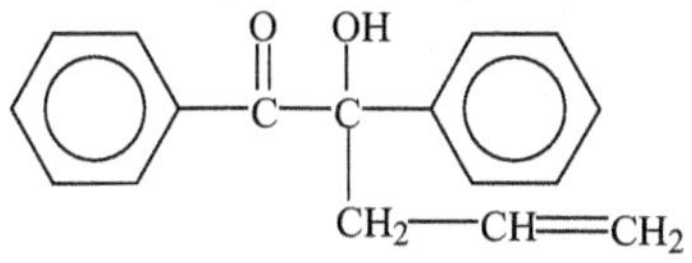	86
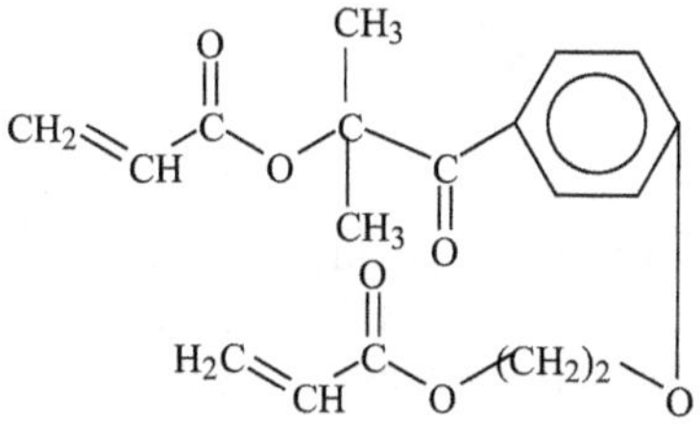	97

More recently, several polymeric photoinitiators with pendant chromophore-borate ion pairs were developed.[101] These materials form active alkyl and tertiary amine radicals upon irradiation.[101] Generally, in free radical mediated polymerizations, photo generated radicals as well as the growing macro radicals are quenched by oxygen. As a result, the rates of polymerizations and molecular weight distributions of the resultant polymers are affected. The formation of tertiary amine in polymerization processes helps to ameliorate the situation. This benefit is somehow offset if the amines are added separately to the formulation by a tendency to yellow and give off volatile toxic materials. [100] It is claimed that the formation of alkyl radicals takes place after electron transfers to the excited chromophores from the borate anions. The alkyl radicals in turn initiate polymerizations of acrylic monomers. Following is an example of polymeric systems containing chromophore-borates as the side chains. Bond cleavage and generation of polymeric tertiary amines and reactive radicals results from irradiation with light of 350 nm wavelength causes. The formation of free radicals can be illustrated as follows[101]:

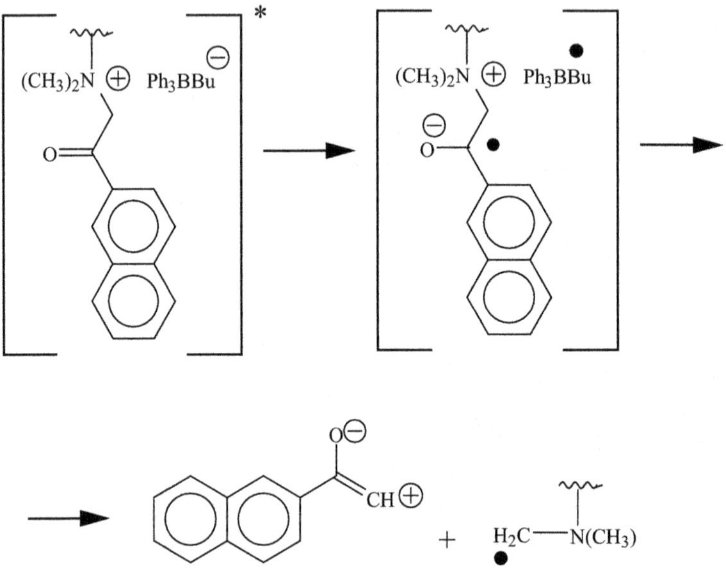

Preparation of polymers from carbazole-containing methacrylic monomers, 2-(N-carbazolyl)ethyl-methacrylate, 6-(N-carbazolyl)hexyl methacrylate, and ll-(N-carbazolyl)undecyl methacrylate was reported.[102] Upon irradiation, the poly(carbazolylethyl methaclylate) are claimed to show greater initiation ability than similar low molecular weight model compounds. This was ascribed to the photoinduced intramolecular charge-transfer interaction. Also, the initiation efficiency of the homopolymers is higher than that of copolymers with methyl methacrylate. This was, however, interpreted to be a result of singlet energy migration of the excited carbazolyl chromophores along the polymeric chains.[102]

There are also several reports in the literature on use of poly(arylsilanes) as initiators in photopolymerization of acrylic monomers.[102-105] The photoinitiating efficiency of silane polymers with thioxanthone side groups was found to depend not upon exiplex formation with tertiary amines but rather upon the micro heterogeneity of the distribution of the partners in the initiation process.[106]

In addition there is a report of a water soluble polymeric (copolymer) photoinitiators.[107] These initiators were synthesized by copolymerizing 2-acryloxy and 2-acrylamido anthraquinone monomers with three water-soluble comonomers: acrylamide, 2-acrylamido-2-methylpropane sulfonic acid and 2-acryloxyethyl trimethylammonium iodide. The polymerization activity correlates well with their measured photo reduction quantum yields in water using triethanolamine as the co synergist [107'] :

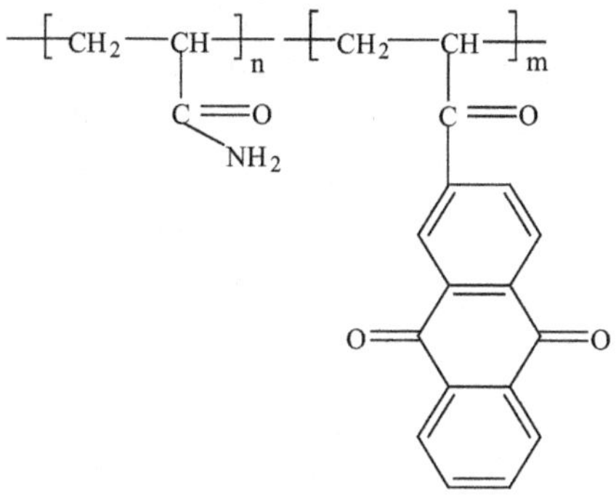

Jiang and Yin [39] reported that they prepared a dendridic macrophotoinitiator that contains thioxanthone and amine coinitiators. The material is a condensation product of 2-(2,3-epoxy)propoxyl thioxanthone and poly(propylene imine):

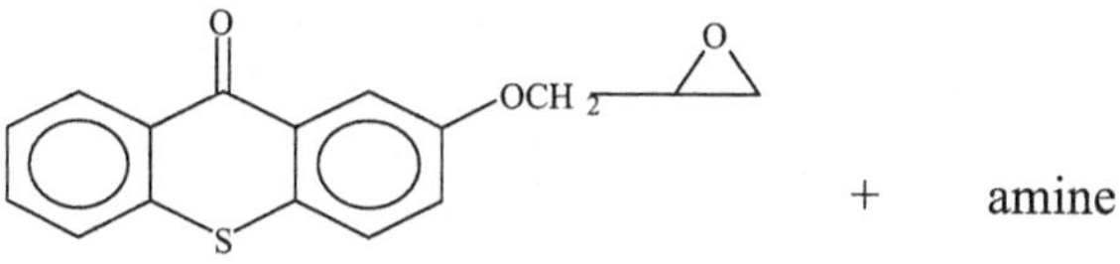

+   amine

It was not explained fully what are the advantages of a dendridic macrophotoinitiator.

2.2.1.7. Photoinitiators Designed for Use with Visible Light

There are many examples where ultraviolet light is undesirable for applications as, for instance, in preparation of photopolymer master printing plates or in photopolymerizations in biological systems, such as dental composites and bone cements. Also, laser processing requires photoinitiators that function in the near-visible, visible or even near infrared light. Syntheses of numerous photoinitiators were reported over the last ten to fifteen years that are capable of initiating polymerizations of acrylic and methacrylic monomers in the presence of visible light. Some of these materials were already mentioned in sections 2.2.5 and 2.2.6. In addition, in many instances, photoinitiators that are commonly used for cures with ultraviolet light were chemically modified with substituents to expand light absorption into the visible region. Among such materials can be found derivatives of thioxanthones,[108] conjugated ketones,[2] benzophenone peresters sensitized with either thiopyrilium salts,[109] or with substituted coumarins,[110] and triazineyl derivatives,[111] as well as coumarins.[112]

There are also compositions that function primarily in visible light. An example is diphenyliodonium salt or methylene blue dye in combination with triphenyl alkyl borate complexes. They were found to be effective with light in the visible spectrum.[113] In addition, titanocene derivatives will photolize directly upon exposure to visible light.[112,113] One such commercially available photoinitiator is Irgacure 784 :

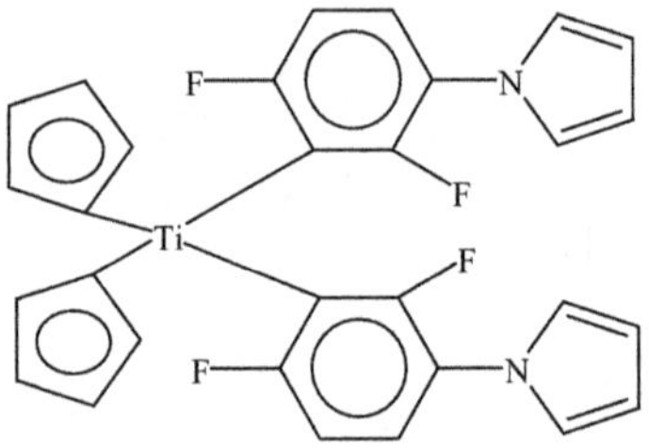

Upon irradiation , the titanocene compounds presumable break up into diradicals This was illustrated as follows [115]:

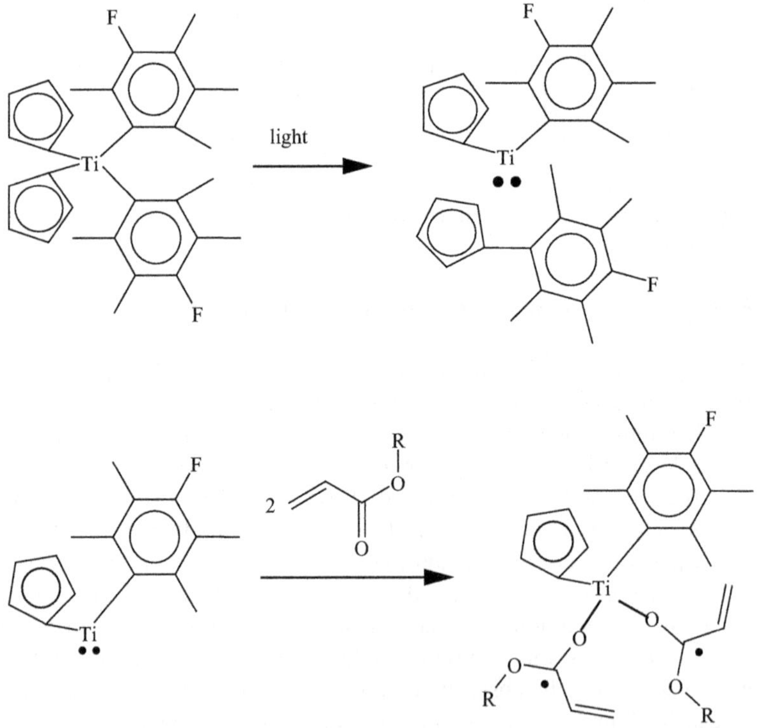

Addition of some amino acids or ketocoumarins is believed to increase the efficiency of initiation.[116]

Also, it was found that hexaarylbisimidazole will initiate polymerizations as a result of irradiation with visible light.[117] The same is true of bisacylphosphine oxde.[118] Other compounds are ketocoumarins that are efficient photoinitiators for acrylic and methacrylic monomers in the presences of amines, phenoxy acetic acid, and alkoxy pyridinium salts.[109,119] It was also shown that free-radical initiation is possible through visible light decomposition of ferrocenium salts in a three component composition, combined with either a hydroperoxide[120] or a epoxide, and a third ingredient, a dicyanobutadiene derivative[121]:

Various other metal salts were also reported as being capable of photo initiating free-radical polymerizations.[122-124] Thus, salts of boron compounds are often used in achieving cures of coatings and ink formulations that are based on acrylates and methacrylates.[125] The aryl alkyl borates that are paired with cationic chromophores can act as co-initiators in either UV or in visible light systems.[126,127] Among them, triphenylbutyliborates are preferred, due to their better stability, compared to substituted di-, trialkyl borates.[128-132] The mechanism of radical chain initiation by triphenylbutylborate salts involves photoinduced electron transfer from the borate anion to a chromophore in the excited state and probably formation of a triphenylbutylborate radical. The radical then undergoes rapid cleavage to triphenylboron and butyl radical. It is believed that the alkyl radicals are the primary initiating species in this reaction.[125]

Until recently, it was generally supposed that in order for the initiators to be useful in visible light, organoborate salts must contain at least one alkyl group linked directly to the boron.[125] Tetraarylborate salts by themselves were thought to be inefficient, because radicals that result from their decomposition are less stable and also more difficult to generate from oxidized borates . Also, in a model study with triphenylbutylborate coinitiators, no evidence of a free phenyl radical was found.[125] It was reported recently, however, that when the reduction potential of the chromophore is near -1.3 V, tetraphenylborate salts become just as efficient as are triphenylbutylborate salts. The following borate salts were reported as being useful in visible light[125]:

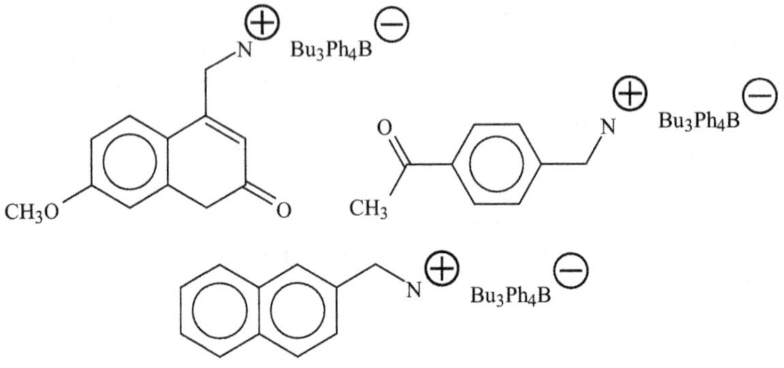

Another complex that was also found to be useful in visible light for free radical initiation is [133,134]:

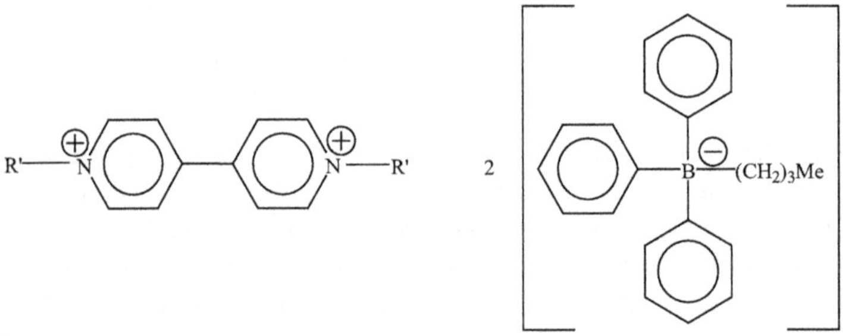

where, R^1 = lower alkyl, $PhCH_2$

Gosh and Mukherjee reported[135] that monochloroacetic, dichloroacetic and trichloroacetic acids readily initiate photopolymerization of methyl methacrylate at 40 °C in visible light when used in the presence of dimethylaniline. The inhibition period decreases with an increase in the number of chlorine substitution of acetic acid. Addition of benzophenone decreases the induction period . They believe that free-radical initiation is a result of a collapse of the exiplex (acid-dimethylaniline) and an electron transfer from the nitrogen to the carbonyl oxygen of the acid.[135]

It was also reported that halogenated xanthene dyes, such as Eosin or Rose Bengal, when combined with electron donors, such as iodonium salts or amines, act as efficient visible-light photoinitiators for the polymerization of acrylic monomers.[87] Such materials were already described in section 2.2.5. Mixtures of these dyes, electron donors, and monomers are active at the emission frequencies of commercial argon and helium-neon lasers and can be used even in the presence of oxygen. During the process the dye usually bleaches so that fading of the color can be related to the degree of polymerization. The accepted photochemical mechanism includes an electron-proton transfer from the donor to the excited dye which gives rise to active donor radicals that can actually initiate chain polymerizations.[87]

Fouaissier and Chesneau[136] reported that an eosine dye and ketocoumarin in the presence of an amine, acetone and iodonium salts initiate photocuring with light greater in wave length that 400 nm., The ketocoumarin (7-diethylamino-3,3'-oxydo-2H-chromen-2-one) was found by them to be a very useful additive to photo curable formulations, either in free-radical or in ionic curing processes.[136] The coumarin derivative used with onium salts can be illustrated as follows [137]:

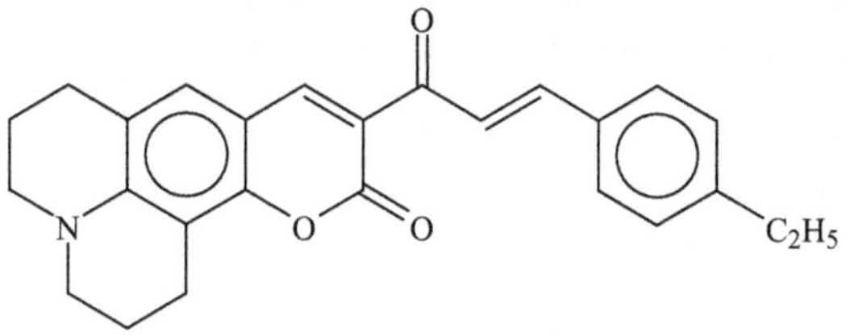

Many other dyes are also capable of initiating polymerizations of acrylic monomers in the presence of visible light.[138-144] Among these are anthraquinones, cyanine, merocyanine, xanthene, oxonol benzopyrans, thiazine and others. Such dyes are photo reduced by various reducing agents, like ascorbic acid, amines, diketones, and toluene sulfonic acid.[144] Photopolymer systems containing xanthene dyes show an increase in their polymerization rates when irradiated with visible light in the presence of aromatic carbonyl compounds, such as benzophenones, benzoins, or benzils.[144]

Also, the polymerization of 2-hydroxyethyl methacrylate in a methanol solution proceeds efficiently when safranine T is used in the presence of tertiary aliphatic amines[145]

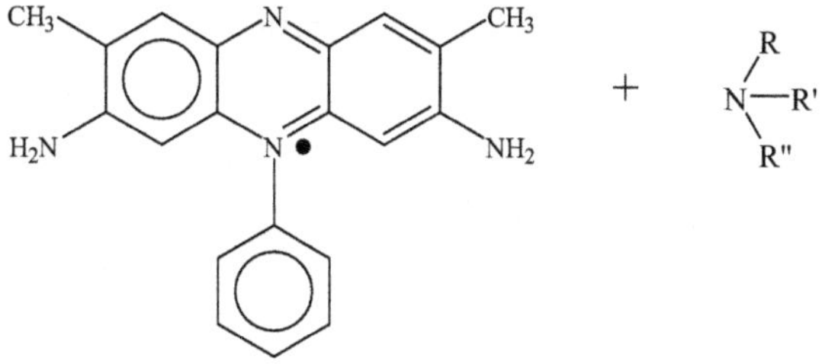

With this composition, the polymerization rates increase with amine concentration, but reach a maximum and then decrease with further increases in the amine concentration.[145] The excited singlet quenching involves a charge-transfer process with rate constants near the diffusion limit. All results indicate that the polymerization involves the interaction of the excited triplet wit amine, because quenching of the excited singlet inhibits the polymerization. [145]

A comparison was made of the activity of xanthene dyes as free-radical polymerization photoinitiators that included succinylfluorescein and some of its halogenated derivatives as well as an ester of Eosin with an *ortho*-benzoyl-α.-oxooxime group (shown below) under different experimental conditions.[146] The polymerization studies were carried out on 2-hydroxyethyl methacrylate with ethylene glycol dimethacrylate as cross linkers using visible light irradiation. N,N,- dimethylaniline was the co-initiator. The polymerization rate obtained with xanthene dyes was found to be similar to or as much as 4.5 times higher (in the case of the Eosin ester) than that reached with Eosin alone. [146]

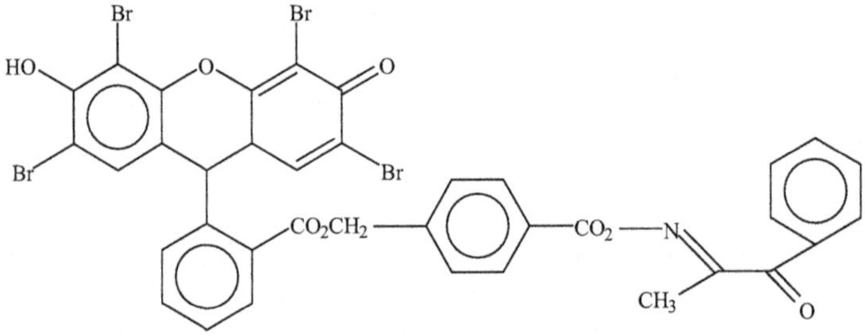

Eosin-benzoyloxyoxime compound

Photo bleaching photoinitiators are useful when it is desirable to touch up and cure, pigmented coatings on metal surfaces in environmentally friendly light.[147] It is important, of course, that such coatings not yellow during aging and that they can be applied to cold surfaces. Photoinitiators based on [bis(2,6-dimethoxybenzoyl)- 2,4,4-trimethylpentylphosphine oxide] and (5,7-diiodo-3-butoxy-6-fluorone) were found to fulfill these requirements.[147] The fluorene compound initiates curing by both free radical mechanism, when used with an amine, and also by a cationic one, when they are exposed to day light.

It was reported [148] that polymerizations employing 450 nm light and bis(2,6-dimethoxy-benzoyl)(2,4,4-trimethylpentyl)phosphine oxide initiator,

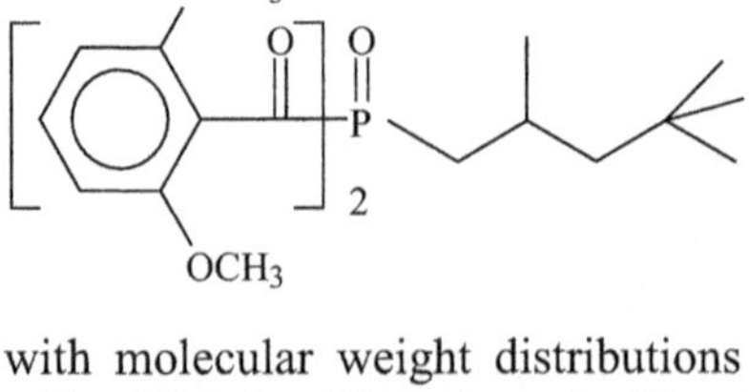

yield polymers with molecular weight distributions somewhat different from those obtained with UV light. There is a significant contribution from higher molecular weight overtones, particularly in styrene polymerizations This is attributed to a phenomenon of end group reinitiating that has the effect of producing greater amounts of high molecular weigh material. Either low radical fluxes or high extents of photoinitiator consumption are suspected to be the major factors in causing the results.[148]

Another photoinitiating system that is effective in visible light that is based on a combination of a squarylium dye[149] and diaryliodonium salt was reported[150] :

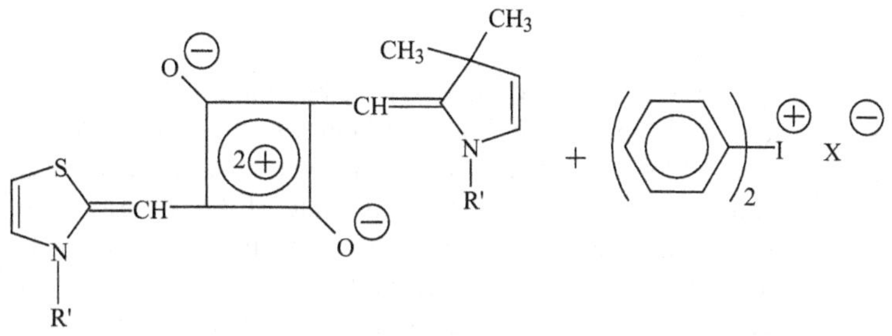

This system is claimed to possess good thermal stability and high photochemical driving force. Upon irradiation, an electron transfer reaction between excited dye and iodonium salt leads to photo bleaching of the squarylium dye and generation of active radicals. The results obtained indicate that squarylium dye and iodonium salt systems will effectively initiate visible light photopolymerization and photocrosslinking of acrylic esters.[150]

The same is true of compositions based on 7-diethyl-amino-3-(2'-N-methylbenzimidazolyl)-coumarin dye and diphenyliodonium hexafluoro-phosphate that will initiate the polymerization of methyl methacrylate in visible light.[151] When exposed to the light, the coumarin dye/iodonium salt undergoes quick electron transfer from the coumarin to the iodonium salt and forms free radicals. The reaction between the two compounds is mainly through the excited singlet state of the coumarin dye. As a result, it shows little sensitivity to O_2. The fluorescence of the coumarin derivative is quenched efficiently by the iodonium salt according to the Stern-Volmer equation.[151] The quenching constant k_q can reach 10^{12} $m^{-1}s^{-1}$. A study of the influence of the concentration of coumarin on the polymerization rate of methyl methacrylate shows that coumarin free-radicals act mainly as chain terminators in the polymerization and have a stronger tendency to terminate the chain reaction.[151]

When a three component visible light initiating system consists of methylene blue, N-methyl-diethanolamine, and diphenyliodonium chloride[152]:

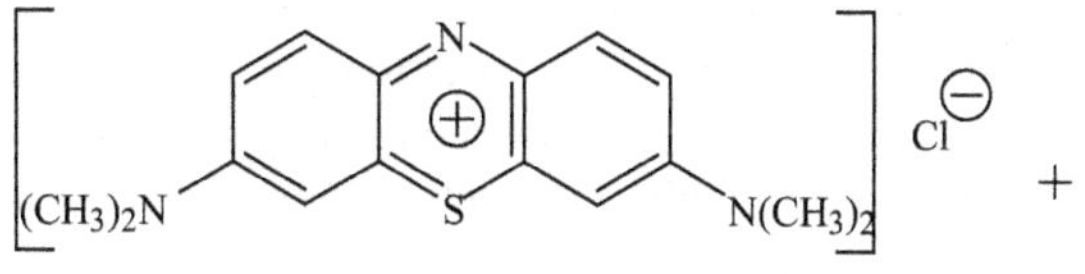

$$+ \qquad \underset{CH_2CH_2OH}{\overset{CH_3}{N-CH_2CH_2OH}} \qquad + \qquad \left(\left\langle \bigcirc \right\rangle \right)_2 \overset{\oplus}{I} \quad \overset{\ominus}{Cl}$$

initiates polymerization of 2-hydroxyethyl methacrylate, there is a significant increase in rate with increasing concentration of either the amine or the diphenyliodonium chloride. It was found, however, that while increasing the amine concentration dramatically increases the rate of dye fluorescence decay, increasing the diphenyliodonium chloride concentration, actually slows consumption of the dye.[152] This suggests that the primary photochemical reaction involves an electron transfer from the amine to the dye. It was concluded, therefore, that the iodonium salt reacts with the resulting dye-based radical (which is active only for termination) to regenerate the original dye and simultaneously produce the initiating phenyl radical derived from diphenylio-donium chloride.[152]

Padon, Kim, and Scranton[152]pesented experimental evidence that in the above tree-component-initiating system comprised of methylene blue, methyldiethanolamine, and a diphenyliodonium salt, the primary reaction involves an electron transfer/proton transfer from the amine to the dye. Interaction between the dye and the iodonium salt is precluded by charge repulsion. The iodonium salt is an electron acceptor and acts to re-oxidize the neutral dye radical back to its original state and allows it to re-enter the primary photochemical process. This reaction also generates initiating phenyl radicals. It appears, therefore, that the iodonium salt has a double role: it regenerates the methylene blue and replaces the inactive, terminating dye radical with an active, initiating radical. Based on the impact of the amine and iodonium concentrations on the fluorescence intensity and the duration of the oxygen-induced retardation period. a mechanism was proposed [152] that includes an oxygen-scavenging pathway in which the tertiary amine radicals formed in the primary photochemical process consume oxygen via a cyclic reaction mechanism.[152]

In addition, Padon et al., also reported[152] that a three-component system consisting of eosin, methyldiethanolamine, and diphenyliodonium salt appears to also produce active radical centers by an electron transfer/proton transfer reaction from the amine to the photoexcited dye. The eosin may be oxidized by the iodonium salt to produce an eosin radicals, which may in tern be reduced by an amine back to the original state. This reaction has a twofold effect: first, it regenerates the original eosin dye, and second, it produces an active amine radical in place of the presumably less active dye radical. Because of the difference in reaction rates for the pair wise reactions, this dark amine-mediated dye regeneration reaction may be the primary source of active radicals in the three-component system.[152]

The initiation mechanism of a similar three-component system, consisting of eosin, methyldiethanolamine, and diphenyliodonium chloride was also studied in the initiation of the polymerization of 2-hydroxyethyl methacrylate.[150] The fastest polymerization occurs when all three components were present The next fastest is a combination of the dye and the amine. The slowest is a combination of the dye with the iodonium salt.[153] In this case, it was also observed that the reaction between eosin and the iodonium salt bleaches the dye more rapidly than when the iodonium salt is included with eosin and the amine. Although a direct eosin and amine reaction can produce active radicals they are formed from the reaction with eosin in the original state. Simultaneously active initiating amine-based radicals are formed[153]

Grabner et al., [34] studied pyridine ketones initiators with laser flash photolysis. The materials were illustrated as follows:

The first compound, 2-hydroxy-2-methyl-1-pyridin-3-yl-propan-1-one was previously found to be water soluble and very reactive.[35] The activity of this compound was found to be comparable to its phenyl analog, 2-hydroxy-2-methyl-1phenyl-propane-1-one. All compounds showed absorption maxima about 410 to 430 nm.

.Neumann et al.[154], reported recently on the polymerization of sodium 4-styrenesulfonate in an aqueous solution photoinitiated with visible light in the presence of the dye safranine (λ_{max} = 532). They observed that the reaction proceeds in the presence of air, but that coinitiators like amines, inhibit the reaction: Also their observations suggest that the monomer tends to aggregate in clusters containing the dye molecules.[154]

An organometallic compound, azidopentaaminocobalt, was shown to be capable of initiating free-radical photopolymerization of N-Vinyl-2-pyrrolidone in the presence of either ultraviolet or visible light.[155] The reaction proceeds at room temperature in an argon atmoslphere.[155] The Co complex is photoreduced through a charge transfer process to the metal. This results in formation of cobaltous ions and azide radicals. The termination step mainly involves a mutual annihilation of growing radicals.[155]

Garcia and coworkers[156] reported a photoinitiating system that exhibits maximum sensitivity at 518 nm. It consists of a sensitizer dye, pyrromethene, and a radical generating reagent 3,3',4,4'-tetrakis(*tert*-butylperoxycarbonyl)-benzophenone. The photosensitization of the benzophenone through the excitation of dye induces high polymerization rates.[156]

There are not many photoinitiators that are designed to be used with infrared light. Some, however, do exist. One such photoinitiator was reported by Li *et al.*,[158] for infrared laser initiation of polymerization of bis(2-acryloyloxyethyl)phthalate. The photoinitiator is 1,3,3,1'.3',3',-hexamethyl-11chloro-10,12-propyletricarbocyanine butylltriphenylborate. [159]

Also, Belfield et al., [159] reported that the fluorene dye,

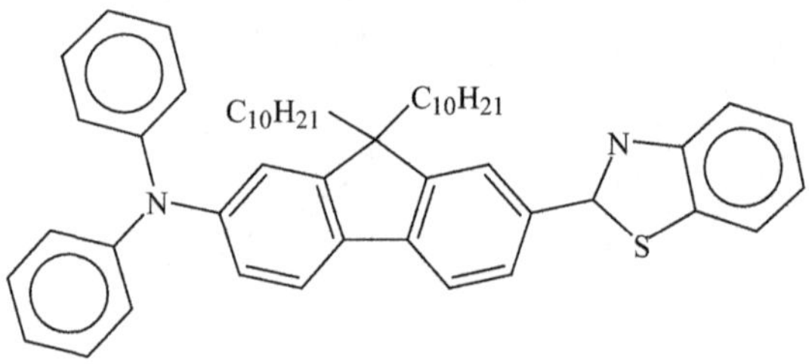

will initiate an acrylate polymerization by irradiation with light of 775 nm wavelength, probably through an electron transfer process.[159]

2.2.1.8. Miscellaneous Free-Radical Photoinitiators

Novikova *et al.*, [160] reported that some pentaaza-1,4-dienes can photo initiate polymerization. Thus, 1,5-Bis(4-methoxyphenyl)-3-methyl-1,4-pentazadiene, 1,5-diphenyl-3-(2-hydroxyethyl)-l,4-pentaza-diene, and 1,5-diphenyl-3-methyl-1,4-pentazadiene were used as photoinitiators of radical polymerization of hexanediol diacrylate and of methyl methacrylate. Photoinitiation abilities of these compounds were compared with those of a commercial photoinitiator, lrgacure 1700. The pentaazodienes showed a high initiation capacity. Also, the activation energy of the polymerization in the presence of the pentaazodiene compounds was lower than that for lrgacure 1700. [160]

Two patents were applied for by Baudin and Jung, for use of three compounds shown below, and similar ones, as photoinitiators that accumulate at the surface of the formulation and ensure surface cures:

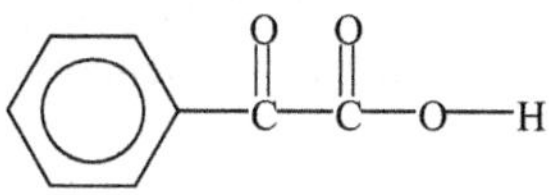

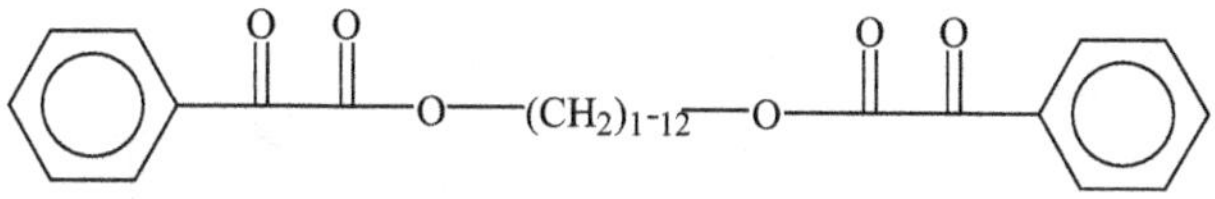

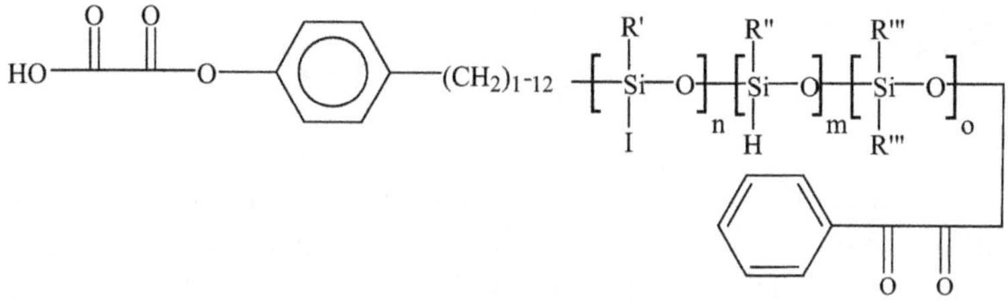

Also, free radical photoinitiation of polymerization of multifunctional acrylate monomers (trimethylolpropane triacrylate and phthalic diglycol diacrylate) was reported to take place by a cationic cyanine dye-borate complex 1,3,3,1',3',3'—hexamethyl-11-chloro-10,12-propylene tricarbocyanine triphenyl-butyl borate. The dye-borate complex was illustrated as follows[161]:

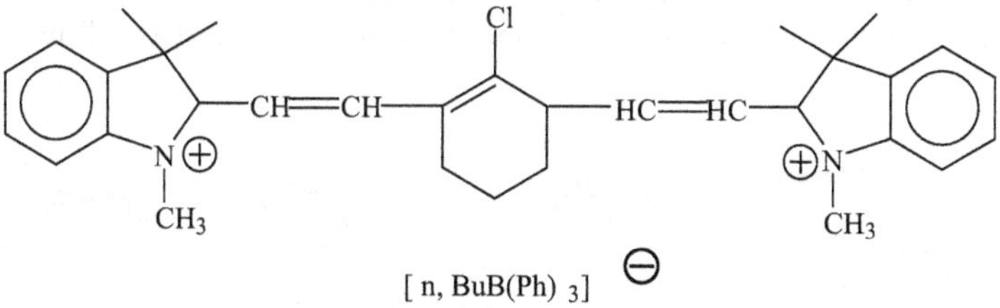

2.2.2. Photoinitiators for Ionic Polymerization

Most of the ionic photoinitiators were developed for use in cationic polymerizations due to practical considerations. These considerations are based on the fact that the anionic photoinitiators are more sensitive to oxygen inhibition than are even the free-radical ones. They are also sensitive to moisture. Some anionic photoinitiators, however, were described in the literature. None appear to be utilized commercially.

2.2.2.1 Anionic Photoinitiators

Kutal, Grutsch, and Yang[162] observed that solutions of ethyl α-cyanoacrylate containing *trans*-$Cr(NH_3)_2(NCS)_4^-$ undergo anionic photopolymerizations when irradiated with visible light. The active initiating species were identified by them as NCS⁻ . More recently, they reported that acyl substituted   ferrocenes,   particularly   benzoylferrocene   and   1,1'-dibenzoylferrocene:

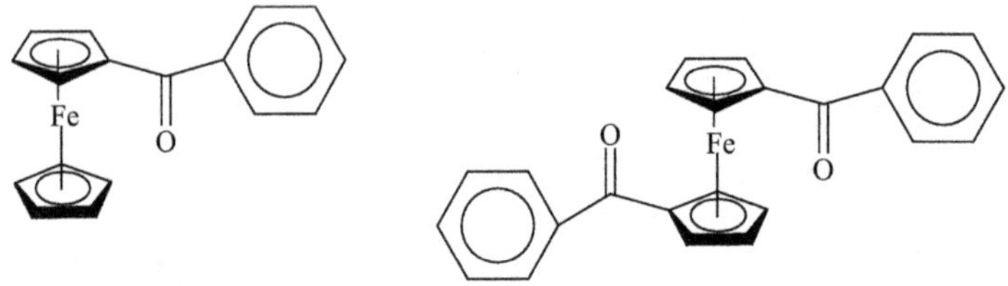

function as photoinitiators for polymerization of α-cyanoacrylate.[163]   These nonionic metallocene complexes absorb strongly light in the ultraviolet and visible regions due to electronic transitions of mixed ligand field/charge-transfer character. Irradiation of these complexes in methanol results in ring-metal cleavage to yield benzoyl-substituted cyclopentadienide carbanions and the corresponding half-sandwich iron^{++} complex. This process occurs with a quantum efficiency that is greater than 0.30 at 546 nm wavelength for 1,1'-dibenzoylferrocenes. By contrast, monobenzoylferrocenes are appreciably less photosensitive. When the solvent is neat ethyl α-cyanoacrylate, the photo generated carbanions initiate rapid anionic polymerization of this electrophilic monomer.[163] The mechanism of carbanion formation is pictured as follows:

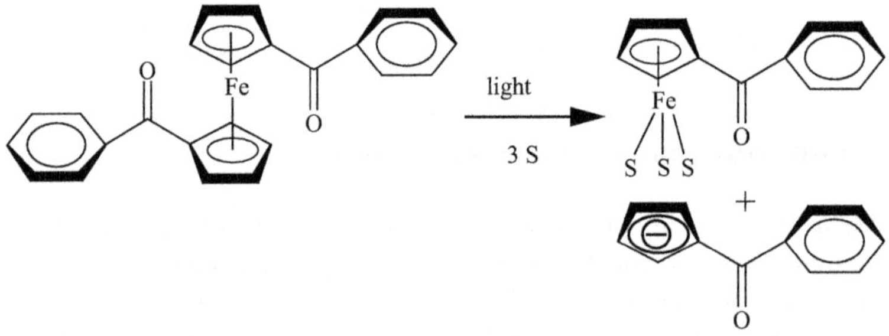

where S represents solvent.

In subsequent papers Kutal et al [164] reported that they found that ferrocene and ruthenocene (FeCp₂ and RuCp₂, where Cp is η^5-C₅H₅) complexes will also photoinitiate anionic polymerizations of α-cyanoacrylate. They suggest that the mechanism of initiation by ferrocene is an attack on the monomer and formation of a radical anion through electron transfer [164]:

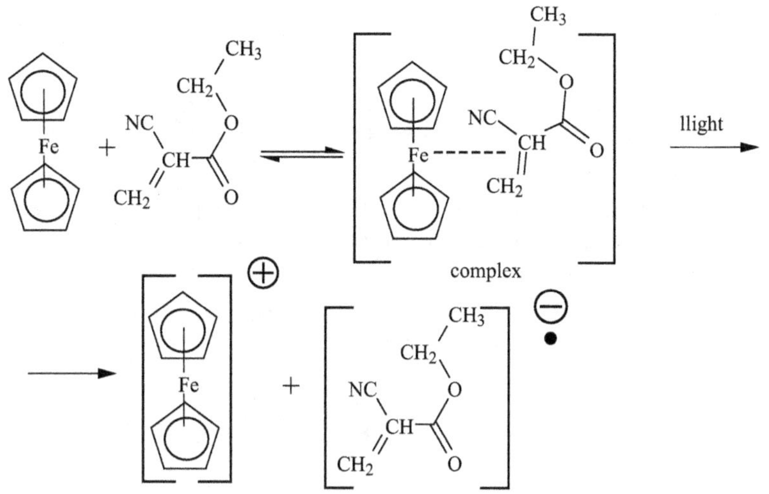

The compounds dissolve readily in the monomer. The electronic spectra of the resulting solutions display a near-UV absorption band that is a charge-transfer-to-solvent (metallocene -*- cyanoacrylate) transition band. Irradiation at this wavelength causes electron transfer or one-electron oxidation of the metallocene to the corresponding metallocenium cation, This is accompanied by a reduction of the ethyl cyanoacrylate to its radical anion and anionic polymerization of the electrophilic monomer. [165]

Jarikov and Neckers[166] carried out anionic polymerization of methyl-2-cyanoacrylate initiated by photoinduced heterolysis of dyes. Crystal Violet leuconitrile and Malachite Green leucohydroxide. The polymerization is accompanied by a color change due to heterolytic cleavage of the dye-leaving group in colorless nitriles. The reaction occurs simultaneously with generation of reactive nucleophiles, producing a colored polymer.

2.2.2.2. Cationic Photoinitiators

There are mainly three types of cationic photoinitiators. The first one consists of aryldiazonium salts. The second one consists primarily of onium salts. Prominent among them are iodonium, sulfonium and selenonium salts. The third type is based on organometallic complexes.

2.2.2.2.1. Diazonium Salts

The photodecomposition of aryldiazonium salts can be illustrated as follows:

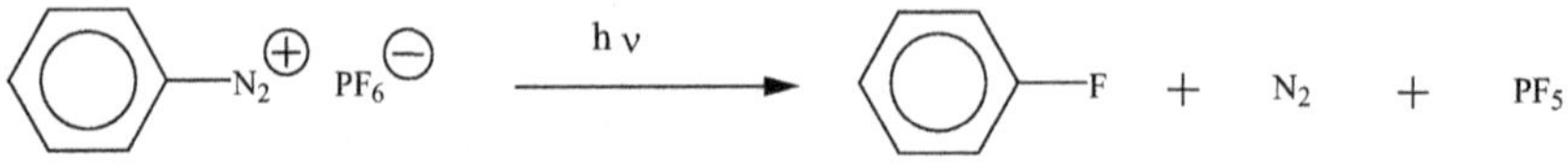

When no hydrogen donors are present the reaction products of the above decomposition initiate cationic polymerizations as follows:

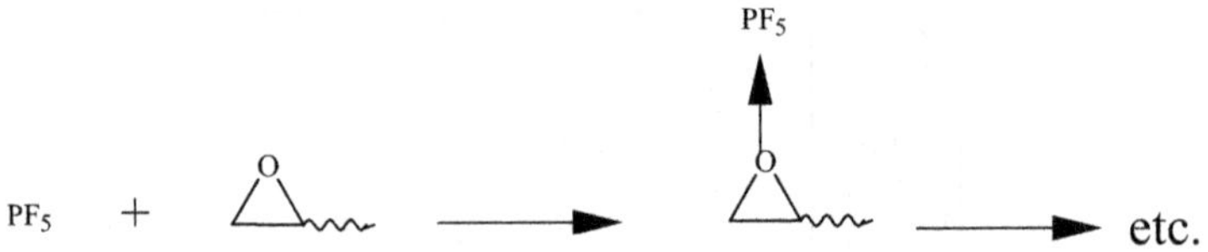

If hydrogen donors are present, however, Lewis and protonic acids form:

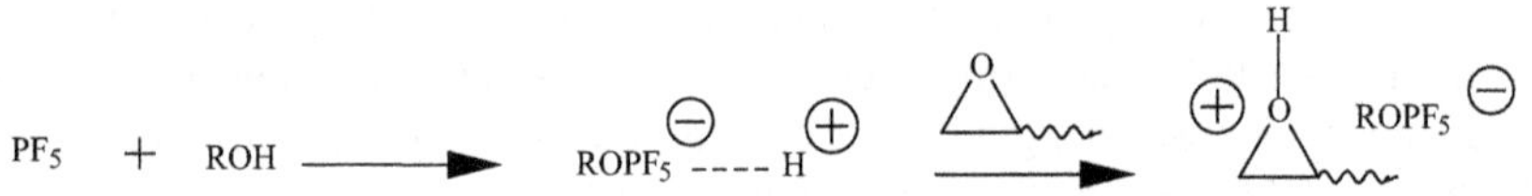

Anions that are low in nucleophilicity, such as SbF_6^-, AsF_6^-, and PF_6^-, are preferred. The overall efficiency of the system depends to a great extent upon the nature of the anion. It is also affected by the substitutions on the phenyl ring of the diazonium salt. The advantage of the systems is low termination rates and ability to obtain cures at room temperature. Also, considerable amount of post curing accompanies these reactions. There are some distinct disadvantages, however, to using diazonium salts in photopolymerizations of epoxides. The formulations tend to gel, unless stored in the dark. In addition, the films also tend to be colored, further yellowing upon aging. Also, use of these initiators causes nitrogen evolution and formation of bubbles or pinholes in thicker films.[158]

2.2.2.2.2. Onium Salts

Currently, the most commonly used photoinitiators for photoinduced cationic polymerizations are onium salts, particularly sulfonium salts, with the general structures shown below. In these illustrations MtX,,-represents a nucleophilic counter ion:

$$Ar_3\text{-}S^+ \!-\!\!-\! MtX_n^-$$

These photoinitiators give high quantum yields on photolysis and are efficient initiators of cationic polymerizations when the irradiation is carried with light in the short UV wavelength region (230-300 nm). Among the most useful cationic photoinitiators are diaryliodonium and triarylsulfonium salts that are used widely for photoinduced cationic crosslinking reactions. Photoinitiators based on onium salt are used in a multitude of practical applications in coatings, adhesives, printing inks, release coatings, stereo lithography, hologram recording, photo curable composites, and microelectronic photoresists.[167]

The general mechanisms of photoinitiation of cationic polymerizations by iodonium, sulfonium, and selenonium salts are similar. These salts can be illustrated as follows.[167, 168]

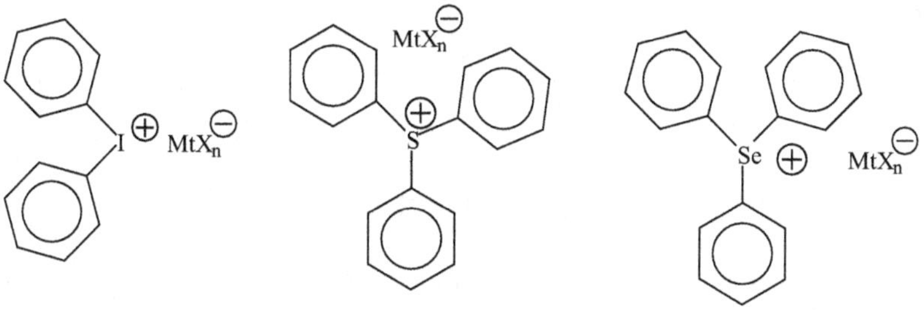

iodonium salt   sulfonium salt   selenonium salt

The photodecomposition reaction of the iodonium salt (in a simplified form) can be illustrated as follows:

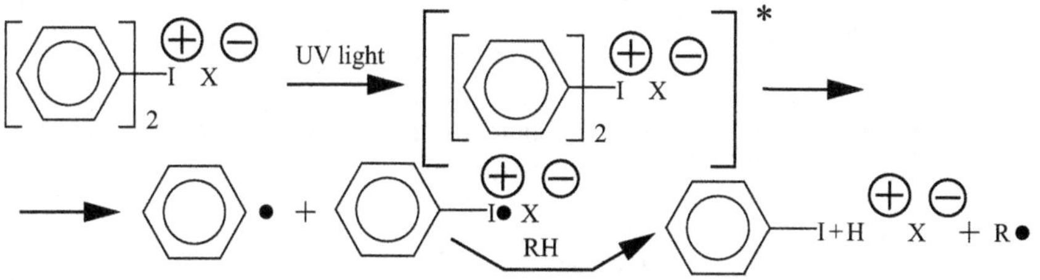

where H^+X^- is the initiating species. These systems generate Bronsted acids of the type $HAsF_6$, HPF_6, and $HSbF_6$. It was demonstrated recently,[169] however, that the mechanism of direct photolysis of iodonium salts is more complex than originally thought and illustrated above. It involves heterolytic as well as homolytic reactions, and the effect of the solvent cage is considered to be important. The overall mechanism for photolysis of diphenyliodonium salts was described by Dektar and Hacker.[170] Homolysis leads to formation of aryl radicals and aryliodonium radical cations, which in the solvent cage recombine to form iodobiaryls. When diffusing outside the solvent cage, the aryl radicals abstracts hydrogen atoms from the solvent to form benzene and other products. The various reaction byproducts were reported to be as follows [169]:

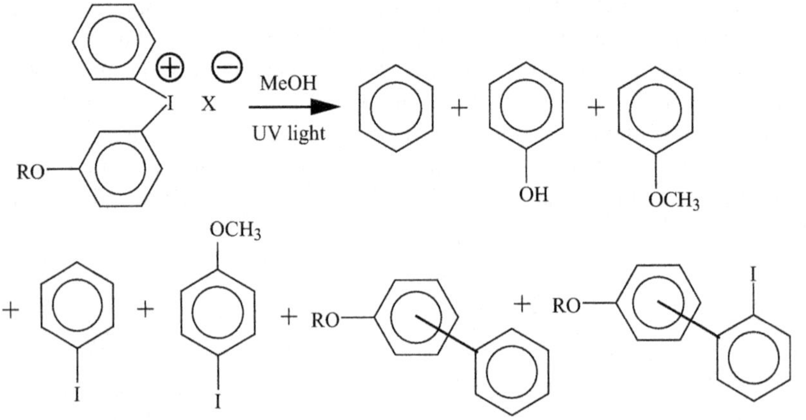

$X = Cl^-; SbF_6^-; B(C_6F_5)_4^-; \quad R = CH_2CH_2CH_2SiMe_3$

The decomposition mechanism of the triaryl sulfonium salt is believed to take place somewhat similarly [171]:

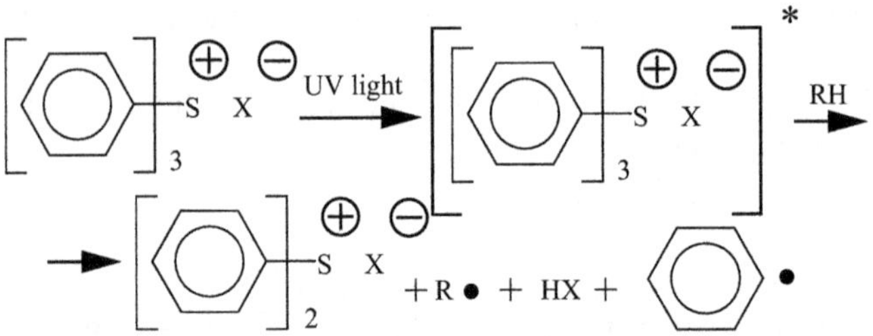

A somewhat different mechanism is visualized for the photo-decompositions of dialkyl phenacylsulfonium salt

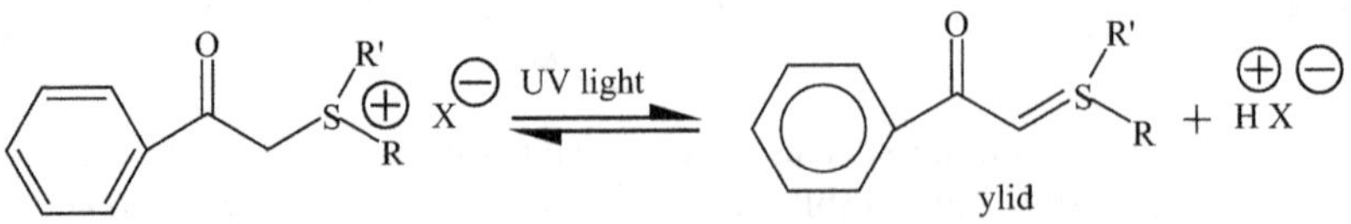

By proper selection of the alkyl groups (R',R) the solubility of this initiator may be tailored to a specific monomer systems.[172,173] The photolysis of this compound is believed to result in a Norrish II type cleavage.[172] The subsequent steps, as proposed by Crivello et al.[172] are as follows: First there is a triplet carbonyl abstraction of hydrogen atoms attached to one of the carbons adjacent to the positively charged sulfur atom. This takes place by way of an intermediate six-membered ring transition state. This is followed by formation of a diradical that is in resonance equilibrium with the sulfur cation-radical An internal electron transfer then takes place within the molecule that results in protonated species. Deprotonation yields a sulfur ylide and a protonic acid. A subsequent protonic shift converts the initially formed ylide to a more stable one[172]:

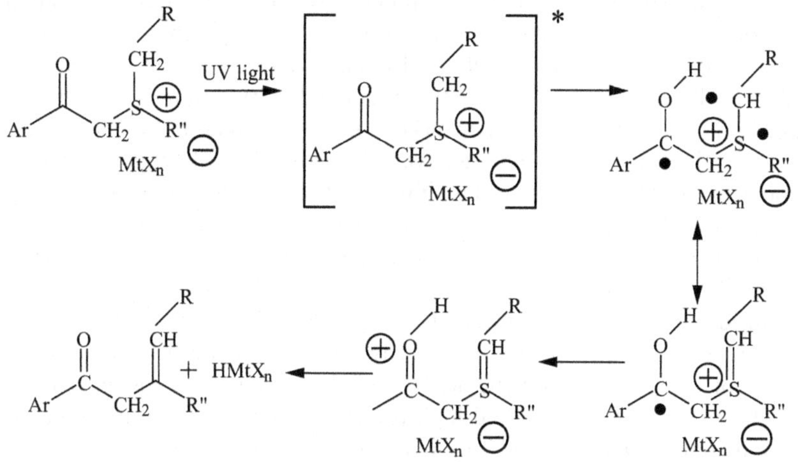

Kinetic studies show that dialkyl phenacylsulfonium salts compare favorably in their initiating efficiency to diaryliodonium and triarylsulfonium salts in the polymerization of epoxides.[172] An induction period was reported in polymerizations of vinyl and 1-propenyl ethers due to termination reactions with the photo generated ylides.[172]

The cure rates by triarylsulfonium salts are also influenced by the size of the anions. Large anions give faster cures. The rates of initiation fall in the following order:

$$SbF_6^- > PF_6^- > AsF_6^- > BF_4^-$$

The rates are also influenced by the temperatures. There may, however, be an optimum. The mechanisms of photoinitiations by triaryl selenonium salts are believed to be generally similar to those by triarylsulfonium salts.[168]

Acceleration of cure rates was reported by Crivello et al.,[175] when simultaneous photo initiated cationic polymerizations of epoxides and vinyl ethers are carried out in the presence of diaryliodonium salts. This, however, results in accelerations of ring-openings in epoxide polymerizations but deceleration of vinyl ether polymerizations. The effects are seen both, in mixtures of the two monofunctional monomers as well as in hybrid monomers that bear both functional groups simultaneously, vinyl ether and epirane groups on the same molecule. The combination of the two mechanisms were proposed to account for these effects. The reversible conversion of alkoxycarbon cation to oxiranium ions results in a two-stage reaction in which first the epoxide, and then the vinyl ether polymerizations take place. Free radical chain induced decompositions of the diaryliodonium salts produce large incremental numbers of carbon cationic species and subsequent acceleration.

In addition to iodonium, sulfonium and selenonium compounds, onium salts of bromine, chlorine, arsenic, and phosphorus are also stable and can act as sources of cation radicals as well as Bronsted acids, when irradiated with light.[173, 174] Performance of diaryl chloronium and diaryl bromonium salts was studied by Nickers and Abu.[176] Also, aryl ammonium and aryl phosphonium,[177, 178] and an alkyl aryl sulfonium salt were investigated.[179] It appears that the general behavior of these materials is similar to diphenyl iodonium and triphenyl sulfonium salts. These are formations of singlet and triplet states followed by cleavages of the carbon-onium atom bonds and in-cage and out of cage-escape reactivity.[180] The anions of choice appear to be boron tetrafluoride, phosphorus hexafluoride, arsenic hexafluoride, and antimony hexafluoride.

One of the difficulties with diphenyl iodonium and triphenyl sulfonium salts is poor light absorption in the near ultraviolet region, at wavelengths longer than 300 nm., where medium and high pressure mercury lamps emit much of their radiation.. One method of overcoming this limitation is to use photosensitizers to expand the spectral region where the onium salts are effective. Use of photosensitizers is a common strategy today to extend the performance of these photoinitiators. The broader spectral sensitivity provided by a photosensitizer permits the capture of a higher fraction of the available light emitted by commercial UV sources. Anthracene and its derivatives are useful for that purpose. It was also reported,[169] for instance, that anthracene is effective as a sensitizer for the onium tetrakis(pentafluorophenyl)borate initiators and that the initiation rate of vinyl ethers cationic photopolymerization increases. This reportedly is accompanied by a decrease in the free energy change for the excited singlet state of anthracene.[169] Jang and Crivello [182] also found that anthracene compounds with electron-donating substituents are very efficient sensitizers for the photopolymerization of vinyl ethers, epoxide, and oxetane compounds in the presence of onium salt initiators. The effects of viscosity on the rate of photosensitization of diaryliodonium salts by anthracene were correlated with the mechanisms of UV-initiated cationic polymerization.[182] As the viscosity increased, the rate of photosensitization decreased in a manner qualitatively described by the Smoluchowski-Stokes-Einstein equation for bimolecular elementary reactions.[182]

Many other electron-rich polynuclear aromatic compounds also act as efficient electron-transfer photosensitizers for both diaryliodonium and triarylsulfonium salt cationic photoinitiators. Examples are polynuclear aromatic compounds like pyrene, and perylene. The drawback, however, to using compounds like pyrene, and perylene is poor solubility in many polymerizable monomers. Also, they tend to volatilize from thin films during processing and are potentially carcinogenic.

A number of structural modifications were carried out to shift the absorption to longer wave lengths.[172] One example is formation of a 4-thiophenoxy derivative of triphenyl sulfonium salt, that acts as a chromophore.[172] Variations of that and other modifications are shown below:

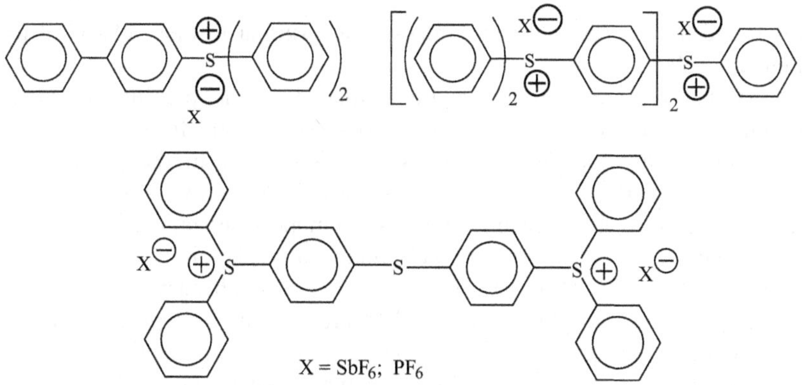

X = SbF$_6$; PF$_6$

This modification does extend the sensitivity to longer wave length ultraviolet light. The compounds were reported to display excellent thermal latency in the presence of various monomer systems and high efficiency as photoinitiators for cationic polymerization. Furthermore, their initiation efficiency was reported to be on par with current commercial triaryl sulfonium salts. [182]

Another approach, described in a Japanese patent,[183] is to attach a photosensitizer, a thioxanthone molecule to the sulfonium salt:

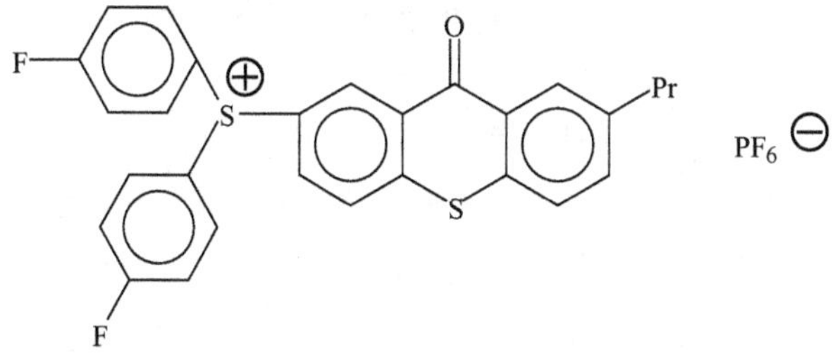

The patent states that photoinitiators composed of quinolinum salts (N-benzylquinolinium hexachloro-antimonate) and sulfonium salts like the one shown above show a maximum molecular extinction coefficient greater that 100 at 360 to 500 nm.[183]

Crivello, Burr and Fouassier reported preparation of an iodonium salt with a fluorenone chromophore as part of the molecule [184] :

It was found that the excited state processes are similar to those observed for fluorenone itself. [184]

When benzoin derivatives are irradiated in the presence of arene diazonium salts or aryl iodonium salts cationic species are formed. [185] These species initiate ring-opening reactions of glycidyl ether in the presence of ketones or aldehydes. The method, however, also forms dioxolane and formation of the dioxolane predominates over homopolymerization.[185]

Preparation of a series of sulfonium salts photoinitiators with the general structure $Ar'S^+CH_3(C_{12}H_{26})SbF_6^-$, where Ar' is phenacyl, 2-indanonyl, 4-methoxyphenacyl, 2-naphthoylmethyl, I-anthroylmethyl or 1-pyrenoylmethyl was reported.[186] Sulfonium salts that contain polycyclic aromatic structures, however, were found to be less effective as cationic photoinitors.[186]

Crivello et al., [187] described several methods for increasing the rates of initiations of cationic photopolymerizations through the use of synergistic reactions with free radical species. He pointed out that although aryl free radicals are also generated by the photolysis of onium salts this is usually ignored, since these species react further to form products that are of little consequence to the cationic polymerization. If, however, such primary free radicals react with monomers or other hydrogen donor compounds (RH) present in the reaction mixture, the secondary free radical species that form may react with onium salts by redox reactions to give cations. This can be shown as follows[187]:

$$Ar^{\ominus}_{\oplus} + RH \longrightarrow Ar H + R\bullet$$
$$Ar_2I^{\oplus}MtXn^{\ominus} + R\bullet \longrightarrow Ar_2I\bullet + R^{\oplus}MtXn^{\ominus}$$
$$R^{\oplus}MtXn^{\ominus} + nM \longrightarrow R(M)_{n-1}M^{\oplus}MtXn^{\ominus}$$

The mechanism shown above involves a three-step radical chain process where aryl radicals are converted to oxidizable radicals (R•) that interact with the onium salt. Subsequently, the cations generated by this process initiate polymerizations, while the reduced onium salts that undergo irreversible decomposition to regenerate aryl radicals. The overall consequence is an acceleration of the initiation process due to an increase in the number of initiating cationic species.[187] In addition, Sangermano and Crivello [187] reported that certain onium salt photoinitiators can be reduced by free radicals produced by the hydrogen abstraction reactions of photoexcited aromatic ketones. The result is efficient photoinitiation in the presence of longer wavelength UV and visible light. [187]

Based on the above, an initiating composition for cationic photopolymerization. with visible and long-wavelength UV light was described by Crivello et al.[188] The structure of the monomers plays a key role in these photosensitization processes. Useful aromatic ketones are camphoquinone, benzyl, 2-isopropylthioxanthone, or 2-ethylanthraquinone. The monomer-bound radicals reduce diaryliodonium salts or dialkyl phenacylsulfonium salts rapidly to form monomer-centered cations. These cations then initiate the polymerization of epoxides, vinyl ethers, and heterocyclic compounds. Onium salts with high reduction potential, however, such as triarylsulfonium salts, do not undergo this reaction.

There are also several mechanisms by which photosensitization of onium salts can take place. Of these, electron-transfer photosensitization appears to be the most efficient and generally is the applicable process. Here, absorption of light by the photosensitizers is followed by formation of *exiplexes* in the intermediate step. This is followed by reductions of the onium salts through electron-transfer and collapse of the *exiplexes*. Subsequent rapid decomposition of the resulting unstable diaryliodonium free radical prevents back electron-transfer. The overall process is irreversible. Generally, due to their lower reduction potentials, diaryliodonium salts are more easily photosensitized by an electron-transfer process than are triarylsulfonium salts. The mechanism is illustrated below as follows:

$$\text{Photosensitizer} + h\nu \longrightarrow [\text{Photosensitizer}]*$$

$$[\text{Photosensitizer}]* + \text{Ar}_2\text{I}^+\text{X}^- \longrightarrow [\text{Photosensitizer} \bullet\bullet\bullet \text{Ar}_2\text{I}^+\text{X}^-]*$$
$$\textit{exiplex}$$
$$[\text{Photosensitizer} \bullet\bullet\bullet \text{Ar}_2\text{I}^+\text{X}^-]* \longrightarrow [\text{Photosensitizer}]*\bullet \text{X}^- + \text{Ar}_2\text{I}\bullet$$

$$\text{Ar}_2\text{I}\bullet \longrightarrow \text{ArI} + \text{AR}\bullet$$

Chen et al.[189] reported efficient photosensitization of onium salts by various compounds containing a carbazole nucleus. Both diaryliodonium and triarylsulfonium salts are photosensitized by such compounds. Thus, the polymer of N-vinylcarbazole was found by them to be an excellent electron-transfer photosensitizer for various onium salts.[189] They also found that poly(9-vinylcarbazole) yields similar results.[189] Poly(2-vinyl carbazole) turned out to be the most efficient photosensitizer among various polymers with carbazole tested.[189] In addition, Chen et al.,[189] concluded that the redox photosensitization by the carbazole molecule or its N-alkylated derivatives occurs predominantly from the singlet excited states. On the other hand, the carbazole derivatives with carbonyl substituents sensitize onium salts via triplet excited states. This follows the Rehm-Weller equation of a photoinduced electron-transfer processes.[1] Hua and Crivello,[190] also found that they were able to enhance of the rates of onium salt photoinitiated cationic polymerizations of epoxides and vinyl ethers through the use of oligomeric and polymeric electron-transfer photosensitizers. Poly (N-vinylcarbazole) was found by them to be a particularly useful photosensitizer. This polymer can be illustrated as follows:

Hua and Crivello found that this polymer functions as an electron-transfer photosensitizer for a wide variety of onium salts, including diaryliodonium, triarylsulfonium, and di-alkylphenacylsulfonium salts. The broadening of the spectral response through the use of these photosensitizers accounts for the observed rate enhancement. In addition, alternating copolymers prepared by the free radical copolymerization of N-vinylcarbazole with vinyl monomers also exhibit excellent photosensitization activity.[190]

Cationic photopolymerization of epoxides and vinyl ethers by direct and sensitized photolysis of onium tetrakis(pentafuorophenyl)borates as photoinitiators was reviewed by Toba.[191] Most of the onium borates are sensitized by anthracene. The rates of polymerization increase with decreasing free energy changes between borates and the excited singlet state of anthracene. The acid generated by the photodecomposition of the onium borates produces nonacidic products upon heating.[191]

Suyama et al.,[192] reported that visible light can induce cationic polymerization of epoxides when sensitized by benzoquinonylsulfanyl derivatives. The cationic polymerizations are initiated by a photoacid generator that was sensitized by these quinones. The spectral analyses suggests that an initial step is a reduction of the quinone derivatives upon irradiation.

Chen and Yagci also reported that salts of allylic compounds will accelerate the initiation by onium salts of cationic polymerizations when irradiated with ultraviolet light.[193] They picture an electron transfer that is followed by a fragmentation mechanism. The intermediate radical cations, formed by reactions of photochemically generated radicals with the allylic salts, undergo fragmentations, producing different radical cations that are capable of initiating cationic polymerizations.[193] This is illustrated on an allyl thiophenium salts :

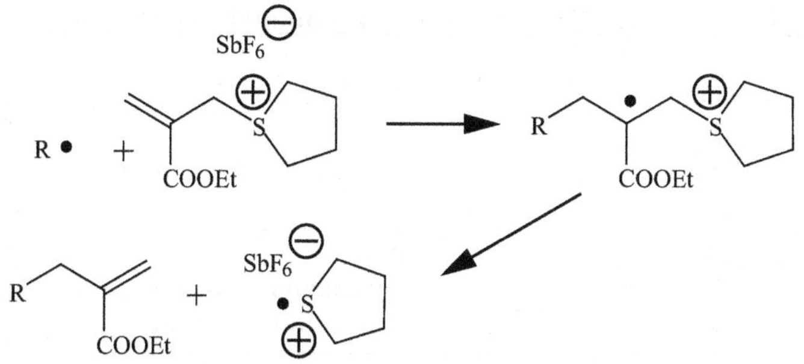

Compounds of this type include allyl sulfonium salts, allyl pyridonium salts, allyloxy pyridonium salts, allylphosphonium salts, etc.[193] The attractive features of this technique include the flexibility of the polymerization conditions that can be tuned to the desired wavelengths.

Crivello *et al.,*[194] described the syntheses of 5-arylthianthrenium salts that form efficient triarylsulfonium salt photoinitiators for cationic polymerization. Through the use of electron-transfer photosensitizers, the response of these photoinitiators can be spectrally broadened to the long wavelength ultraviolet-visible light regions of the spectrum. The results obtained suggest that. 5-arylthianthrenium salts are potential replacements for now available triarysulfonium salt photoinitiators in many applications. [194]

In addition, Crivello et al.,[195] observed marked acceleration of the onium-salt initiated photopolymerizations of epoxides and vinyl ether monomers in the presence of small quantities of phenothiazine or its derivatives, bearing a wide variety of functional groups. Particularly efficient photosensitizer was reported to be 10-allylphenothiazine-5,5'-dioxides. Moreover, the ability to readily change the wavelength of absorption of these compounds in a very predictable manner makes this class of compounds very valuable for photosensitization in the long wavelength UV-visible regions.[195] Phenothiazine molecule can be illustrated as follows:

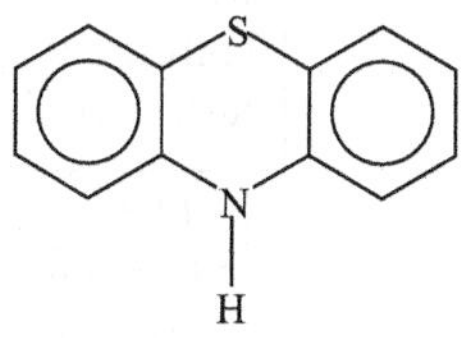

phenothiazine

Sun *et al.,*[196] reported a significant accelerating effect by free-radicals on the polymerization of cycloaliphatic epoxy siloxanes in the presence of diaryliodonium salts. The optimum ratio of a free-radical initiator to the iodonium salt being 1:2. [196] Some commercially available onium slats are listed in Table 2.10.

Table 2.10. Examples of Some Onium Salt Photoinitiators

Chemical Structure	Compound	λ max nm
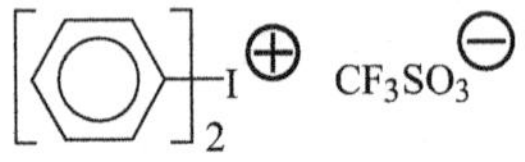	diphenyliodonium triflate	193
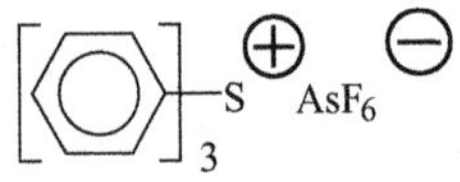	triphenylsulfonium hexafluoroarsenate	227
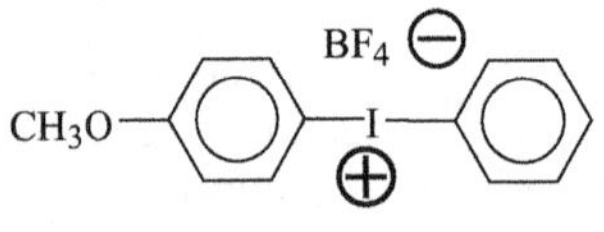	4-methoxydiphenyliodonium tetrafluoroborate	246
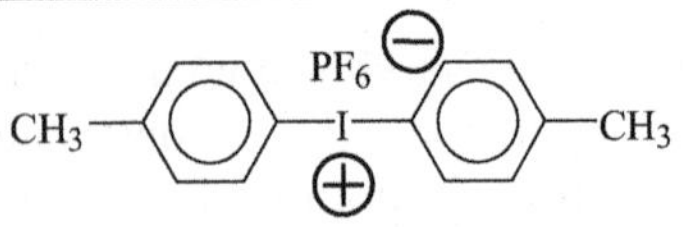	4,4'dimethyldiphenyl-iodonium hexfluorophosphate	237
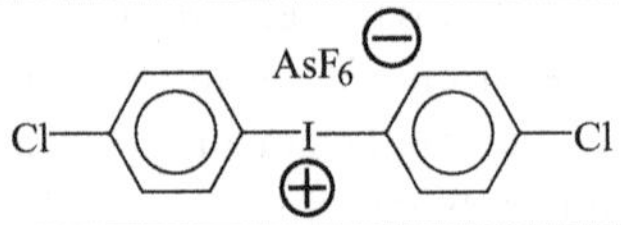	4.4'dichlorodiphenyl iodonium hexafluoroarsenate	240
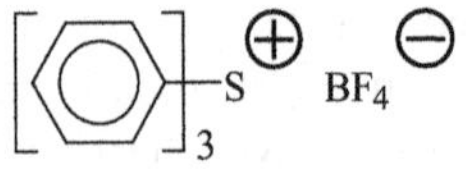	triphenylsulfonium tetrafluoroborate	230
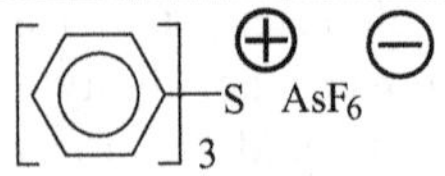	triphenylsulfonium hexafluoroarsenate	230
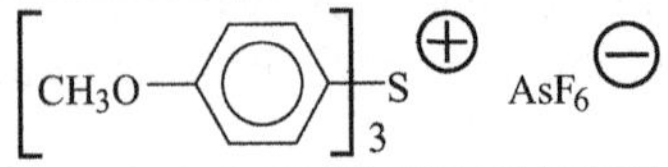	tri-*p*-methoxyphenyl-sulfonium hexafluoroarsenate	225 280
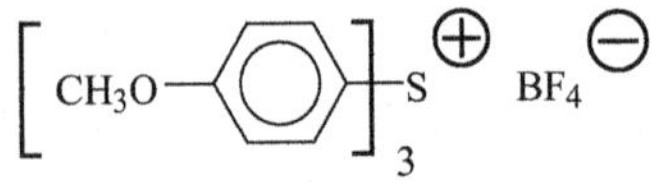	tri-p-methoxyphenyl-sulfonium tetrafluoroborate	243 278
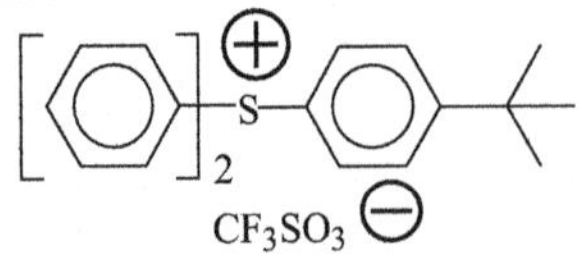	(4-*tert*-butylphenyl)diphenyl-sulfonium triflate	238
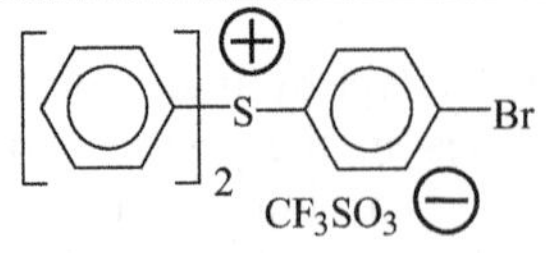	(4-Bromophenyl)-diphenylsulfonium triflate	243

Table 2.10. (Continued)

Chemical Structure	Compound	λ max nm
	(1-naphthyl)diphenyl-sulfonium triflate	299
	diphenyliodonium-decane ether hexafluoroantimonate	
	diphenylsulfonium-phenyldecane ether hexafluoroantimonate	
	dialkyl phenacyl sulfonium hexafluoroantimonate	
	triphenylselenonium tetrafluoroborate	238 266 275
	triphenylselenonium hexafluoroantimonate	257 275 266
	[4-*tert*-butoxycarbonyl-methoxy)-naphthyl diphenyl-sulfonium triflate	233

a from various sources, including commercial brochures

Development of new sulfonium salts was patented by Herlihy. [40,43] These materials are claimed to be useful as photoinitiators for surface coatings, printing inks and varnishes. They were illustrated in two ways as is shown on the next page. In the first illustration R_1 = O,CH_2, S,C=O,$(CH_2)_2$, NH, N-alkyl ; R_{4-7} = H; R_{8-11} = H, Hydroxy, or alkyl; or R_9 and R_{11} form a fused ring system with the benzene rinds to which they are attached. R_{12} that can be a direct bond or it can be an O or a CH_2.

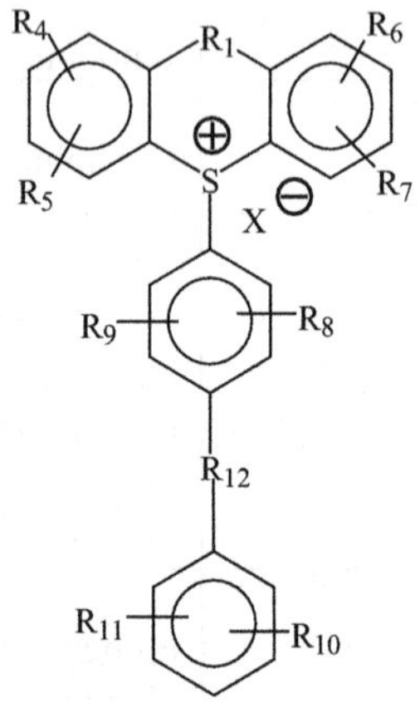

The second illustration also includes thioxanthone based compounds [40]:

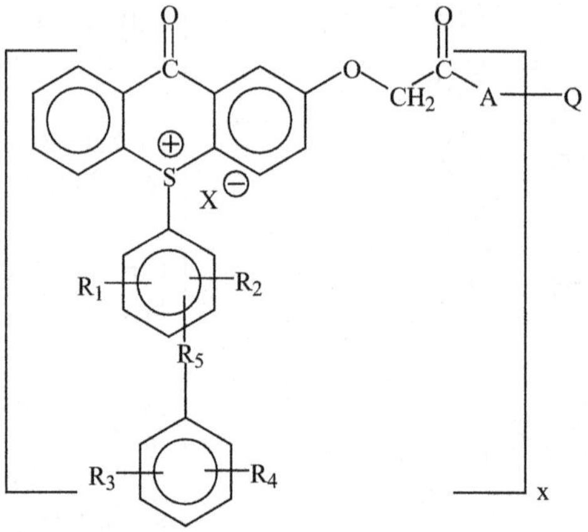

where, A is a direct bond or [O-(CH$_2$)$_2$CO], or(CH$_2$CH$_2$) and similar compounds and Q is a residue of polyhydroxy compounds; R$_{1-4}$ are H, hydroxy, or alkyl, and R$_1$ and $_R$3 are joined to form a fused ring system; R5 is an O, CH$_2$ or a direct bond.

Crivello, Jiang, and Ma [182] also described the preparation of a number of 5-arylthrenuiuim salts. The synthesis was shown as follows:

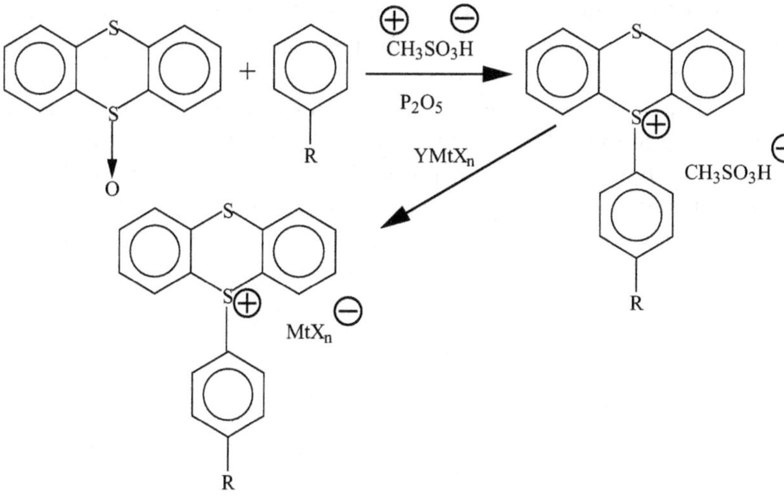

where R represents various substituents like OCH_3, or alkyl groups on the benzene ring, or a phenyl ether group. It was reported that 5-arylthianthreneiium salts exhibit photoinitiating activity that compares well with diaryliodonium and triarylsulfonium salts as cationic photoinitiators. In addition, the spectral response of these photoinitiators can be readily broadened into the longer wavelength UV light through use of electron-transfer photosensitizers. [182]

Norcini et al., [197] disclosed preparation and use of sulfonium salts I and II shown below. The compounds are 4,4'-bis(thianthrenium-9-yl)diphenyl ether dihexafluorophosphate and 9-[4-(2- hydroxyethoxy)-phenyl]thianthrenium hexafluorophosphate):

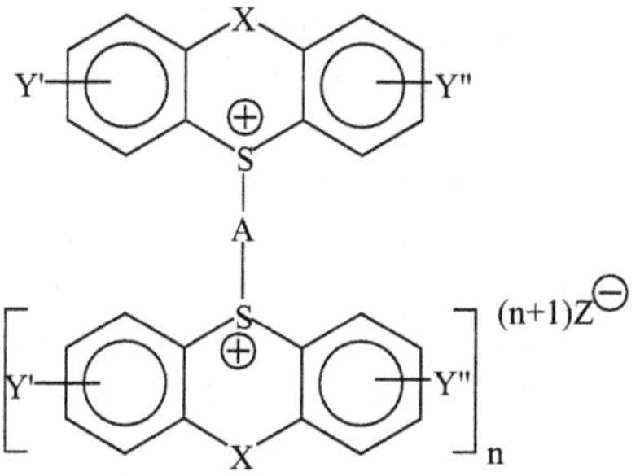

compound I 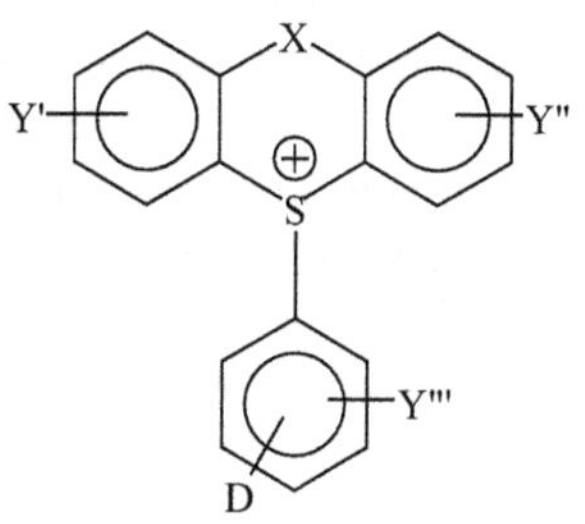   compound II

These sulfonium salts are claimed to exhibit good solubility in the formulations without the drawbacks of release of undesired compounds (e.g. benzene). This is claimed to be useful in food packaging. For compounds I and II, n = 1 or 2; X = S, O, CH_2, CO, single bond, N-R (R - H, or alkyl or aryl); Y' and Y'' = H, C_1-C_6 linear or branched alkyl, cycloalkyl, O-alkyl, hydroxy, halogen, S-alkyl, S-aryl; $Z^{\ominus}$ = MQ_p (M is B, P, As or Sb; Q is F, Cl, Br, I, or perfluorophenyl; p = 4—6). A is compound III, carrying two or three sulfonium salt units, while m = 1 or 2; R^1-R^9 are single bonded H, halogen atoms (F, Cl, Br, 1), or nitro moeties.[197]

compound III

Preparation of three allylic triphenylphosphonium salts, allyltriphenyl-phosphonium, methallyltriphenyl-phosphonium, and ethoxy carbonyl allyltriph-enylphosphonium salts with hexafluoroantimonate counter anions was reported.[198] The materials are capable of initiating cationic polymerization of oxiranes, such as cyclohexene oxide, vinyl ethers and N-vinylcarbazole. Initiation of polymerization of theses monomers by these salts requires free radicals that must be generated simultaneously and photochemically with light of λ_{max} > 350 nm.[198] The mode of action of ethoxy carbonyl allyltriphe-nylphosphonium salts in promotion of the cationic polymerization seems to be based on an addition-fragmentation mechanism. Accordingly, a free radical adds to the carbon-carbon double bond and fragmentation of the adduct radical results in the formation of a reactive onium radical cations.[198]

Sipani and Scranton [199] carried out comprehensive investigations of cationic photopolymerizations of phenylglycidyl ether initiated by diaryliodon-ium hexafluoroantimonate and also by (tolylcumyl) iodonium tetrakis(pentafl-uorophenyl) borate. The studies included dark-cure experiments in which the polymerizations were monitored in the dark after illumination for predetermined periods of time. The results show similar reaction kinetics for both salts. Different results, however, were obtained in the dark-cure experiments. Thus, the borate salts exhibit a higher polymerization rate at a given time, and higher limiting conversions (76%) than does the antimonate salt. [199]

Other onium salts described in the literature include tripenylpyrylium tetrakis(perfluorophenyl)gallate and antimonite, reported by Nickers *et al.*, [200] They are claimed to be efficient photoinitiators for the cationic polymerization of epoxy compounds.[200] The gallium salt can be illustrated as follows:

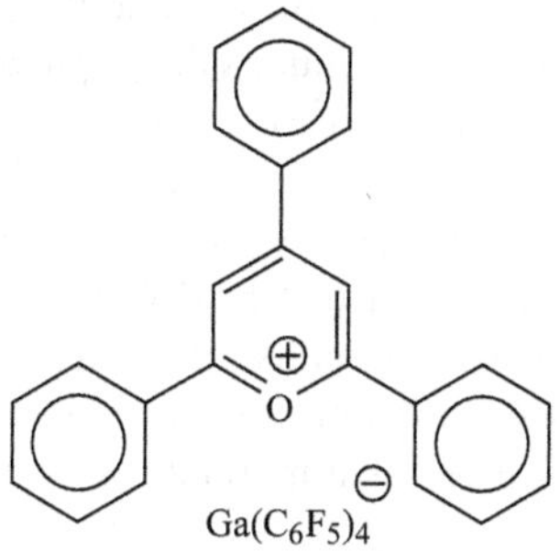

Triphenylpyrylium tetrakis(perfluorophenyl)gallate.

Also, a European patent was applied for that describes cationic photoinitiators based on 2,4,6-triarylpyrylium salts with nonnucleophilic anions and electron donors.[201] It is claimed that the photoinitiators rapidly photo crosslink mixtures of cycloaliphatic epoxides. Micro encapsulation of the salts in polystyrene was found to increase thermal and photo stability of these photoinitiators.[200] The encapsulation material can be dissolved in monomers when polymerization is desired.

Similarly, cyclopropenium salts were also reported to act as initiators for photopolymerization of glycidyl phenyl ether[200]:

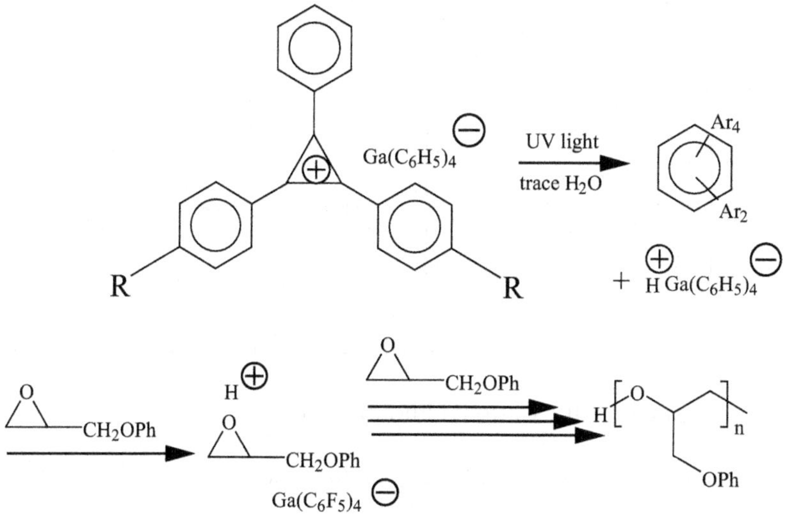

In a review artice,[202] Schnabel, points out that pyridinium-type salts containing an N-ethoxy group also belong to the family of onium salts and are useful as photoinitiators for the polymerization of oxiranes and vinyl ethers. The initiation is accomplished by direct or indirect (sensitized) photolysis of the salts at wavelengths of visible light. As a useful tool, addition free radical initiating species enlarges the versatility of pyridinium salts as photoinitiators, [202]

The photo-acid generation and the photosensitivity of three kinds of bis(trichloroethylene)triazine derivatives were reported.[203] From the measurements of absorption, fluorescence and fluorescence quenching spectra and acid generation it was concluded that acid generation has the nature of a photo-electronic transfer process. Results also indicate that the localized excitation and emission is most important for the photo-acid generation and that intramolecular charge transfer effect is important for extension of the absorption region. [203]

Endo and coworkers, [204] investigated nonsalt type latent initiators, like phenols. Phenols initiate polymerizations of epoxides with difficulty, even at elevated temperatures, probably due to insufficient acidity and relatively higher nucleophilicity of the dissociated phenoxide ion that acts as a polymerization termination. This, however, can be overcome by the substituent on the benzene ring. Hino and Endo report that some substituted phenols, like difluorophenol do act as nonsalt type photoinitiators. [204] They propose the following mechanism to explain the mechanism of the reaction:

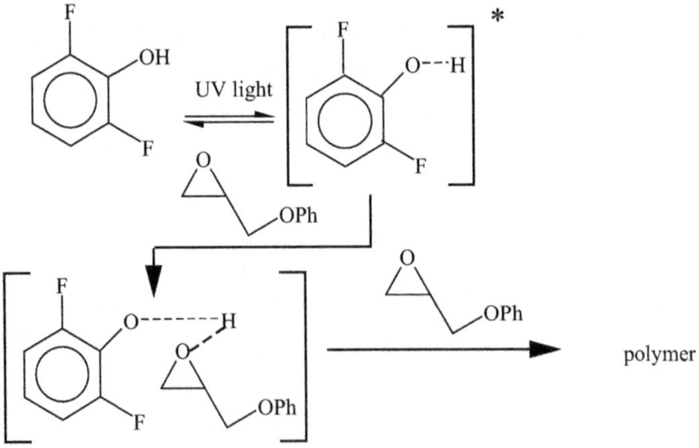

Suitable substituents are electron withdrawing CF_3 and F groups. [204]

2.2.2.2.3. Organometallic Cationic Photoinitiators

Iron-arene complexes, with a general formula as shown below:

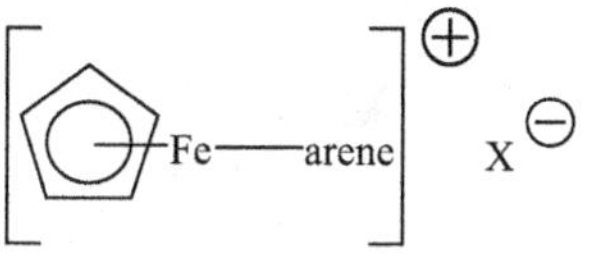

where, arene is benzene, toluene, naphthalene, or pyrene, and X = PF_6^-, SbF_6^-, AsF_6^-, BF_4^-, $CF_3SO_3^-$ are effective cationic photoinitiators for the polymerization of epoxy resins as well as aromatic dicyanate esters.[207] The data, based on time-resolved laser spectroscopy, suggests that [CpFe(η^4-arene)(η^2-ligand)]X, an intermediary, is reduced. Subsequently it undergoes relatively rapid displacement of arene by incoming nucleophilic ligand to form [CpFeL$_3$]X, where L is equal to an η^2 - ligand. [207]

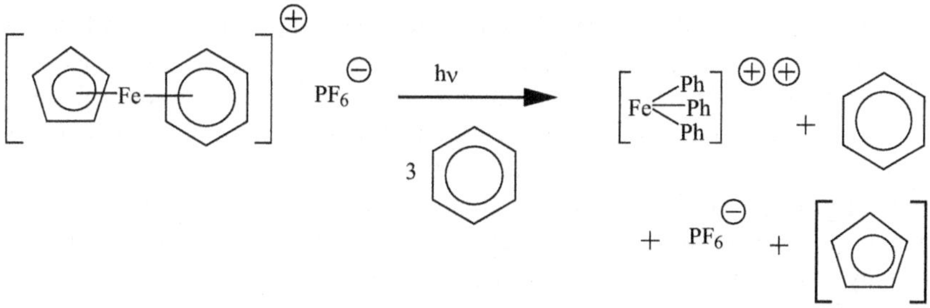

A photochemical reaction was suggested that actually results in formations of a number of different products. They can be shown as follows[207]:

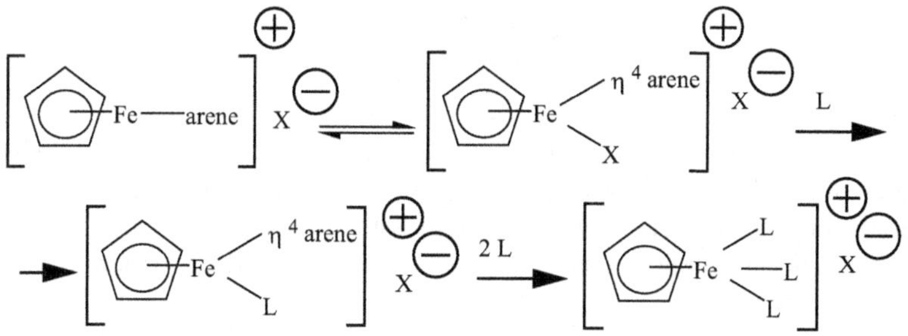

Phenylglycine derivatives were found to act as co-synergists for the iron-arene complexes when used in conjunction with dyes and amines.[204] It should be pointed out, however, that at this time the exact identity of the photo generated species that do the initiation is still uncertain. A commercially available photoinitiator of this type is Irgacure 261, (η^5-2,4-cyclopentadienyl)-1,2,3,4,5,6---[(1-methyl ethyl)-benzene]iron hexafluorophosphate[274] :

A comparison was made of the photo initiating capability of two photoinitiators, bis [4(diphenylsulfonium)-phenyl sulfide bis-hexafluoroanti-monate (Cyracure UV16974) and (lrgacure 261) in initiation of polymerization of 1,2-epoxy-6-(9-carbazolyl)-4-oxahexane.   Higher initiation rates were claimed for photopolymerizations initiated by Cyracure UV16974 than by lrgacure 261.[208]

Kutal *et al.*,[209] reviewed the chemistry of several iron (II) metallocenes that   are   effective   photoinitiators   for   ionic   polymerization   reactions. Photoexcitation of ferrocene and 1,1-dibenzoyl -ferrocenes in solutions of ethyl-α-cyanoacrylate produces anionic species that initiate the polymerization of electrophilic monomers. Irradiation of C_5H_5-Fe (η^6 - arene) in epoxide containing media generates several cationic species capable of initiating ring-opening polymerizations. It was concluded that iron(II) metallocenes exhibit a diversity of photoinitiation mechanisms. [209]

Wang *et al.*, [210]   prepared a ferrocene tetrafluoroborate photoinitiator for cationic polymerizations of epoxides:

They observed two main absorptions above 300 nm. These are λ_{max} = 355 nm and 620 nm. The second one is in the visible region. This shows that the material can well match commercial high pressure mercury arc lamps. They also report that optimized photosensitivity in epoxide is 41.8 mJ/cm^2 when irradiated by 365 nm. UV light.

Also, Wang, Huang and Shi, [211] studied the photosensitivity of an epoxide   system   with   several   different   arene-cyclopentadienyl   iron tetrafluoroborate salts. Among them, two compounds, described as, [C_5H_5Fe-2,5-(CH_3)$_2$-COMe-C_6H_3]BF$_4$ and [CH_3CO-C_5H_4Fe-1,4-(CH_3)$_2$-C_6H_5]BF$_4$, show two main absorption peaks at 370 and 450 nm. The salts are suitable for light curing reactions, when irradiated with high pressure mercury lamps. [211]

Preparation of a novel cationic photoinitiator [Cp-Fe-diphenylether]$^+$ PF$_6^-$ was reported. [212] It was claimed that polymerizations of epoxy resins and cycloaliphatic epoxides can be photo initiated by this compound effectively. Addition of benzoyl peroxide to this photoinitiator promotes the photopolymerization of epoxy oligomers. Post curing of the epoxy system was observed.

An aluminum containing photoinitiation catalyst was reported by Hayase and coworkers[213] It consists of the aluminum compound and o-nitrobenzyl triphenylsilyl ether. The o-nitrobenzyl triphenylsilyl ether photo decomposes to form triphenylsilanol:

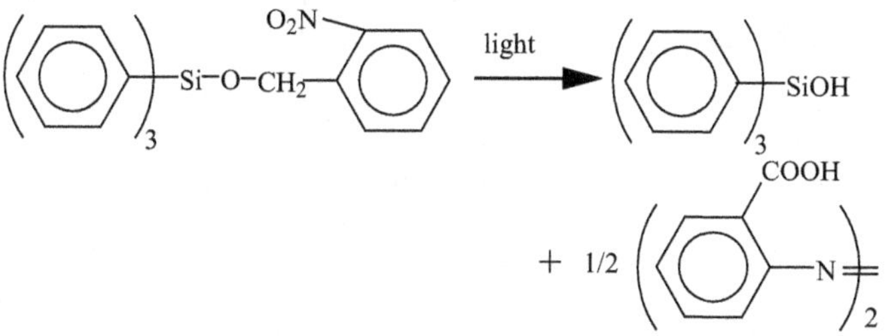

The formation of the dicarboxylic acid intermediate may be via a hemiacetal intermediate. In any event, it appears that an aluminum compound originally complexes with the silyl ether (not shown above) and releases it as a result of irradiation. This initiates the polymerization. The oxygen of SiOH behaves as a donor and coordinates with aluminum to form an AI-O-Si linkage that acts as an acceptor. Because of this coordination, the H of SiOH is polarized to form H$^{\oplus}$ that polymerizes epoxies.[213]

2.2.2.3. Photoinitiators Capable of Initiating Simultaneously Cationic and Free-Radical Polymerizations

An ultraviolet light photoinitiator, diphenyliodonium 9-acridinecarbo-xylate shows different absorption and fluorescence profiles and photochemical properties when irradiated with near-UV light. The anion absorbs the radiation and sensitizes the photolysis of the iodonium cation and formation of a cationic photoinitiator. At the same time, the free radicals thus formed initiated polymerization of vinyl monomers The structure of ion pairs influences the rate and efficiency of the intra-ion-pair electron transfer and the polymerization. [214]

A Japanese patent[215] reports phosphine oxide sulfonium borate salts as photoinitiators. The compounds show good storage stability and compatibility. The materials are illustrated below. Both boron salts shown in the illustration are claimed to be effective in initiating either free radical and cationic polymerizations:

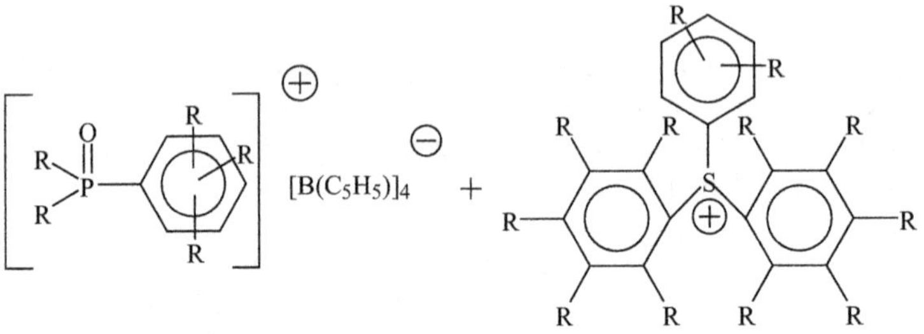

or

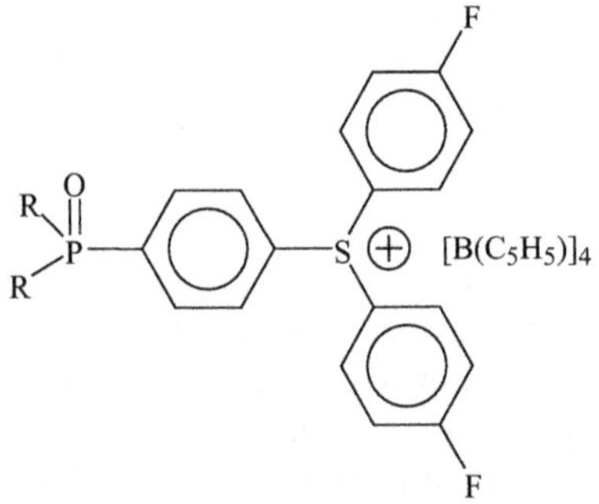

2.2.3. Photoinitiators for Two-Photon-Induced Photoinitiation

In response to increased demands for three-dimensional photolithography, three-dimensional spatial resolution of a two-photon absorption was adopted for three-dimensional microstructure formation and microlithography[216]. Unlike single photon absorption, whose probability is linearly proportional to the incident intensity, the two-phototon-induced process depends on both spatial and temporal overlap of the incident photons and takes on a quadratic (nonlinear) dependency on the incident irradiance. Because the probability of a two-photon initiation process is proportional to the square of the incident light intensity, photoexcitation is spatially confined to the focal volume. Using a dye, 5,7-diiodo-3-butoxy-6-floorene in the presence of an aryl amine and an acrylic monomer Belfield *et al.*,[217] initiated free-radical polymerizations by direct excitation and an electron transfer process using light of 773 nm. from a Ti-sapphire laser. This excitation wavelength is well beyond the linear absorption spectrum of this dye. Single-photon visible photoinitiators are used in combination with chromophores that possess high two dimensional cross sections. Belfield and Schafer also reported a two-photon-induced polymerization of a thiol-ene system using a near infrared light source and isopropylthioxanthone photoinitiator.[216]

Mortineau et al.,[219] developed a new initiator to be used with Nd-YAG micro laser for two-photon induced polymerizations. The material is described as a symmetrically conjugated ketone with terminal amino groups.

A patent was issued to Campston et al., [220] on chromophores for two-photon or higher-order absorbing optical materials for generation of reactive species for polymerization. These compounds are described as generally including a bridge of π-conjugated bonds connecting electron donating groups or electron accepting groups.

Preparation of two novel two-photon polymerization initiators, 10-ethyl-3-E-(4-(N,N'-di-n-butyl-amino)styryl)phenothiazine-5 and 10-ethyl-3,7-E,E-bis(4-(N,N'-di-butylamino)styryl)phenothiazine-6 was reported. [221] These initiators were found to exhibit good single-photon fluorescence. [221]Synthesis, structures, and properties of two additional two-photon photopolymerization initiators was reported. [222] The compounds are diphenyl-(4-{2-[4-(2-pyridin-4-ylvinyl)phenyl]vinyl}phenyl)amine and 9-(4-{2-[4-(2-pyridin-4-ylvinyl)-phenyl]vinyl}phenyl)-9H-carbazole. Both compounds were found to be good two-photon-absorbing chromophores and effective two-photon photopolymerization initiators.

References

1. J. Guillet, "Polymer Photophysics and Photochemistry", Cambridge University Press, Cambridge, 1985; R.W. Bush, A.D. Ketley, C.R. Morgan, and D.G. Whitt, *J. Radiation Curing*, **1980**, *7*(2), 20; A.R. Gutierrez, and R. J. Cox, *PolymerPhotochemistry*, **1986**, *7*, 517

2. L.A. Gatechair and A. Tiefenthaler, Chapter 3, *Radiation Curing of Polymeric Materials,* C.H. Hoyle and J.F. Kinstle, editors, ACS Symposium Series, #417, Washington D.C 1990; A. Hayachi, Y. Goto, and M. Nakayama, Japan, Kokai 0,129.337 (1989)

3. N.J. Turo, *Molecular Photochemistry*, W.J. Benjamin/Cummings, New York, 1978; b. C. Decker, T. Bendaikha, D. Decker, and K. Zahouily, *Am. Chem. Soc. Polymer Preprints*, **1997**, *38* (1), 487

4. G. Terrones and A.J. Pearlstein, *Macromolecules* **2001**, *34*, 3195-3204; *ibid.*, **2004**, *37*, 1565-1575; J. Hutchison and A. Ledwith, *Polymer*, **14**, (1973)

5. S.P. Pappas and A.J. Chattopadhyay, *J. Am. Chem. Soc.,* **1973**. *95*, 6484 H.G. Heine, *Tetrahedron Letters*, **1972**, *47*, 4755

6. H.G. Heine, H.J. Resenkranz, and H. Rudolph, *Angew. Chem., Int. Ed. Engl.* **11**, 974 (1974)

7. J. Huang, D. Feng, D. Zhu, and K. Geng, *Geofenzi Cailiao Yu Gongcheng*, **1994**, *10*, 23 (From Chem. Abstr.)

8. J.E. Christensen, A.F. Jacobine, and C.J.V. Scanio, *Radiation Curing,* **1981**, *8*, 1; .J.-P. Fouassier, *Photopolymerization, Photoinitiation, and*

Photocuring, Hanser/Gardner, Cincinnati, 1995; G. Berner, R. Kirchmayer, and G. Rist, *J. Oil Chem. Assoc.,* **1978**, *61*, 105

9.   S.P. Pappas and L.H. Carbolm, *J. Polymer Sci., Chem Ed.,* **1977**, *15*, 1381

10.   C.S. Colley, D.C. Grills, N.A. Besley, S. Jockusch, P. Matousek, A.W. Parker, M. Towrie, N.J. Turro, P.M.W. Gill, and M.W. George, *J. Am. Chem. Soc.* **2002**, *124*(50), 14952-14958; N.S. Allen, M. Edge. F. Catalina, T. Corrales, M. Blanco-Pina, and A. Green, *J. Photochem. Photobiol. Chem Ed.,* **1997**, *103*, 183; N.S. Allen, *Photochemistry*, **1997**, *38*, 381

11.   R. Kirchmayr, G. Berner, and G. Rist, *Farbe und Lack*, **1980**, *86* (3), 224

12.   G. Berner, J. Puglist, R. Kirchmayr, and G. Rist., *J. Rad. Curing*, **1979**, *61* (2), 2

13.   G. Berner, R. Kirchmayr, and G. Rist., *JOCCA*, **1978**, *61*,105

14.   A. Borer, R. Kirchmayr, and G. Rist, *Helv. Chim. Acta*, **1978**, *61*, 305

15.   D. Ruhlmann and J.P. Fouassier, *Eur. Polymer J.*, **1992**, *28*, 1063

16.   V. Lemee, D. Burget, J.P. Fouassier, H. Tomioka, *Eur. Polym. J.* **2000**, 36(6), 1221-1230

17.   W.A. Green and A.W. Timms, *Proc. Rad. Techn,* **1992**, 3

18.   X. Allonas, G. Grotzinger, J. Lalevee, J.P. Fouassier, and M. Visconti, *Eur. Polym. J.* **2001**, 37(5), 897

19.   R.M. Williams, I.V. Khudyakov, M.B. Purvis, B.J. Overton, N.J. Turro, *Phys Chem. B* **2000**, 104(44), 10437-10443; I.V. Khudyakov, N. Arsu, S. Jockusch, and N.J. Turro, *Designed Monomers and Polymers*, **2003**, *6*(1),

20.   N.S. Allen, N.G. Salle, M. Edge. T. Corrales, M. Shah, F. Catalina, and A. Green, *J. Photochem. Photobiol. Chem. Ed.,* **1997**, *103*, 185 ; N.S. Allen, M. Edge, F. Catalina. C. Peinado, A. Green, *Trends Photochem. Photobiol.* **1999**, *5,* 7;

21.   S.P. Pappas, *Radiation Curing*, **1981**, *8*, (3), 28

22.   G.A. Delzenne, U. Landon, and H. Peters, *Eur. Polymer J.*, **1970**, 933

23.   W. Rutsch, G. Berner, R. Kirchmayr, R. Husler, G. Rist, and N. Buhler, in *Organic Coatings – Science and Technology* (G.D. Parfitt and A.V. Patsis, *eds.*), Vol 8, Dekker, New york, 1986

24.   K. Dietliker, M. Rembold, G. Rist, W. Rutcsh, and F. Sitek, Rad-Cure Europe 87, Proceeding of the 3rd Conference Association of Finishing Processes SME, Dearborn, MI, 1987

25.   G. Rist, A. Borer, K. Dietliker, V. Desobry, J.P Fouassier, and D. Ruhlmenn, *Macromolecules*, **1992**, *25*, 4182

26.   V. Desobry, K. Dietliker, R. Husler, L. Misev, M. Rembold, G. Rist, and W. Rutsch, *Symp. Am. Chem. Soc., Ser., 274*

27.   D. Ruhlmann and J.P. Fouassier, *Europ. Polym. J.*, **1991**, *27*, 991; *ibid.*, **1992**, *28*, 287

28.   R. Liska and B. Seidl, *Am. Chem. Soc., Polymer Preprints*, **2004**, *45* (2), 31

29. J.E. Baxter, R.S. Davidson, H.J. Hageman, G.T.V. Hakoort, and T. Overeen, *Polymer*, **1988**, *29*, 1575

30. T. Majama, W. Weber, and W. Schnabel, *Makromol. Chem.*, **1991**, *192*, 2307

31. J.E. Baxter, R.S. Davidson, and H.J. Hageman, *Eur. Polymer J.*, **1988**, *24*, 419

32. C. Decker, D. Decker, and F. Morel, Chapt 6 in "Photopolymerization, Fundamentals and Applications"A.C.S. Symposium Series #673, American Chemical Society, Washington, D.C., 1996; C. Decker, *Polym. Intern.*, **1998**, *45,* 133; C. Decker, K. Zahouily, D. Decker, T. Nguyen, T. Viet, T. *Polymer* **2001,** 42(18), 7551

33. E. Beck, C. Kandzia, B. Pranti, M. Lokai, P. Enenkel, E. Keil, K. Menzel, Ger. Offen. DE 19,660,562, (1998)

34. R. Liska, B. Seidl, and G. Grabner, *Am. Chem. Soc., Polymer Preprints*, **2004,** *45* (2), 75

35. R. Liska and D. Herzog, *J. Polymer Sci., Part A, Polymer Chem.*, **2004**, *42*, 752-764

36. C.E. Hoyle, S. Jonsson, M. Shimose, J. Owens, P.Sundell, *Am. Chem. Soc. Symp. Ser.* **1997**, 673; T.B. Cavitt, B. Phillipps, C.L. Mguyen, C.E. Hoyle, S. Jonsson, and K. Viswanathan, A.C.S. Polymer Preprints, **2001**, *42*(2), 709; T.B. Cavitt, B. Phillips, C.K. Nguyen, C.E. Hoyle, S. Jonsson, K. Viswanathan, *A.C. S. Polymer Preprints*, **2001**, *42*(2), 709; C.W. Miller, S. Jonsson, C.E. Hoyle, C. Haraldsson, T. Haradlsson, and L. Shao, *RadTech '98 North America Conf. Proc.*, 182, **1998**; C.W. Miller, L. Viswanathan, S. Jonsson, C. Nason, W.-F. Kuang, D. Yang, B. Kess, C. Hoyle, *Am .Chem. Soc. Polymer Preprints*, **2001**, *42*(2), 811; S. Jonsson, V. KalyanaramanK. Lindgren, S. Swami, and L.-T. Ng, *Am. Chem. Soc. Polymer Preprints*, **2003**, *44*(1), 7; C. Nguyoen, C.E. Hoyle, T.B. Cavitt, and S. Jonsson, *Technical Conference Proceedings, UV and EB Technology Expo, and Conference, NC., US*, **2004**, May 5

37. C.W. Miller, C.K. Nguyen, K. Viswanathan, Johnson, T. Ah, M. Cole, C.E, Hoyl., L.Shao, D. Hill, W. Xia, S. Jonsson, *Am. Chem. Soc. Polym. Prepr.* **1999,** 40(2), 936; C. K. Nguyen, R. S. Smith,T. B. Cavitt, C. E. Hoyle, S. Jonsson, C. W. Miller, and S. P. Pappas, *A.C.S. Polymer Preprints* **2001***, 42(2),* 707

38. C.K. Ngyuen, R.S. Smith, T.B. Cavitt, C.E. Hoyle, S. Jonsson, C.W. Miller, and S.P. Pappas, *Am. Chem. Soc. Polymer Preprints*, **2001**, *42*(2), 707

39. X. Jiang and J. Yin. *Macromolecules*, **2004**, *37*, 7850

40. S.L. Herlihy, WO Patent 03 33,429 (April 2003)

41. A. Ledwith, *Pure Appl. Chem.*, **1977**, *49*, 431

42. 224. T.B. Cavitt, B. Phillips, C. Daniels, and C.E. Hoyle, *Proceeding of the International Waterborne, High-Solids, and Powder Coatings Symposium,*

2002, 29th, 431; T.B. Cavitt, B. Phillips, C.E. Hoyle, C.K. Nguyen, V. Kalyanaraman, S. Jonsson, *ACS Symposium Series*, **2003**, *847*, 41

43. S.L. Herlihy, PCT Int. Appl. WO 03 72,567 (Sept 4, 2003) and **WO** 03 72,568 (Sept. 4, 2003)

44. N.S. Allen, F. Catalina, B. Moghaddam, P.N. Green, W.A. Green, *Eur. Polymer J.*, **1986**, *22*, 691

45. J. Hutchison, M.C. Lamber, and A. Ledwith, *Polymer*, **1973**, *14*, 259

46. L. Cokbaglan, N. Arsu., Y. Yugei, S. Jockusch, and N.J. Turro, *Macromolecules*, **2003**, *36*, 2649

47. M. Aydin, N. Arsu, Y. Yagci, S. Jockusch, and N. J. Turro, *Macromolecules* **2005**, *38*, 4133—4138

48. C. Devadoss, R.W. Fessendem, and W. Richard, *J. Phys. Chem.*, **1991**, *95*(19), 7253

49. H. Block, A. Ledwith, and A.R. Taylor, *Polymer*, **1971**, *12*, 271

50. X. Feng, *Makromol. Chem., Makromol. Symp.*, **1992**, *63*, 1

51. *Radiation Curing Science and Technology,* S.P. Pappas (ed.), Plenum, New York, 1992

52. *Radiation Curing in Polymer Science and Technology*, J.P. Fouassier and J.F. Rabek (eds.), Elsevier, London, 1993; a. Corrales, T.; Peinado, C.; Catalina, F.; Neumann, M. G.; Alien, N. S.; Rufs, A. M.; Encinas, M. V. *Polyiner* **2000**, 41(26), 9103-9109

53. D.Ruhlmann and J.P. Fouassier, *Europ. Polymer J.*, **1991**, *27, 991; ibid., **1992**, *28*, 287, *ibid.*, **1992**, *28, 591; 254. X. Allonas, J. Lalevee, and J.P. Fouassier, *ACS Symposium Series* **2003**, *84*, 7 140—151;

54. N.S. Alien, N.G. Salleh, M. Edge, M. Shah, C. Ley, F. Morlet-Savary, J.P. Fouassier, F. Catalina. A.Green, S.I. Navaratnam, B.J. Parsons, *Polymer* **1999**, *40*(15), 4181; N.S. Allen, , N.G. Salleh, M. Edge, T. Corrales, M. Shah, F. Catalina, and A. Green, *J. Photochem. Photobiol. Chem. Ed.*, **1997**, *103*, 185; N.S. Allen, G. Pullen, M. Shah, M. Edge, I. Weddell, R. Swart, and F. Cataline, *Polymer*, **1995**, *24*, 4665

55. L. Angioline, C. Caretti, E. Carlini, J.P. Fouassier, and F. Morlet-Savary, *Polymer*, (1995). *36*, 4055; Y. Goto., E. Yamada, N. Nakayama, and K. Tokumara, *J. Polymer Sci.*, **1988**, *26*, 1671

56. N.S. Allen, F. Catalina, P.N. Green, W.A. Green, *Eur. Polymer J.*, **1986**, *22*, 793

57. K.S. Padon, D. Kirn, and A.B. Scranton, *Am.Chem.Soc. Polymner Preprints*, **2001**, *42*(2),705

58. N.S. Allen, F. Cataline, J. Luc-Gardette, W.A. Green, P.N. Green, W. Chen and K.O. Fatinikun, *Eur Polymer J.*, **1988**, *24*, 435 59

59. C. Valderas, S. Bertolotti, C.M. Previtali, and M.V.Encinas, *Journal of Polymer Science, Part A: Polymer Chemistry,* **2002**, *40*(16), 2888-2893

60. M. V. Encinas, A.M. Rufs, Bertolotti and C.M. Previtali, *Macromolecules* **2001,** *34*, 2845-284761.

61. T. Urano and E. Hino, J. *Photopolym. Sci. Technol.* **2000,** 13(1), 97-101
62. Bradley, G.; Davidson, R. S. *Reel. Trav. Chim. Pays-Bas* **1995,** *114,*528.
63. Paczkowski, J.; Kucybala, Z. *Macromolecules* **1995,** *28,* 269
64. X.X. Ren, J. Su, H.B. Yin, and J.Y. Feng, *J. Photoplym. Sci. and Technol.,* **1997,** *37,* 329
65. Lemee, V.; Burget, D.; Fouassier, J-P., *Trends Photochem. Photobiol.* **1999,** *5,* 63-77
66. R. Liska, G. Ullrich, D. Herzog, P. Burscher, and N. Moszner, *Am. Chem. Soc., Polymer Preprints,* **2004,** *45* (2), 69; R. Liska, *Journal of Polymer Science, Part A: Polymer Chemistry* **2004,** *42*(9), 2285-2301
67. S. Morlet, J.P. Fouassier, T. Matsumoto, K. Inomata, *Polymers for Adv. Technol.,* **1994,** *5,* 56
68. P. Gosh, P.K Sengupta, and N. Mukherjee, *J. Polymer Sci., Chem. Ed.,* **1979,** *17,* 2119
69. S. Jockusch, H.J. Timpe, H. Fischer, and W. Schnabel, *J. Photochem. Photobiol. A: Chem.,* **1992,** *63*
70. S. Shimizu, S. Fumya, and T. Urano, Japan. Kokai # 6,344,651 (1988)
71. M. Rodriguez, L.Garcia-Moreno, O.Garcia, R.J. Sastre, *Polymer* **2000,** 41(21), 7871-7875
72. G. Rist, A. Bohrer, K. Dietliker, V. Desobry, J.P. Fouassier, and D. Ruhlmann, *Macromolecules,* **1992,** *25,* 4182
73. K. Dietliker, M. Rambold, G. Rist, W. Rutsch, and F. Sitek, *Radcure Europe Technical Paper,* 3-37, **1987,** SME ed. Dearborn, MI
74. K. Dietliker, *Chemistry and Technology of UV and EB Formulation for Coatings, Ink and Paints,* Vol.III, *Photoinitiators for Free Radical and Cationic Polymerization,* SITA Rechnology Ltd., London, **1991**
75. M.V. Encinas, J. Garrido, and E.A. Lissi, *J. Polymer Sci., Chem. Ed.,* **1985,** *23,* 2481 (1985); M. Mitry, R. McCartney, M.R. Amin, *PCT Int. Appl. WO 00 06,613* (Cl. C08F2/46), 10 Feb 2000, US Appl. 94,742, 31 July 1998
76. N.S. Allen, F. Catalina, P.N. Green, and W.A. Green, *Eur. Polymer J.,* **1986,** *22,* 347; C. Liu, Y. Cui, C. Chen, Q. Gao, *Huaxue Yanjiu* **1999,** *10*(4), 23 Q. Gao, R. Li, G. Yang, X. Yu and F. Du, *Ganguang Kexue Yu Guang Huaxue* **2001,** 19(2), 116-121 (from Chem. Abstr.); S. Jonsson, D. Yang, K. Viswanathan, E. Shier, C.E. Hoyle, K. Belfielf and K. Lindgren, *A.C.S. Polymer Preprints* **2001,** *42(2),* 703; C.W. Miller, E.S. Jonsson, C.E. Hoyle, K. Viswanathan, and E.J. Valente, *J. Phys. Chem. B* **2001,** 105(14), 2707; S. Joensson, D. Yang, K. Viswanathan, E. Shier, K. Lindgren, K. Lindgren, and C.E. Hoyle, *ACS Symposium Series,* **2003,** *847,* 76
77. F. Catalina, P.N. Green, and W.A. Green, *Eur. Polymer J.,* **1986** 22, 871; A. Costela, L. Garcia-Moreno, J. Dabrio, R.J. Sastre, *Polym. Sa., Part A:*

Polym. Chem. **1997,** *35*(17), 3801; A. Costela, L. Garcia-Moreno, J. Dabrio, R.J. Sastre, *Recent Res. Dev. Phys. Chem.* **1999,** 3(Pt. 1), 111-131

78. C. Wang, Y. Wei, K. Simone, H. Gruber, and J. Wendrinsky, *A.C.S. Polymer Preprints,* **1997,** *38* (2), 217

79. R. Liska, *J. Polymer Sci., Part A, Polymer Chemistry*, **2002,** *40* (10), 1504

80. S.P. Pappas and R.A. Asmas, *J. Polymer Sci., Polymer Chem.,* **1982,** *20,* 2643

81. G. Le Bassi and F. Broggi, and A. Revelli, , *RadTech.* **1998,** *88*, 16082.

82. M. Koehler, and J. Ohngemach, *RadTech* , **1988.** 150

83. P.N. Green, *Polym. Paint Resins,* **1985,** *175*, 246

84. F. Cataline, R. Corrales, C. Peinado, N.S. Allen, W.A. Green, and A. Timms, *Eu. Polymer J.*, **1993,** *29*, 125

85. N.S. Allen, F. Catalina, C. Peinado, R. Sastre, J.L. Mateo, and P.N. Green, *Eur. Polymer J.*, **1987** *23*, 985 86.

86. V.B. Ivanov and E. Yu. Khavina, *Vysokomolekulyarnye Soedineniya, Seriya A i Seriya B*, **2002,** *44*(2), 2084

87. K.D. Ahn, K.J. Ihn, and I.C. Kwon, *J. Macromol. Sci. - Chem.,* **1968,** *A23* (3), 355

88. K. Padon, D. Kim, M. El-Maazawi, and A.B. Scranton, *ACS Symposium Series* **2003,** *847*, 15

89. A. Wrzyszczynski, M. Pietrzak, and J. Paczkowski, *Macromolecules* **2004,** *37,* 41-44

90. A. Tsuchida, N. Masuda, M. Yamamoto, and Y. Nishijima, *Macromolecules*, **1986,** *19*, 1299

91. Y. Hua and J.V. Crivello, *Macromolecules* **2001**, 34(8), 2488; Y. Hua and J.V. Crivello, *ACS Symposium Series*, **2003,** *847*, 21992. M. AbdAlla, *Makromol. Chem.* **1991,** *192*, 277; N. Angoli, C. Carlini, M. Tramontini, and A. Altomare, *Polymer,* **1990,** *31*, 212

93. W.Y. Chiang and S.C. Chan, U.S. Patent # 4,935,535 (1989)

94. N.S. Allen, S.J. Hardy, A.F. Jacobine, D.M. Glaser, F. Catalina, S. Navaratnam, and B.J. Parsons, *J Photochem. Photobiol. A; Chem.*, **1991,** *62*, 125; R. West, A.R. Wolff, and D.J Peterson, *J. Radiation Curing*, **1986,** *13*, 378

95. L. Aiigiolini, C. Luigi, D.L. Caretti, and E. Salatelli, *Macromol. Chem. Phys.* **2000,** 201(18), 2646- 2653

96. W. Rutsch, G. Hug, and M. Koehler, Eur. Patent # 495,751 (1992)

97. A. Ajayaghosh, R. Francis, S. Das, *Eur. Polymer J.* **1993**, *29,* 63; L. Angiolini, D. Caretti, C. Carlini, N. Lelli, and P.A. Rolla, *J. Appl. Polymer Sci.*, **1993,** *48*, 1163

98. F. Catalina, C. Peinado, J.L. Mateo, P. Bosch, and N.S. Allen, *J. Photochem. Photobiol. A: Chem.,* **1989,** *47*, 365

99. F. Catalina and C. Peinado, *J. Photochem. Photobiol. A: Chem.*, **1992**, *67*, 255

100. W. Bauemer, H. Koehler, and J. Ohngemach, *Radcure Technical Paper*, **1986**, 4-43,

101. A.M. Sarker, K. Sawabe, B. Strehmel, Y. Kaneko, and D.C. Neckers *Macromolecules* **1999**, *32* ,5203

102. E.-S. Du, Z.-C. Li, W. Hong, Q.-Y. Gao, F.-M. Li, *J. Polym. Sci., Part A: Polym. Chem.* **2000,** 38(4), 679

103. H.G. Woo, L.Y. Hong, S.Y. Kim, S.H. Park, S.J. Songand and S.H. Ham, *Bull. Korean Chem. Soc.,* **1995**, *16*, 774.

104. H.G. Woo, L.Y. Hong, S.Y. Kim, S.H. Park, S.J. Songand and S.H. Ham, *Bull. Korean Chem. Soc.,* **1995**, *16*, 1056.

105. D. Yuecesan, H. Hostoygar, S. Deniziigil and Y. Yagci, *Angew. Makromol. Chem.,* **1994**, *221*, 207; Y.A. Yagci, V. Onen, M. Harabagiu, C. Pinteala, C. Cotzur and B. C. Simonescu, *Turk. J. Chem.,* **1994**, *18*, 101

106. X. Coqueret and L. Pouliquen, *Macromol. Symp.* **1994**, *87*, 17

107. F. Catalina, C. Peinado, M. Blanco, T. Corrales, and N.S. Alien, *Polymer.,* **2001** (Pub. **2000**), *42*(5), 1825

108. W. Fischer, V. Kvita, H. Zweifel and L. Fiedler, French Patent # 8,108,981 (1981)

109. T. Yamaoka, Y. Nakamura, K. Koseki, and T. Shirosaki, *Polymers for Adv. Technol.* **1990**, *1*, 287

110. F. Morlet Savarty, J.P. Fouassier, T. Matsumoto, and K. Inomata, *Polymers for Adv. Technol.,* **1994**, *5*, 56

111. G. Pawloski, F. Erdman, and H. Lutz, Eur. Patent # 332,042

112. K. Inomata, H. Sawada, T. Matsumoto, Y. Minoshima, and O Nakachi, Japan Kokai Tokkyo Khoho, J. Pat. 04,225,002 (1992)

113. M. Li, H. Chen, X. Wang, H. Song, and E. Wang, *J. Photopolym. Sci. and Technol.,* **1994**, *7*, 199

114. B. Klingert, A. Roloff, B. Urwler, and J. Wirtz, *Helv. Chim. Acta,* **1988**, *71*, 1858

115. H. Angerer, V. Desobry, m. Riediker, H. Spahni, and M. Rembold, *Radcure Asia,* 461

116. F. Kaneko and M. Kaji, Japanese Patent # 02,245,003 (1992)

117. W. Yang, Y. Yang, J. Wang, C. Zhang and M. Yu, *J. Pholopolym. Sci. and Technol.,* **1994**, *7*, 187

118. K. Dietliker, G. Hug, R. Kaeser, M. Kohler, U. Kolzak, D. Leppard, L. Misev, G. Rist and W. Putsch, *Radtech'94 North Am. VVIEB Conf. Exhibit. Proc..* **1994**, *2*, 693

119. W.G. Herkstroeter and S. Farid, *Photochem.* **1986**, *35*, 71

120. M. Reidiker, K. Meier, and H. Zweifel, Eur. Patent #186,626 (1986)

121. S. Imahashi and A. Saito, Japan Kokai #0,202,562 (1990)

122. B. Robert, R.A. Bartolini, and D.L. Ross, *J. Chim. Phys.,* **1985**, *82*, 361

123. M.C. Gonzalez, M.R. Feliz, and A.L. Capparelli, *J. Photochem, Photobiol. A: Chem.*, **1992**, *63*, 149

124. V.F. Plyusnin, E.P. Kuxnetzova, E. P. Khmelinski, V.P. Grinvin and V.N. Kirchenko, *J. Photochem, Photobiol. A: Chem.*, **1992**, *63*, 289

125. R. Popielarz, A.M. Sarker, and D.C. Neckers, *Macromolecules,* **1998**, *31*, 951

126. S. Chatterjee, P. Gottschalk, P.D. Davis, G.B. Schuster, *J. Am. Chem. Soc.* **1988**, *110,* 2326

127. S.Chatterjee, P. Gottschalk, P.D. Davis, M.E. Kurz, B. Saurwein, X. ang, and G.B. Schuster, *J. Am. Chem. Soc.* **1990,** *112,* 6239

128. P. Gottschalk; D.C. Neckers, G.B. Schuster, U.S. Patent 4 842 980,1989

129. P. Gottschalk, U.S. Patent 4 874 450, 1989

130. K. Kawamura and Y. Okamoto, U.S. Patent 4 971 891, 1990

131. M. Koike and N. Kita, U.S. Patent 4 950 581, 1990

132. P. Gottschalk, D.C. Neckers, and G.B. Schuster, U.S. Patent 5 051 520, (1992)

133. Y. Toba and Y. Usui, Jpn. Kokai Tokkyo Koho J.P. # 09,328,487 [97,328,487] (1997); Y. Toba, *J. Polymer Sci., Part A:Polym. Chem.* **2000**, *38*(6), 982

134. P. Gosh and G.S. Mukherjee, *Eur. Polymer. J.*.**1986**, *22*, 103

135. J.P. Fouassier and E. Chesneau, *Makromol. Chem., Rapid Commun*, **1988**, *9*, 223

136. T. Corrales, F. Catalma, C. Peinado, N.S. Alien, A.M. Rufs. C. Bueno, and M.V. Encinas, *Polymer* **2002**, *43*(17), 4591

137. U.S. Patent 4,921,827 (1990 to Mining and Manufacturing Co.)

138. D.F. Eaton, *Topics Curr. Chem.*,**1990**, *156*, 199

139. J. Nakazato, k. ito, S. Wakumoto, and S. Yamauchi, Japan. Kokai #02,245,003 (1990)

140. O. Valdes-Aguilera, C.P. Pathak, J. Shi, and D.C. Neckers, *Macromolecules*, **1992**, *25*, 541

141. H.J. Timpe, E. Kuestermann and H. Bottcher, *Eur. Polymer J.*, **1991**, *27*, 429

142. H. Gan, X. Zhao, and D.G. Whitten, *J. Am. Chem. Soc.*, **1991**, *113*, 9409

143. S.K. Lee, and D.C. Neckers, *Chem. Mater.*, **1991**, *3*, 852

144. T. Yamaoka, Y.C. Zhang, and K. Koseki, *J. Appl. Polymer Sci.* **1989**, *38,* 1271

145. C.M. Previtali, S.G. Bertolotti, M.G. Neuman, I.A. Pastre, A.M. Rufs, and E.M. Encinas, *Macromolecules*, **1994**, *27*, 7454

146. F. Amat-Guerri, R. Mallavia, R. Sastre, *Trends Photochem. Photobiol.* **1999**, 5, 103-116

147. B.P. Howell., A. de Raaff, T. Marino, *ACS Symp. Ser.* **1997**, 673

148. M.T.L. Rees, G.T. Russell, M.D. Zammit, and T.P. Davis, *Macromolecules* **1998**, *31*(6),' 1763

149. H. Nagasaka and K. Ota, Japan Kokai 04,106,548 (1992)

150. H. Yong, W. Zhou, G.Liu, L.M. Zhen, E.Wang, *J. Photopolym. S(1. Tecniol.* **2000,** 13(2), 253-258

151. G. Gso and Y. Yang, *GaofenziXuebao* **2000,** (1), 125-128. (From Chem Abstr.)

152. K.S. Padon and A.B. Scranton, *J. Polym. Sci.., Part A: Polym. Chem.* **2000,** *38*(11), 2057

153. K.S. Padon and A.B. Scranton, *J. Polym. Sci., Part A: Polym. Chem.* **2001,** 39(5), 715-723

154. M.G. Neumann, C.C. Schmitt, and H.M. Maciel, *J. Phys. Chem. B* **2001,** 105(15), 2939-2944

155. C. Billaud, M. Sarakha, and M. Bolte, *J. Polym. Sci., Part A: Polyn. Chem.* **2000,** *38*(21), 3997-4005

156. O. Garcia, A.Costela, I. Garcia-Moreno, and R. Sastre, *Macromolecular Chemistry and Physics* **2003,** 204(18), 2233-2239

157. D. Burget, C. Grotzinger, and J.P. Fouassier, *Trends in Photochemistry and Photobilogy,* **2001,** *7,* 71; Allonas, X.; Fouassier, J.P.; Kaji, M.; Miyasaka, M., *J. Photopolym. Scl. Technol.* **2000,** 13(2), 237-242; b. Allonas, X.; Fouassier, J.P.; Kaji, M.; Miyasaka, M.; Hidaka, T., *Polymer* **2001,** 42(18), 7627-7634

158. B. Li, S. Zhang, L. Tang, and Q. Zhou, *Polymeric Materials Science and Engineering,* **2001,** *84,* 139

159. K.D. Belfield, J. Liu, K.J. Schafer and S.J. Andrasik, *Am. Chem. Soc. Polymer Preprints,* **2001,** *42*(2), 713

160. O.O. Novikova, V.G. Syromyatnikov, L.F. Avramenko, N.P. Kondratenko, T.M. Kolisnichenko, and M.J.M. Abadie, *Materials Science* **2002,** *20*(4), 19

161. S. Zhang, B. Li, L. Tang, R. Yang, and Q. Zhou, *A.C.S. Polym. Prepr.* **2000,** 41(2), 1104-1105

162. C. Kutal, P.A. Grutsch, and D.B. Young, *Macromolecules,* **1991,** *24,* 6872

163. Y. Yamaguchi and C. Kutal, *Macromolecules,* **2000,** *33,* 1152

164. Y. Yamaguchi, C. Kutal. *Molwgr. Ser. Int. Conf. Cooid. Chem.* **1999,** 4, 209-214; C. Kutal, Y. Yamaguchi, W. Ding, X. Li, C.T. Sanderson, and I.J. Amster, A.C.S. Polymer Preprints, **2001,** *42*(2), 719

165. C.T. Sanderson, B.J. Palmer, A. Morgan, M. Murphy, D.A. Diuhy, T. Mize, I.J. Amster, and C. Kutal, *Macromolecules* **2002,** 35(26), 9648-9652

166. V. Jarikov and V. Nickers, *Macromolecules,* **2000,** *33*(21) 7761

167. J.V. Crivello, *J. Polym. Sci., Part A: Polym. Chem.* **1999,** 37(23), 4241

168. D.R. Randall (ed.), *Radiation Curing of Polymers,*CRC Press, Boca Raton, 1987

169. H. Gu, W. Zhang and D.C. Neckers, *Am. Chem. Soc. Polymer Preprints,* **2000,** *41* (2), 1266

170. J.L. Dektar and N.P. Hacker. *J. Org. Chem.* **1990,** *55,* 639; J.L. Dektar and N.P. Hacker, *J. Org. Chem.,* **1991,** *56.* 1838

171. J.V. Crivello and S. Kong, *J. Polyn. Sci., Part A: Polym. Chem.* **2000**, *38*(9), 1433-1442

172. J.V. Crivello and S. Kong, *Macromolecules*, **2000**, *33*, 825

173. J.P. Fouassier, D. Ruhlmann, Y. Takimoto, M. Harada, and M. Kawabata, *J. Imaging Sci. Technol.* **1993**, *37*, 208 ; J.V. Crivello, *Adv. Polymer Sci.* **1984**, *62*, 1

174. S.P. Pappas, *Progr. Organic Coatings*, **1985**, *13*, 35

175. J.V. Crivello, S. Rajaraman, W.A. Mowers, and S. Liu, *Macromol. Symp.* **2000**, 109-119

176. D.C. Nickers and A. Abu, *Macromolecules*, **1984**, *17*, 2468

177. E.O. Alonso, L.J. Johnson, J.C. Scaiano, and V.G. Toscano, *J. Can. Chem.*, **1992**, *70*, 1784

178. T. Takato, K. Takuma, and T. Endo, *Makromol. Chem. Rapid Commun.*, **1993**, *14*, 203

179. F.D. Saeva. D.T. Breslin, and H.R. Lun, *J. Am. Chem. Soc.*, **1991**, *113*, 5333

180. F. Lohse and H. Zweifel, *Adv. In Polymer Sci.*, **1986**, *78*, 62 ; S.I. Schlesinger, *Photog. Sci. Eng.* **1974**, *18*, 387

181. M.-S. Jang and J.V. Crivello, *Am. Chem. Soc. Polymer Preprints*, **2003**, *44*(1), 15; Z. Gomurashvili and J. V. Crivello, *A.C.S. Polymer Preprints* **2001**, *42(2)*, 755; *ACS Symposium Series*, **2003**, *847*, 231; Z. Gomurashvili and J. V. Crivello, *Macromolecules*, **2002** , *35*, 2962

182. J.V. Crivello, F. Jiang, and J. Ma, *Am. Chem. Soc. Polymer Preprints*, **2003**, *44*(1), 13

183. N. Taniguchi and M. Yokojima, *Jpn. Kokai Tokkyo Koho,* Japan Patent 10,182,711(98 182,711) (Jul 1998) (from Chemical Abstracts)

184. J.V. Crivello, D. Burr, and J.P. Fouassier, *Macromol. Sc;i. Pure and Appl. Chem.* **1944**, *A31*, 677

185. H.J. Timpe, C. Meckel, and H. Baumann, *Makromol. Chem.*, **1986**, *187*, 187

186. J.V. Crivello and S. Kong, *J. Polyn. Sci., Part A: Polym. Chem.* **2000,** 38(9), 1433-1442

187. J. V. Crivello, *Polymer Preprints* **2001**, *42(2),* 773; b. M. Sangermano and J. V. Crivello, *ACS Symposiym Series*, **2003**, *847*, 242

188. J. V. Crivello and M. Sargenmano, *J. Polyn. Sci., Part A: Polym. Chem.* **2001**, *39*(3)m 343

189. Chen, Yamamura, and K. Igarashi, *J. Polym. Sci., Polym.Chem. Ed.*, **2000**, *3811*, 90

190. J.V. Crivello and M. Sangermano, *Polymer Preprints* **2001**, *42(2),* 783 : *J. Polym. Sci., Part A.-Polym. Chem.* **2001,** 39(3), 343-356

191. Y. Toba, *Trends in Photochemistry and Photobiology*, **2001**, *7*, 31

192. K. Suyama, W. Qu, M. Shirai, and M. Tsunooka, *Chemistry Latter*, **2003**, *32*(1), 92

193. Aysen Onen and Yusuf Yagci, *Macromolecules* **2001**, *34,* 7608-7612

194. J.V. Crivello, J-q. Ma, and F. Jiang, *J. of Polymer Sci., Part A: Polymer Chemistry* **2002**, 40(20), 3465-3480

195. Y. Hua and J. V. Crivello, *Macromolecules* **2001**, 34, 2488-2494

196. F. Sun, J. Liu, and Y. Huang, *Fushe Yanjiu Yu Fushe Gongy Xuebao*, **2002**, 20(4), 255 (from Chem. Abstr. 138: **369221p**)

197. G. Norcini, A. Casiraghi, M. Visconti, G. Li Bassi, and Giuseppe, **PCT Int. Appl. WO 03 8,404** (Jan. 30, 2003)

198. L. Atmaca, L. Kayihan, and Y. Yagci, *Polymer* **2000,** *41*(16), 6035

199. V. Sipani and A.B. Scranton, *J. Photochemistry and Photobiology*, **2003**, *159* (2), 189

200. E.Y. Komarova, K. Ren, and D. C. Neckers, Am. Chem. Soc., *Polymer Preprints* **2001,***42(2),* 737; H. Li, K. Ren, and D.C. Noickers, *Macromolecules* , **20001**, *34,* 8637

201. R.B. Frings and G.F. Grahe, **Eu. Pat. Appl. EP** 1,262,506, (Dec. 2, 2002)

202. Schnabel, Wolfram, *Macromol. Rapid Commun.* **2000**, 21(10), 628- 642

203. Y. Jiang, L. Yue, X. Zhang, S. Wu. *Gaofen-zi. Xuebao* **2003,** (4), 495- 499 (From *Chem. Abstr.*)

204. T. Hino and T. Endo, *Macromolecules*, **2004**, *37*, 1671

205. J.P. Fouassier, *et al.*, *J. Polymer Sci., A. Chem.*, **1988**, *26*, 1021

206. X. Allonas, J.P. Fouassier, M. Kaji, and M. Miyasaka,, *J. Photopolym. Scl. Technol.* **2000**, 13(2), 237-242; X. Allonas, J.P. Fouassier, M. Kaji, M. Miyasaka, and T. Hidaka, *Polymer* **2001**, 42(18), 7627-7634

207. J.P. Fouassier, F. Morlet-Savary, K. Yamashita, and S. Imahashi, *Polymer*, **1997**, *38*, 1415; V. Jakubek, A.J. Lees, S.J. Fuerniss, and K.I. Papathomas, *Am. Chem. Soc. Polymer Preprints,* **1997**, *38* (1), 185; N.S. Allen, *Photochemistry*, **1999**, *30*, 331

208. Lazauskaite, Ruta; Budreckiene, Ruta; Grazulevicius, Juozas V.; Abadie, Marc J. M. , *J. Prakt. Chem.*, **2000**, 342(6), 569-573

209. C. Kutal, Y. Yamagushi, W. Ding, C.T. Sandesrson, X. Li, G. Gamble, and I.J. Amster, *ACS Symnposiium Series*, **2003**, *847*, 332

210. T. Wang, Y.-l. Huang, L.-J. Ma, and S.-J. Shi, *Ganguang Kexue Yu Guang Huaxue*, **2003**, *21* (1),46 (From Chem. Abstr. 138:**385782e**)

211. T. Wang, Y.-l. Huang, and S.-J Shi, *Gaodeng Xuexiao Huazue Xuebao*, **2003**, *24* (4), 735 (From Chem. Abst., 139:**101452t**)

212. L. Ma, T. Wang, L. Zhang, S. Yu, and Y. Huang, *Fushe Yanjiu Yct Fushe Gongyi Xuebao* **2004**, *22*(4), 219—223

213. S. Hayase, Y. Onishi, S. Suzuki, and M. Wada, *Macromolecules*, **1985**, *18*, 1799

214. J.-H. He, R. J.-C. Zhan., *J Polym. Set. Part A: Polym. Chem.* **1999**, 371241, 4521-4527

215. N. Taniguchi, M. Wada, M. Yokoshima, *Jpn. Kokai Tokkyo Koho* Japanese Patent 09,278,814, Oct 1997
216. K.D. Belfield, J. Liu, K.J. Schafer, and J. Andrasik, *Am. Chem. Soc., Polymer Preprints*, **2001**,*42* (2),713
217. K.D. Belfield, X. Ren, E.W. Van Stryland, D.H. Hagan, V. Dubikovski, and E.J. Meisak, *J. Am. Chem. Soc.*, **2000**, *122*, 1217
218. K.D. Belfield and K.J. Schafer, *ACS Symposium Series*, **2003**, *847*, 464
219. C. Morteneau G. Lemercier, C. Andraud, I. Wang, M. Bouriau, and P.L. Baldeck, *Synthetic Materials*, **2003**, *138*, 353
220. B. Campston, M. Lipson, S.R. Marder, and J.W. Perry, U.S. Patent 6,608,228 (Aug. 19, 2003)
221. Y. Tian, M. Zhang, X. Yu, G. Xu, Y. Ren, J. Yang, J. Wu, X. Zhang, X. Tao, S. Zhang, M. Jiang, *Chemical Physics Letters* **2004**, 388(4-6), 325
222. Y.-X. Yan, X.-T. Tao, Y.-H. Sun, W.-T. Yu, G.-B. Xu, C.-K. Wang, H.-P. Zhao, L.-X. Yang, X.-Q. Yu, X. Zhao, and M.-H. Jiang, *Bulletin of the Chemical Society of Japan* **2005**, *78*(2), 300—306.
223. S. Bader, Eur. Patent #793,030,784
224. 3M Co., Eur. Patent #0,369,645
225. R. Patel, Eur. Patent #97,012
226. E. Chesneau and J.P. Fouassier, Ang. Makromol. Chem., 1985, 135,41
227. S. Imahashi, A. Saito, and k. Yamashita, Eur. Patent #3,918, 105

Chapter 3

Chemistry of Photocurable Compositions

3.1. Basic Background of Photocurable Materials

It is common to define *photocuring* as a process of rapid conversion of specially formulated, usually liquid solventless compositions into solid films by irradiation with ultraviolet or visible light. This is used to form films of varnishes, paints, adhesives, and coatings for paper, plastic, wood, metal surfaces, optical fibers and compact discs. The chemistry is also being used in the imaging areas, like preparation of printing plates, photoresists, in graphic arts, in microelectronics, and in stereo- and microlithography. In addition this technology is also utilized in dental restorative fillers, fiber-optic treatment, aspherical lenses for CD applications, and in preparations of some contact lenses where acrylate and methacrylate compositions are efficiently crosslinked. [1] Typical *photocuring* reactions are very rapid, lasting from sub-second to minute time scales.

A large proportion of the curing reactions described above are carried out with light in the ultraviolet region. This is due to the fact that ultraviolet light is more energetic and, therefore, more efficient in rupturing chemical bonds (see Chapter 1). On the other hand, the use of visible light has many attractions, such as safety and ease of handling. So, lately, curing with visible light is receiving attention as well.

Majority of the compositions of photocrosslinkable materials in commercial use today consist of mixtures of prepolymers or oligomers combined with di- or polyfunctional monomers and photoinitiators. [2] All three are necessary to carry out fast polymerizations with the available light sources and to form films with desired properties Such compositions can be divided into two groups: (1) those that cure by free-radical mechanism and (2) those that cure by an ionic (mostly cationic) one. There are also some compositions that cure simultaneously by both mechanisms. The photoinitiators and sometimes also photosensitizers are included into the reaction mixtures, because the materials used by themselves alone do not absorb sufficient energy in the areas of high emission of the commercial light sources for photopolymerization. Photoinitiators that match as closely as possible the output energy of the light sources and also have high extinction coefficients are, therefore, added. The mono- and polyfunctional monomers usually serve both as solvents and as co-reactants.

In search of the optimum quantity of the photoinitiators for these photo-reactive compositions, it is important to know that the rate of cure is related to the intensity of the absorbed radiation. It is actually not proportional to the concentration of the initiator in the reaction mixture. In ideal systems, the theory does state that the rate of a photopolymerization should be proportional to the square root of the photoinitiator concentration. Unfortunately, however, light curable formulations or light crosslinkable compositions behave far from ideal. This deviation from ideality can be blamed on many factors. These include the average functionality of the composition and the viscosity of the medium. They also include efficiency of reactions with monomers by the initiating species produced, the quantum yield of the photoinitiators, and the presence or absence of inhibiting agents, such as oxygen.

Several publications suggest that an optimum concentration of a pbotoinitiator should be such that the optical density or absorbance of the reacting composition should be 0.434 [2] to yield an optical density at which maximum radiation is absorbed. This number has meaning, however, only for ideal systems, cured by monochromatic radiation. With commercial sources of radiation that consist of 20 to 40 emission lines this does not apply.[3] Also, it is important to know that variations in the concentrations of a photoinitiator within various locations of the film that is being photocured, dramatically affect the locations where the radiation is absorbed, such as at the surface, throughout the material, or at some other location.

Many compositions also contain additives for various purposes. Thus, for instance, to increase storage life of an acrylic composition that crosslinks by free-radical mechanism, some trace amounts of a copper compound, like a copper xanthate might be added, or small quantities of hydroquinone combined with monomethyl ether of hydroquinone or quinone might be added. Pigmented coatings will also, of course, contain pigments. Many white coatings may contain TiO_2, (rutile or anatase) or, zinc oxide or zinc sulfide. The colored coatings will contain colored pigments and, or dyes. Other additives may be flow or wetting agents and additional materials, like, for instance, an internal lubricant.

3.2. Compositions that Cure by Free-Radical Mechanism

The physical properties of light cured polymers are generally governed by the chemical structures of all reactive species, including monomers, oligomers, and prepolymers that form the backbones of the three dimensional films. Many acrylic modified prepolymers and oligomers were developed to meet various end uses. The chemical structures can be quite diverse and the functionality of the prepolymers can range from two to eight. The prepolymers can vary in molecular weights from 500 to 3000, but some even higher. The limiting factor is the viscosity. Also, they must contain at least two reactive groups but often contain more than two. The monomers used usually range in functionality from one to four. Often, the higher the functionality of the monomer the tighter is the crosslinking density of the finished material.

There are two distinct types of compositions that cure by free-radical mechanism. One type utilizes typical free-radical chain growth polymerization reactions of compounds with carbon to carbon double bonds for accomplishing the cure process, while the other one employs thiol-ene reactions. [3] In the first one, in order to acquire high reactivity, the unsaturations are preferably of the $CH_2 = CH-$ or $CH_2 = C<$ type, carrying activating groups of the following nature: -COO, -CON<, -NCO-, etc. The choice of the monomers that function as reactive diluents and solvents is based upon considerations of their solvent power, their reactivity, their volatility, and their influence on properties of the finished film. Useful resin systems that fall into this category include unsaturated polyesters/styrene compositions and multifunctional acrylates or methacrylates. The unsaturated polyesters/styrene compositions are relatively slow curing and are used primarily in wood finishing. The multifunctional acrylates, on the other hand, are fast curing, usually faster than methacrylates and are, therefore, widely used.

The thiol-ene type of cure is based on free-radical additions of thiols to unsaturated compounds. The unsaturated compounds are often, though not necessarily, allylic functionalized resins, such as allyl ethers and allylic urethanes and ureas. [3] The thiol-ene systems find application in poly(vinyl chloride) based flooring and in gaskets.

3.2.1. Photocuring by Free-Radical Chain Growth Polymerization

Light curable composition that fall into this category, as stated above, consists of appropriate photoinitiators, monomers, oligomers, and may or may not contain some stablilizers. The formulations may also contain various additives for desired performance, such flow agents, improved wetting, and others. Gatechair and Tiefenthaler [4] studied the depth of cure and observed that a coating with an optical density of 0.43 or below yields a fairly uniform absorption of radiation throughout the film. Under ideal conditions one might anticipate uniform crosslinking throughout the coating. Oxygen, however, reacts with the free-radicals. A high concentration of free-radicals on the surface may prevent oxygen diffusion into the film. By varying the concentration of the photoinitiator Gatechair and Tiefenthaler [4] demonstrated a dramatic effect on where the radiation is absorbed within a coating and where the initiating radicals are formed.

3.2.1.1. Monomers Used in Photocuring

Numerous monomers are available commercially for radiation curing. Many commercially available monomers are prepared by esterification of polyhydroxy compounds. Each supplier, and there are many of them, offers at least several acrylic and methacrylic esters of mono and polyfunctional hydroxy compounds. Table 3.1 list some mono-, di-and, trifunctional compounds available commercially for ultraviolet light curable compositions. An increase in the functionality of the monomers usually speeds up the reaction . This, however, can result in early gelation or vitrification of the crosslinked structure and interferes with complete conversion. Introduction of heterocyclic structures with oxygen into the chemical structure of the monomer was found to cause a substantial acceleration of the polymerization rate. Such reactions are claimed to proceed to near completion. [5]

Many monofunctional monomers will not yield a three-dimensional network. Monomers with a functionality of two or more are often required for his purpose and attainment of desired properties in a coating. Among the many possible examples of a commercial difunctional acrylic monomer, is 1,4-butylene glycol diacrylate. It is prepared from 1,4-butanediol and can be illustrated as follows:

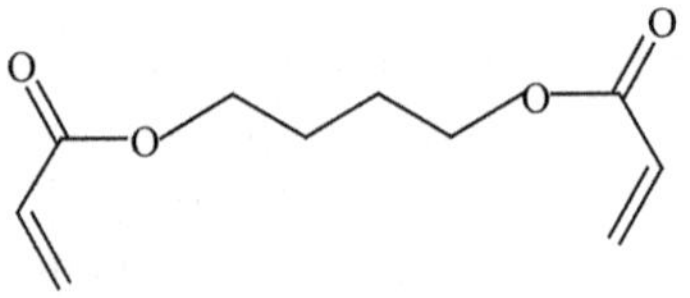

Examples of a trifunctional monomer are acrylate or methacrylate esters of glycerol, pentaerythritol, trimethylolpropane, or triethylolpropane and others. One such material is shown below:

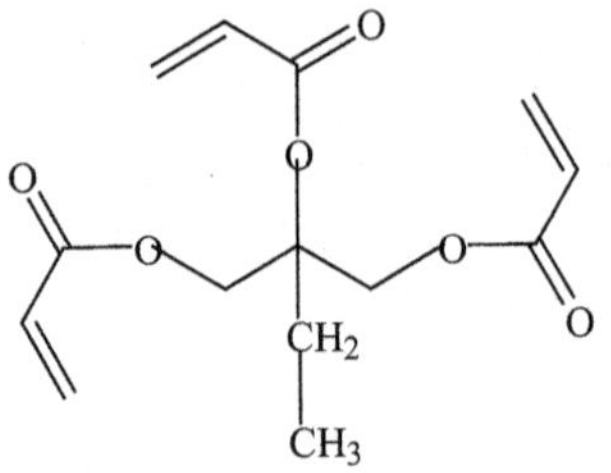

trimethylolpropane triacrylate

An example of a tetrafunctional monomer is pentaerythritol tetraacrylate. It can be illustrated as follows:

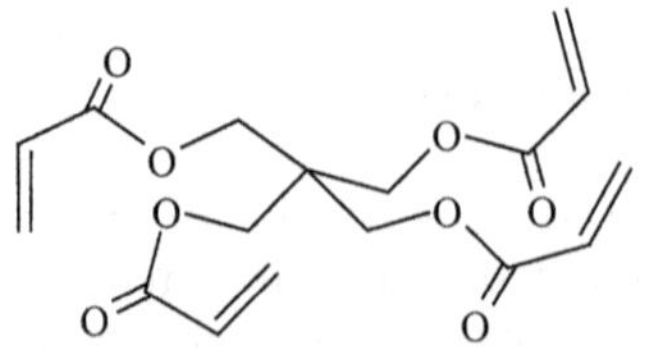

pentaerythritol tetraacrylate

Table 3.1. Common Commercially Available Monomers

Monomer	Structure
Hydroxyethyl acrylate	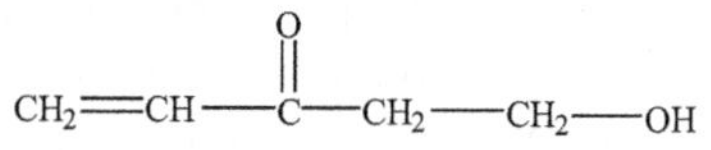 $CH_2=CH-\overset{\overset{O}{\|\|}}{C}-CH_2-CH_2-OH$
N-vinyl pyrrolidone	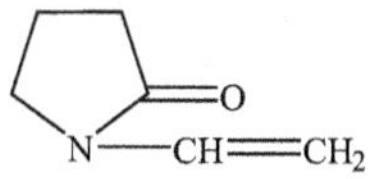 (structure: pyrrolidone ring with $=O$, $N-CH=CH_2$)
Hexanediol diacrylate	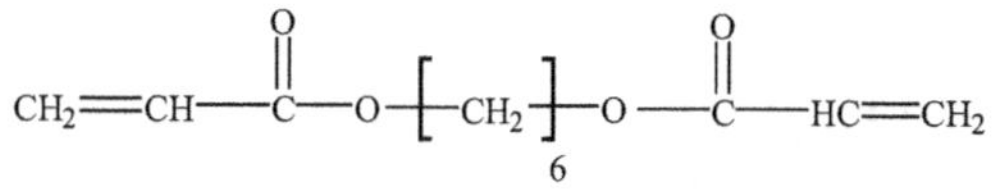 $CH_2=CH-\overset{\overset{O}{\|\|}}{C}-O-[CH_2]_6-O-\overset{\overset{O}{\|\|}}{C}-HC=CH_2$
Diethyleneglycol diacrylate	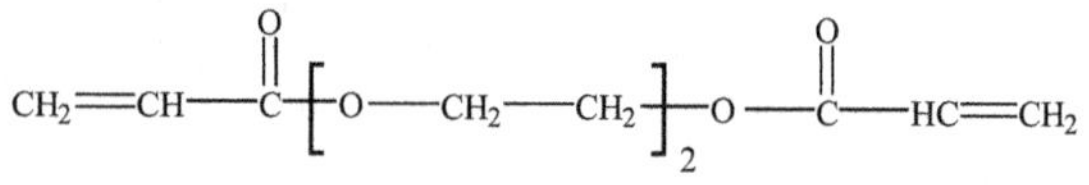 $CH_2=CH-\overset{\overset{O}{\|\|}}{C}[O-CH_2-CH_2]_2-O-\overset{\overset{O}{\|\|}}{C}-HC=CH_2$
Glycerol triacrylate	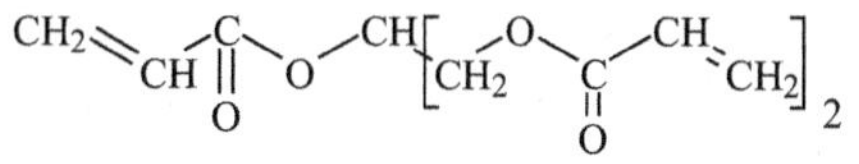 $CH_2{=}CH-\overset{\overset{C}{\|}}{\underset{O}{}}O-CH[CH_2-\overset{O}{\underset{O}{C}}-CH-CH_2]_2$
Trimethylolethane triacrylate	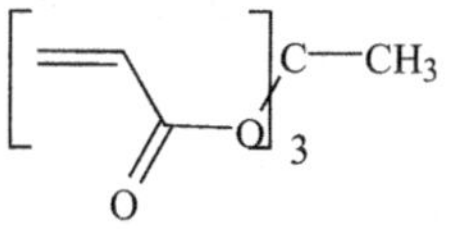 $[\;\;=\;\;\overset{}{\underset{O}{}}-O]_3 C-CH_3$
N-ethyloxalidone acrylate	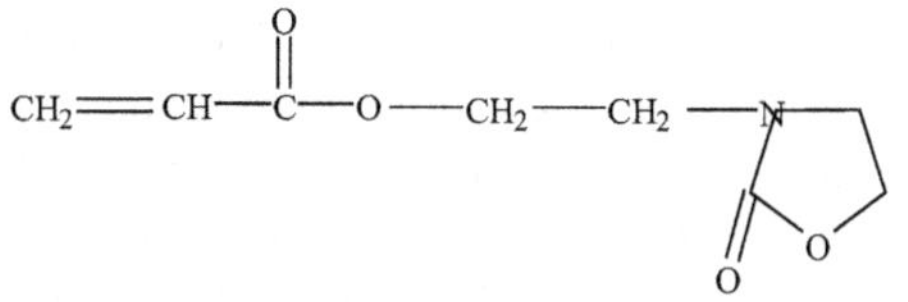 $CH_2=CH-\overset{\overset{O}{\|\|}}{C}-O-CH_2-CH_2-N$ (ring with O, $=O$)
Neopentylglycol diacrylate	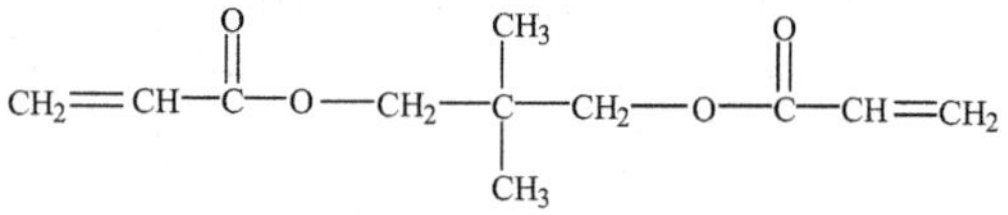 $CH_2=CH-\overset{\overset{O}{\|\|}}{C}-O-CH_2-\overset{\overset{CH_3}{\|}}{\underset{CH_3}{C}}-CH_2-O-\overset{\overset{O}{\|\|}}{C}-CH=CH_2$
Decamethylenediol diacrylate	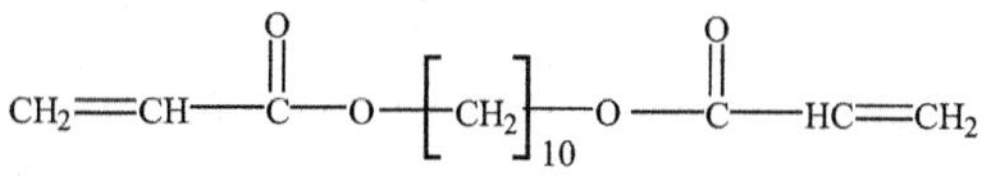 $CH_2=CH-\overset{\overset{O}{\|\|}}{C}-O-[CH_2]_{10}-O-\overset{\overset{O}{\|\|}}{C}-HC=CH_2$
Pentaerythritol triacrylate	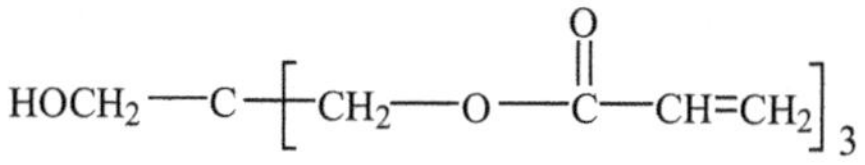 $HOCH_2-C[CH_2-O-\overset{\overset{O}{\|\|}}{C}-CH=CH_2]_3$

[a] From various sales brochures

Andrzejewska observed that the presence of heteroatomes in the side chains of acrylic monomers has an accelerating effect on the photopolymerization. [6] Decker reported [5,7] that some monofunctional monomers exhibit enhanced reactivity, undergo complete conversion, and become completely insoluble when exposed to UV light in the presence of photoinitiators. These monomers display polymerization kinetics that rivals those of polyfunctional acrylic monomers. These monomers were placed by him in the order of their photocuring rates as follows,

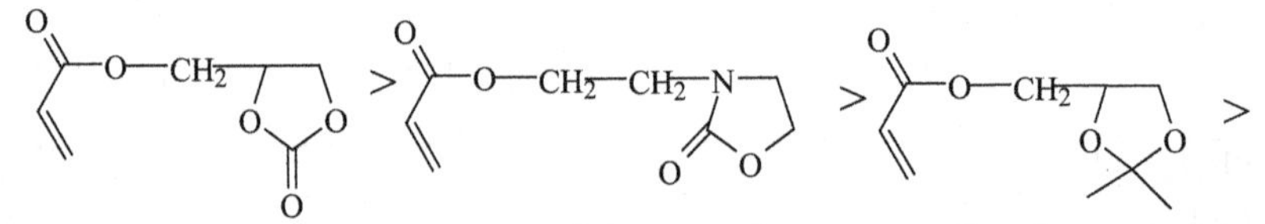

cyclic carbonate acrylate   oxazolidone acrylate   dioxalane acrylate

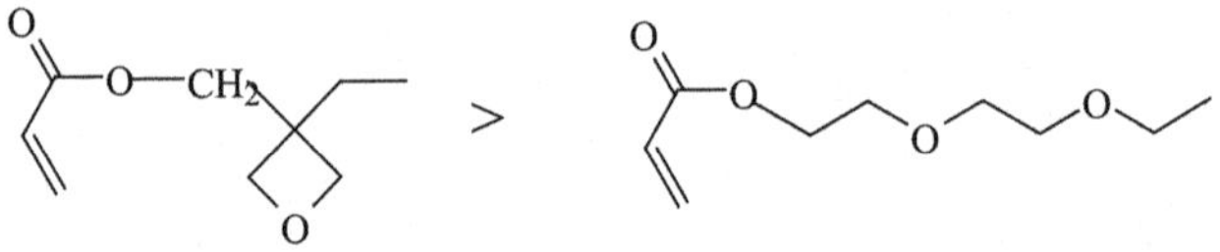

oxetane acrylate   ethyl diethylene glycol acrylate

Thus, incorporation of certain functional groups, like carbonates, carbamates, cyclic carbonates, or oxazolidones into acrylate or methacrylate monomers greatly enhances their reactivity. The exact cause of the enhanced reactivity of such monomer has not been fully understood. Some theories have been proposed. Decker suggested that this reactivity may be due to the presence of labile hydrogen groups in their chemical structures. That may results in enhanced hydrogen abstraction and chain transferring during the crosslinking process. [5,7] The enhanced chain transferring creates branches and radical sites for crosslinking. Jansen et al., [8] however, attributed the increased polymerization rates of these monomers to increases in the dipole moments, greater than 3.5 debyes

Bowman et al., [9] concluded that hydrogen abstraction alone does not fully account for the number of crosslinks that form and the polymerization kinetics of these monomers. Instead, they attributed the enhanced reactivity to electronic effects as well. To carry out electronic and resonance studies on monomers, they synthesized several monomers with various substituents on an aromatic end group:

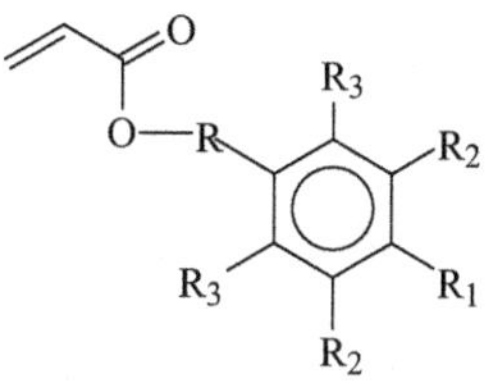

where R substituents can be hydrogen, fluorine, and others. Actually, three possible mechanistic theories for the enhanced reactivities were studied by Bowman et al..[10] They include hydrogen bonding, hydrogen abstraction, and electronic and resonance effects. The mechanical properties were studied for carbonate and carbamate incorporated monomers. Their results show that these materials, upon polymerization, exhibit crosslink formation that exceeds the initiator concentration by almost ten fold. Such a degree of crosslinking can not be exclusively due to hydrogen abstraction. In addition, fluoro substitution of the aromatic ring in the above shown monomer at the para and ortho positions causes a two-fold increase in the chain propagation rate and a ten-fold increase in the chain termination rate compared to the unsubstituted monomers.[10]

Subsequently, Jansen et al.,[11] reported a systematic study of the effects of molecular structures on the photoinitiated polymerization of monofunctional acrylic monomers including those reported by Decker[7,8] (shown above). Their data shows that the rate of polymerization of such monomers is dominated by two factors. These are hydrogen bonding and Boltzmann-average dipole moments. They reported that the two are complementary to each other and are the factors that influence the speed of polymerization.[11] Also, a direct correlation was observed between the rate of polymerization and the calculated Boltzmann-average dipole moments. These correlation hold for pure monomers, mixtures of monomers, and even for mixtures of monomers with inert solvents.[11] Also, their studies suggest that the propagation steps in the polymerization are influenced by hydrogen bonding, while the dipole moments influence the termination rate constants.[11] The fastest among them include monomers reported by Decker[14,15] as well as some additional ones. Based on their data, the monomers can be placed in the following order of reactivity rates[11]:

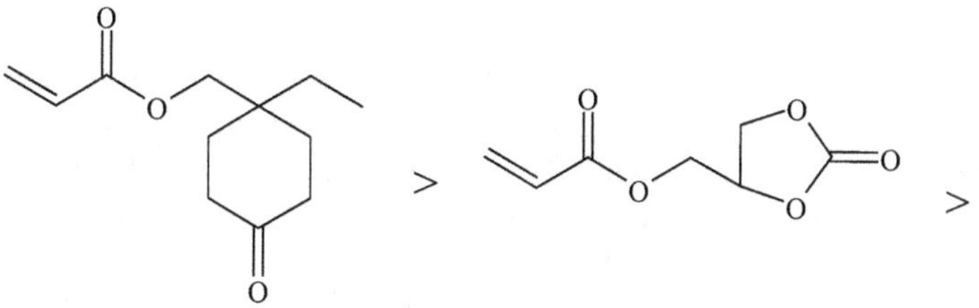

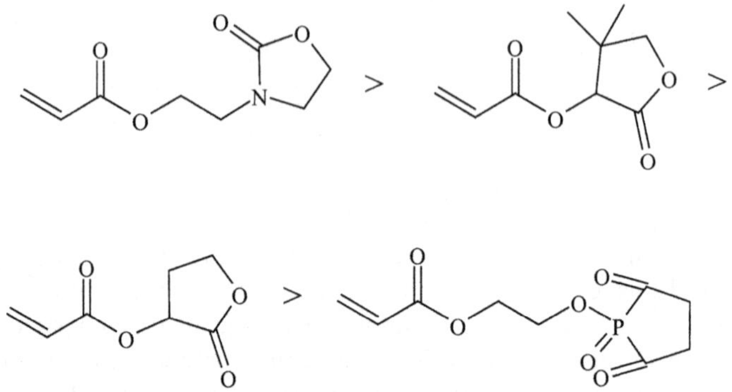

Recently, Bowman *et al.*, [12] reported another investigation of the contribution of dipole moments to monomer reactivity, as well as the claim that the dipole moments of monomers correlate with reactivity of monomers above the threshold of 3.5 debyes. Following monomers were investigated by them:

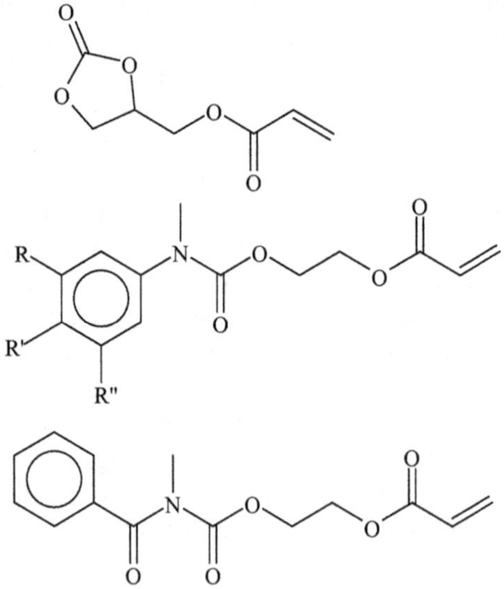

where, R', R'', R''' = H, F, OCH_3. The results of this investigation indicate that the dipole moments are not a significant factor in enhancement of reactivity of such monomers. Also, apparently the correlation between threshold value of 3.5 debyes and enhanced reactivity does not seem to hold universally. [12]

      The influence of hydrogen bonding on the polymerization rates of several acrylates and methacrylates was studied. [13] by correlating the extent of such bonding with the reaction rates. A direct relationship was found. Both, hydrogen bonding and polymerization rate decrease with increasing temperature. This is different from conventional acrylic polymerization. Evidence was presented that hydrogen bonding influences the termination process by reducing polymer radical mobility. [13] Dias and coworkers [14] found that maximum rates of photoinitiated polymerizations of acrylates are strongly influenced by the ability to form hydrogen bonds. In cases where the hydrogen bonds are in the vicinity of the acrylate moiety higher rates are observed. This is explained by preorganization due to such bonding of the molecules. Also it was shown that in cases where hydrogen bonds are present, anti-Arrhenius behavior is observed. [14]

The relationship between polymerization of hydroxyalkyl acrylates and chemical structure was investigated by examining of aggregation (orientation) of monomer and polymer induced by hydrogen bonding. It was found that the presence of the OH group significantly enhances polymerization rate of hydroxyalkyl acrylates. [15] Interestingly, these monomers show almost the same curing rate as multifunctional monomers. The relative degree of hydrogen bonding is proportional to the rate of polymerization. [15]

Novel, self initiating monomers, based on vinyl acrylate and derivatives were reported recently. [16] These monomers were illustrated as follows:

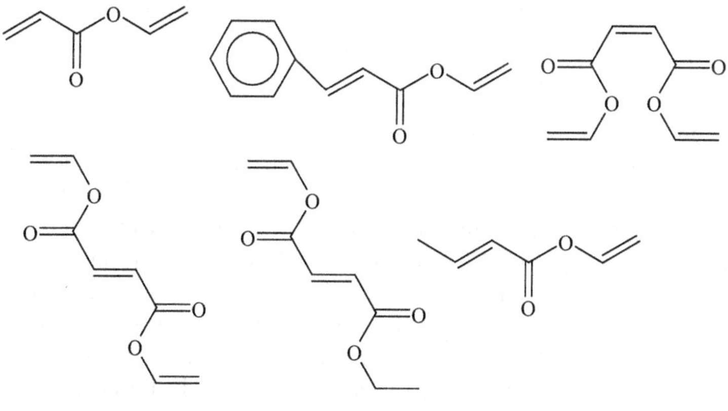

There are definite attractions for monomers that can be used without the aid of initiators. Such monomers are maleimides. The monomers based on vinyl acrylate are also capable of self initiation. The vinyl ester itself, however, is too volatile for practical use and its initiation of polymerization is slower than obtained with the traditional photoinitiators. When the acrylate group is replaced by crotonate, cinnamate, fumarate, or maleate chromophores, these monomers copolymerize readily with thiol and vinyl ether monomers and initiate free-radical polymerization upon direct excitation in the absence of any added photoinitiator. [16]

3.2.1.2. Oligomers and Prepolymers

Pretot, *et al.*, [17] describe in a European patent new reactive diluents, to be used with acylphosphine oxide initiator and new alkyd coating compositions. The new reactive diluents are best illustrated by their chemical structures as follows:

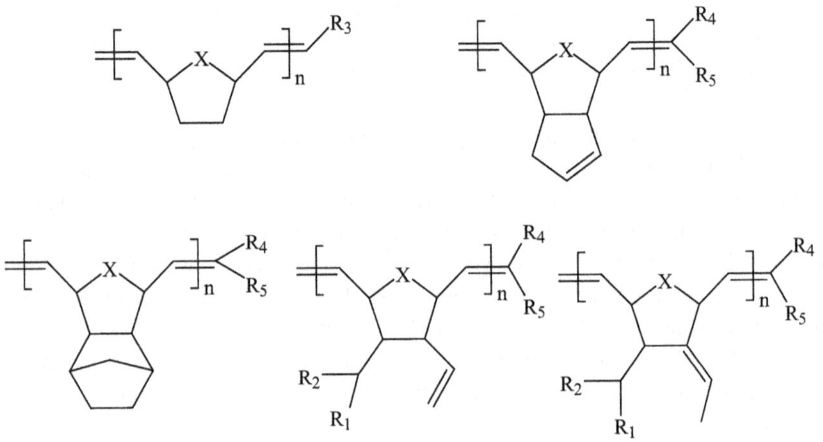

where R_1 and R_2 are H, hydroxy, cyano, halogen, vinyl, formyl, or other compounds; R_3 = methacryloyloxy-Me or phenyl, para substituted by vinyl, hydroxy, halogen, cyano, C_{1-12} alkyl; R_4 = H, 1-naphthyl, 2-naphthyl, biphenyl, anthracenyl, phenyl or phenyl para substituted by vinyl or methacryloyloxy, or a substituted phenyl residue $C_6H_4CH_2W$ (along with R_3); W = methacryloyloxy, or an aliphatic residue, such as CH_2YA, where Y = direct bond or OC_{1-12} alkylene, where the alkylene linker is linear or branched and A = hydroxy. R_5 is H or C_1-C_6 alkyl; n = 1-10; X= CH_2, CO, O, NR, S, or substituted methylene. [17]

Many oligomers that are used in industrial formulations are polyurethane acrylates that are prepared from urethane prepolymers, where the excess isocyanate groups are reacted with compounds like hydroxyethyl or hydroxypropyl acrylates or methacrylates. They can be also epoxy resins, polyethers or polyesters that were esterified by reactions with acrylic or methacrylic acids. The following illustration shows the types of oligomers often used. This illustration only shows a functionality of four. Much higher functionality, however, is possible. On the other hand, some might only have a functionality of two or three. Also, while acrylate groups are shown in the illustration, these can be methacrylate groups instead.

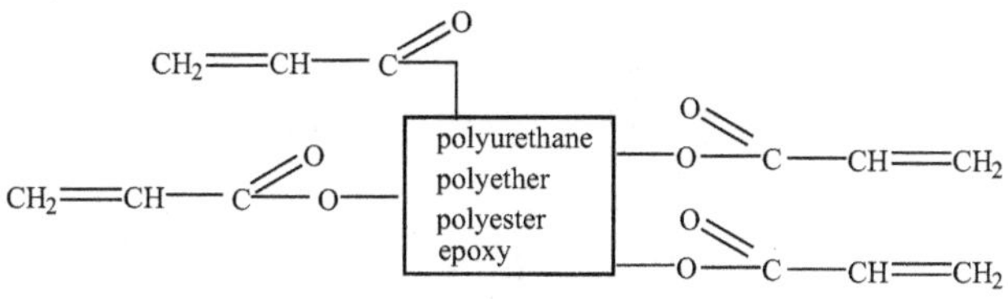

$$CH_2{=}CH-C{\overset{\displaystyle O}{<}}$$

$$CH_2{=}CH-C{\overset{\displaystyle O}{<}_O}-$$

| polyurethane |
| polyether |
| polyester |
| epoxy |

$$-O-C{\overset{O}{\diagup\!\diagup}}-CH{=}CH_2$$

$$-O-C{\overset{O}{\diagup\!\diagup}}-CH{=}CH_2$$

As stated earlier, the reactivity of the light curable compositions and the physical properties of the finished films depend to a great extent on the chemical structure of the oligomers and their functionality. In addition the viscosities of the reaction mixtures affect both the propagation and the termination constants. Also, the molecular weights of the prepolymers control the kinetic chain lengths Presence of aromatic groups in the chemical structures of the oligomers result in formations of films that are hard and tend to be glossy with good scratch resistance. The presence of aromatic groups, however, affects the initiation rates, because aromatic groups absorb in the ultraviolet region and act as inner filters. Aromatic groups can also prevent complete crosslinking due to vitrification.

Decker [5] suggested that the reactivity of light curable acrylic resins can be increased by introducing labile hydrogens in the chains of the oligomers. This was confirmed by demonstrating higher curing rates for an amino-polyester acrylate than for a polyester-acrylate. [5]

Light cured compositions that contain acrylated polyurethanes have a general reputation of possessing good abrasion resistance, toughness, and flexibility. Furthermore, these properties can be varied by picking either aliphatic polyisocyanates, or aromatic ones, depending upon need. Usually, prepolymers derived from aromatic isocyanates yield better adhesion to the substrate than do the aliphatic ones.

Preparation of acrylated polyurethane was illustrated in one disclosure as follows [17]:

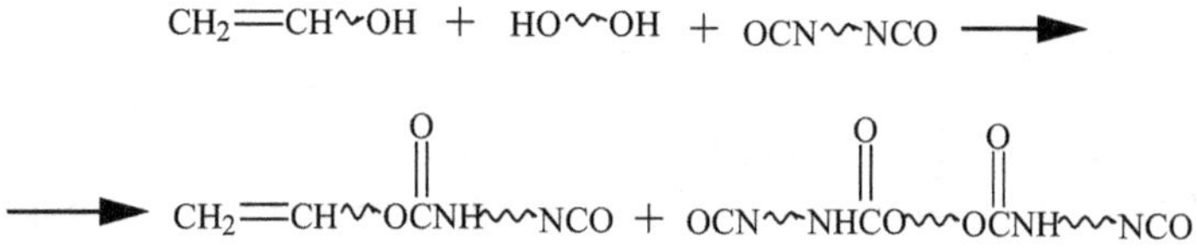

$$CH_2{=}CH{\sim}OH \ + \ HO{\sim}OH \ + \ OCN{\sim}NCO \longrightarrow$$

$$\longrightarrow \ CH_2{=}CH{\sim}O\overset{O}{\overset{\|}{C}}NH{\sim}NCO \ + \ OCN{\sim}NH\overset{O}{\overset{\|}{C}}O{\sim}O\overset{O}{\overset{\|}{C}}NH{\sim}NCO$$

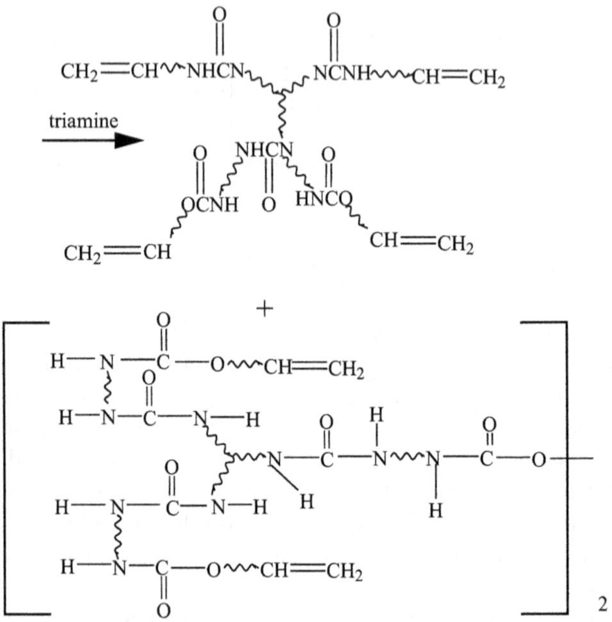

in place of a triamine a triol can also be used. The above shown material was reported to have been prepared from hydroxyethyl acrylate, isophorone diisocyanate, various diols and two trifunctional monomers. It was reported that molecular weights of the diol reactants were the most significant variables in the control of the mechanical properties of the finished films. The higher molecular weight diols yielded films with lower tensile strengths and moduli while giving increased elongation. The selection of the diluent monomers also contributed significantly to adjusting the mechanical properties of the cured films. [17] Linear prepolymers would form, of course, if difucntional amine or a glycol are used in place of trifunctional ones. In fact, such a material is described by McConnel and Willard as follows [18]:

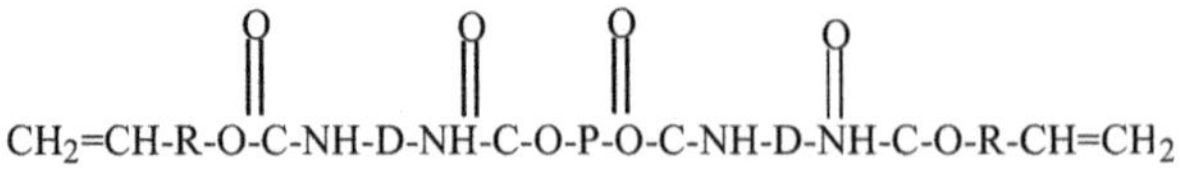

in this illustration R is a hydroxy alkyl acrylate, D stand for the inside components of a diisocyanate, like toluene diisocyanate or isophorone diisocyanate, and P is a polyol. One might also picture such materials as follows:

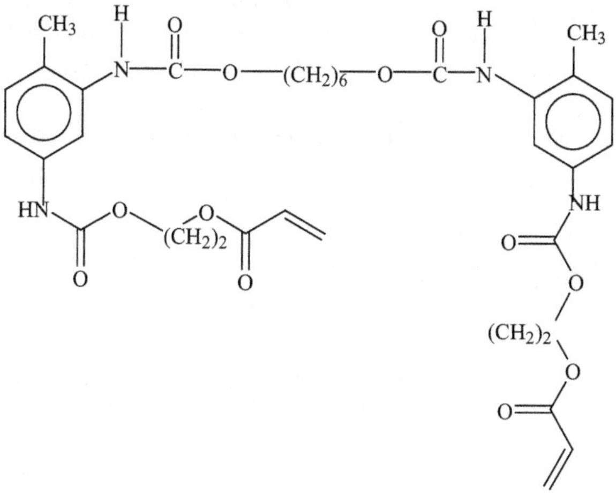

or

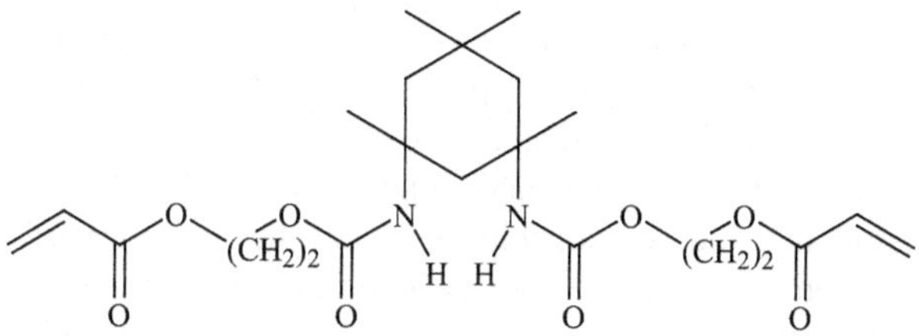

A polyurethane acrylate coating is described in a Japanese patent that contains an ultraviolet light absorbing component, a benzothriazole containing polyester, for more efficient ultraviolet light curing. [19] The component, methylenebis(2-hydroxy-3-2H-benzotriazolyl-5-hydroxyalkylcarbonyloxyethyl) -benzene is shown as follows:

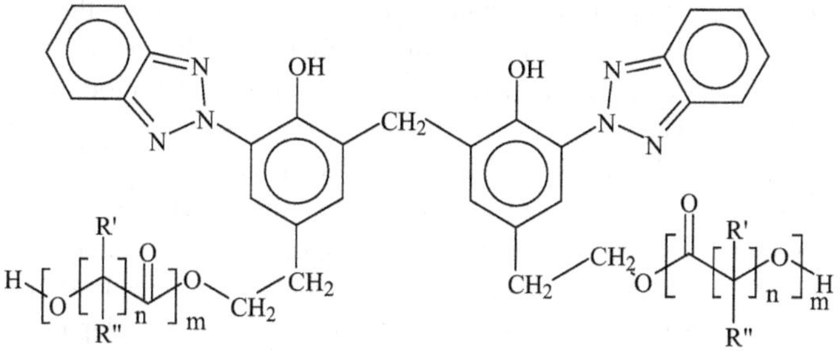

where R',R'' = H,C_{1-10} alkyl; n,n' = 4-8; m,m' = 0-20. This material, shown above is combined in a reactive mixture with isophorone diisocyanate, polycaprolactone diol, hydroxyethyl acrylate copolymer, tripropylene glycol diacrylate to form a polyurethane. A photoinitiator (Irgacure 184) that was mentioned in the patent initiates the crosslinking reaction upon irradiation with UV light. The product is claimed to form a flexible, weather resistant, crosslinked coating. [19]

Another benzotriazine containing polyester was illustrated as follows[19]:

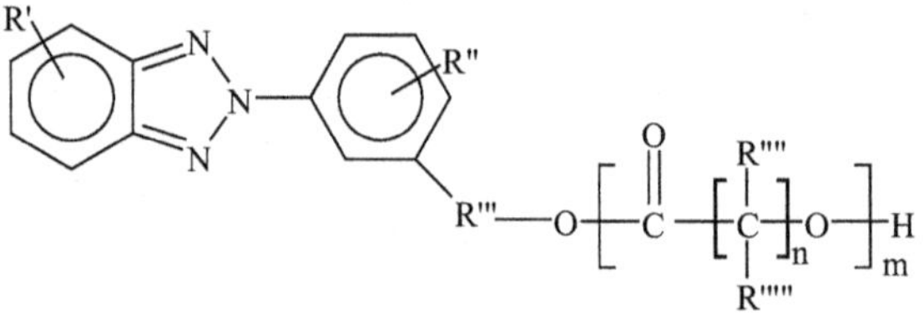

where R' = H, halo; C$_{1-10}$ is alkyl ; R", R"", R""" = H. A sample formulation was described as containing a mixture of hexamethylene diisocyanate-pentaerythritol triacrylate adduct, pentaerythritol triacrylate, and furfuryl acrylate. After UV curing, the material was described as having good scratch resistance, adhesion, transparency, and water resistance. [19]

In studying the relationship between the structure of the urethane acrylates and their performance as finished films, McConnell and Willard [18] used three different diisocyanates in the preparation of the oligomers. These were toluene diisocyanate, isophorone diisocyanate and dicyclohexylmethane-4.4'-diisocyane. They concluded that generally toluene diisocyanete yielded films with the higher tensiles but lower modulus and elongation than does isophorone diisocyane. This was attributed to the more rigid structure of toluene diisocyanate. Also, decreases in molecular weight, weight per urethane linkage, and weight per double bond all gave increases in tensile strength and modulus and decreases in elongation. This, of course, is related to the crosslinking density and the ratio of hard and soft segments in the molecule. The hard segments contribute to higher tensile strength and higher moduli, but also to lower elongation. The reverse is true for soft segments. [18]

Variations in the type of diisocyanates used and well as in the type of diols used can affect the speed of cure. Thus linear polyether linkages in the polyols used are claimed to increase the cure speed of these materials. [18]

Acrylated dimers and trimers of diisocyanates are described in other formulations. These, for instance, were illustrated in another Japanese patent as follows [20]:

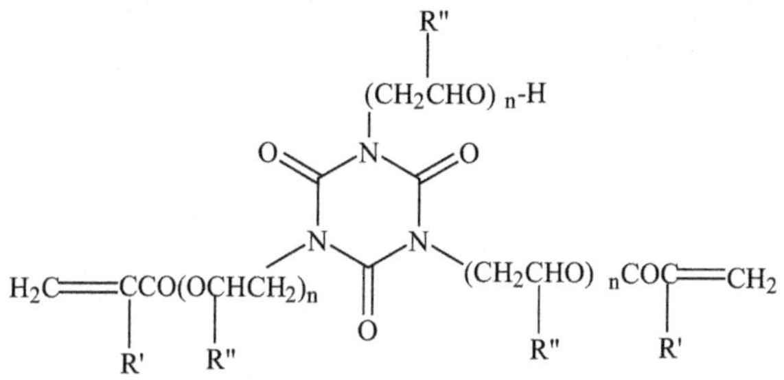

Such urethane acrylates can be formed, according to the patent from polyisocyanates and from isocyanuric acid derivatives. In the illustration R' = H, Me, and R" = H, and n = 1 to 20.[20]

There is some interest in water dispersible light curable coatings and binders. Thus, Heischkel et al, [21] reported preparation of a water based radiation curable polyurethane acrylate coating. Some of the water soluble or dispersible monomers are: diethylaminoethyl acrylate, butanediol monoacrylate, vinyl caparolactam, and propylene glycol monoacrylate.

The oligomers that are based on unsaturated polyesters are often maleic anhydride or fumaric acids based polyesters combined with styrene or vinyl toluene. The relative high reactivity of such mixtures is attributed to the formation of polymerizable charge-transfer complexes between styrene and maleate or fumarate double bonds. Additional unsaturation can also be introduced in a number of ways, including esterification with acrylic or methacrylic acids of the residual hydroxy groups. [22] .Network structures result upon photocuring. Glass-reinforced unsaturated polyesters were reported to cure upon exposure to ultraviolet light when benzoylphosphine oxides are employed as the photoinitiators [22] Also, it was reported that curing with visible light can be achieved when irradiated in the presence of a three-component initiating system based on eosin, a benzoyl oxime ester, and a tertiary benzylamine, [5] or fluorescene-dimethyl ethanol amine. [7] An acrylated unsaturated polyester can be illustrated as follows:

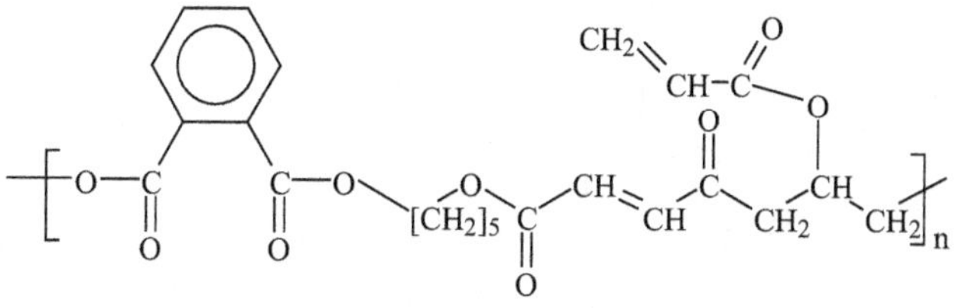

Many modifications of the above shown material are possible. For instance, one might replace some of the acrylate esters with fatty acid ester for internal plasticization. Also, some polyesters are prepared with chlorendic anhydride in place of phthalic anhydride [23]:

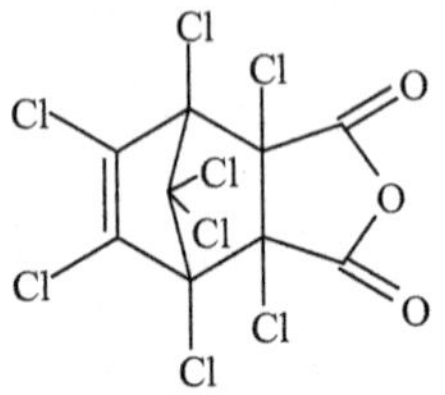

chlorendic anhydride

One advantage of using polyesters based on chlorendic anhydride is that upon UV light irradiation the compound undergoes multiple carbon-chlorine bond ruptures and yields both chlorine and carbon free radicals. [23] This results in greater photocuring rates. An acrylated polyester based on chlorendic anhydride might have the following structure:

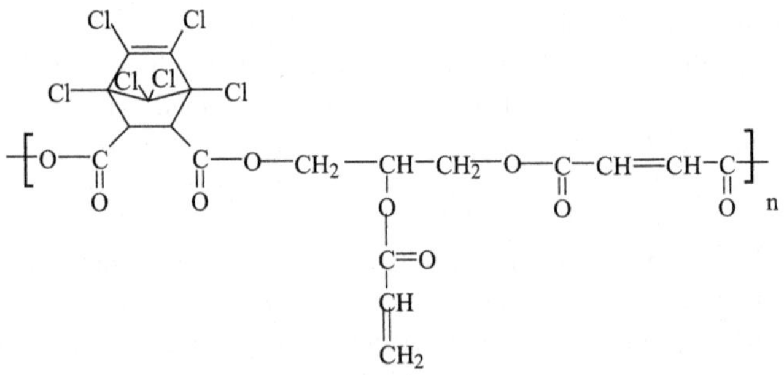

When acrylated epoxy resins are used, the oligomers may simply be reaction products of the diglycidyl ether of Bisphenol A, or one of its higher analogs, with acrylic or methacrylic acids. The preparation of a typical acrylated epoxy oligomer can be illustrated as follows:

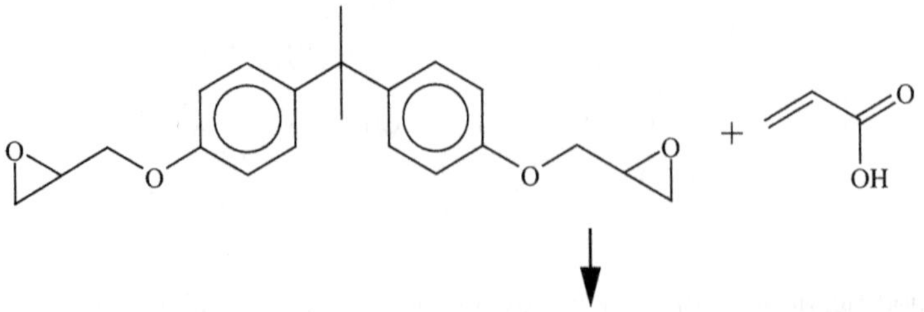

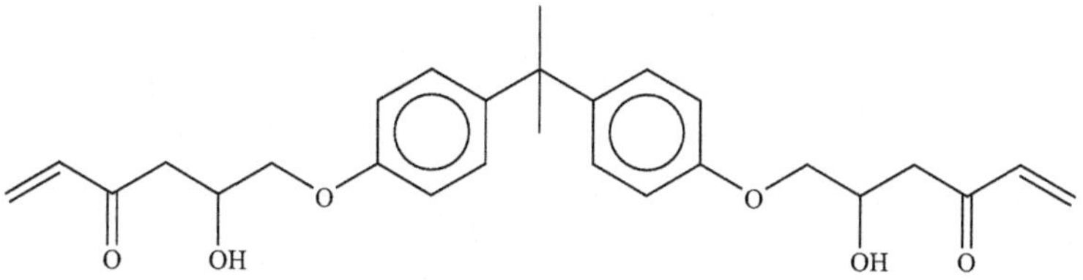

These acrylated epoxy resins crosslink very rapidly by free-radical mechanism in tested UV curable compositions.

Nishikubo and coworkers[24] described the synthesis of dimethacrylate oligomers by addition reaction of epoxy compounds with active esters.

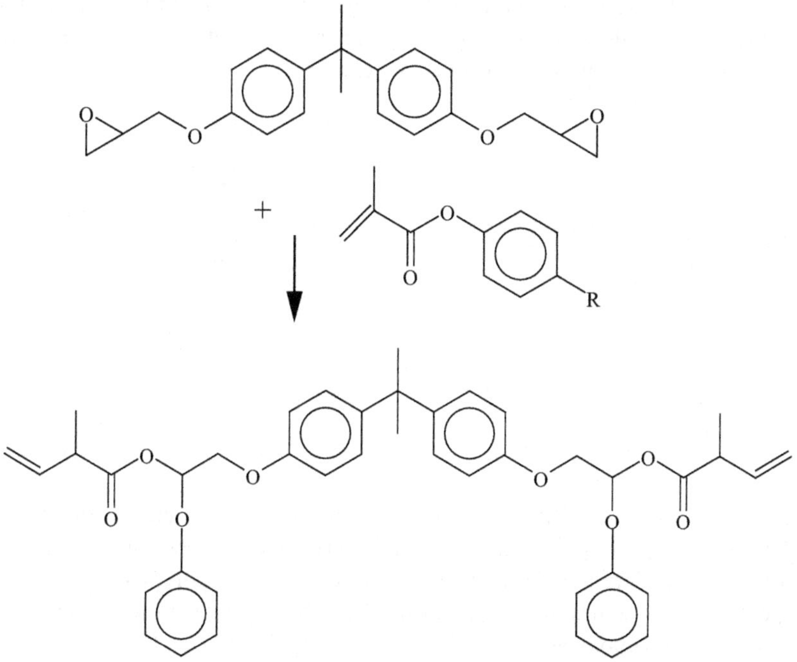

where R represents hydrogen, methyl, nitro, chloro, or other groups. The catalyst used was tetraammonium bromide and the reaction was run at 120 °C.

Yang and Schaeffer [25] pointed out that the photocuring industry is interested in liquid oligomers that can form highly flexible films. They suggest using elastomeric liquid polybutadiene with acrylic or methacrylic esters. Such oligomers can be illustrated as follows [25]:

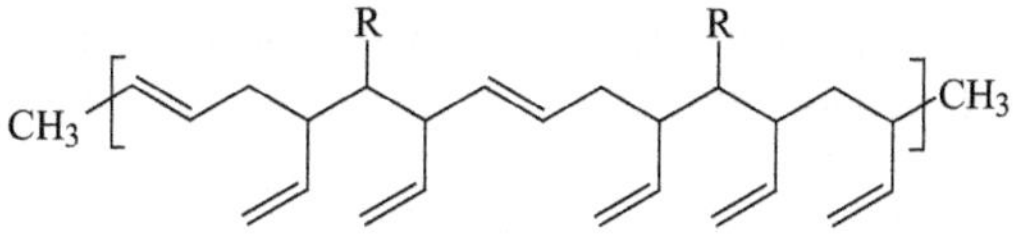

Oligomer A

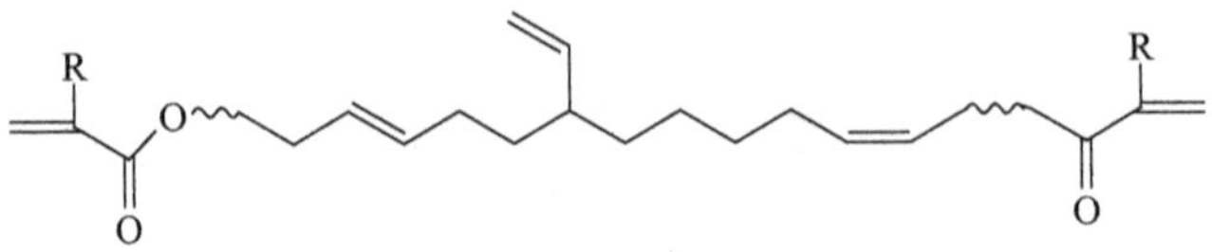

Oligomer B

In oligomer A, R is an acrylate, a methacrylate, or an amine group, while in oligomer B, R it is a hydrogen or a methyl group. [25]

Gummeson [26] reported that the triazine ring has six reactive sites that can be utilized to prepare materials with a range of unsaturated functionality for light curing. This allows formation of a number of acrylated prepolymers. An acrylated melamine resin was illustrated as follows [26]:

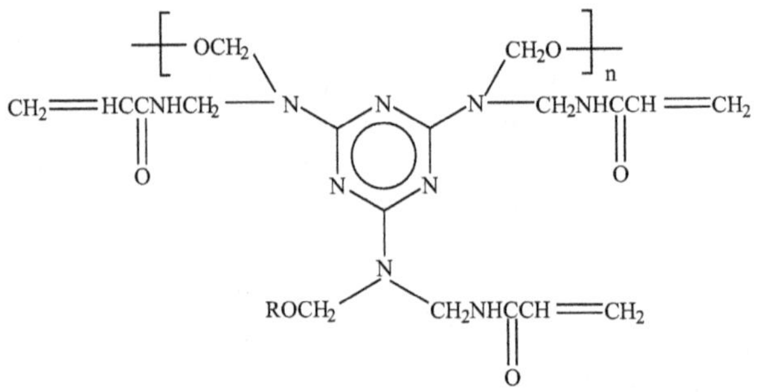

Gummeson's data shows, however, that these materials do not cure as rapidly as do the epoxy acrylate prepolymers.[26] Their shrinkage, however, is lower than that of the epoxy acrylates. Also, he reported that these prepolymers yield films that possess better adhesion to the substrate and better abrasion resistance than films formed with epoxy acrylates. [26]

A prepolymer that contains calixarene derivatives was reported. The material can also include methacrylate, vinyl ether, propargyl ether, oxetane, oxirane, or spiro ortho ester groups. [25] Such materials were synthesized by reacting calixarenes with methacrylic acid derivatives, or with vinyl ether compounds. Reactions with propargyl bromide, oxetane derivative, epibromohydrin, and spiro ortho ester derivatives were also carried out. The resulting calixarene derivative, containing photoreactive groups were claimed to exhibit excellent thermal stability and high photochemical reactivity. [25] The synthesis was illustrated as follows:

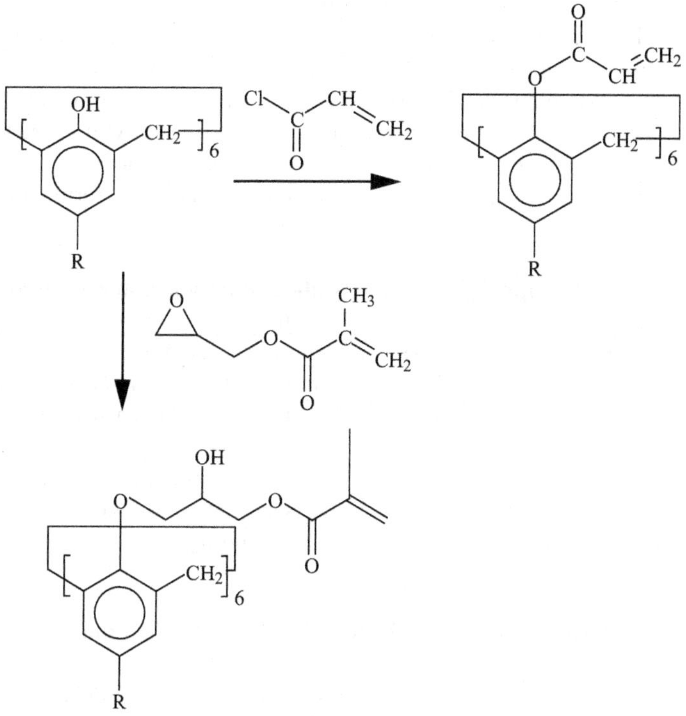

Kou *et al.*, [27] reported using a hyperbranched acrylated aromatic polyester as a modifier in UV curable epoxyacrylate resin. The material is compatible with the epoxy-acrylate resins. They found that the photopolymerization rate of the resin is promoted by this modifier. Also, the shrinkage of the resin was reduced. At the same time, the tensile, flexural, compressive strength, and thermal properties of the ultraviolet light cured films are greatly improved. [27]

Krongauz and Chawla, [28] reported on the effect of aromatic thiols on kinetics of acrylate radical photopolymerization in the presence and absence of photoinitiators. They observed that aromatic thiols at concentrations, < 0.5% (-0.05 M), can accelerate radical photopolymerization. Initiation of radical photopolymerization. by some aromatic thiols in the absence of conventional photoinitiators was also observed. On the other hand, there was also an unexpected inhibition of photopolymerization at higher concentration of the aromatic thiols due to chain transfer. Also, a ground state charge-transfer complex formation between thiols and benzoin based photoinitiators was detected. [28]

In a study of polymerization kinetics of photocurable compounds based on an epoxy acrylate oligomer and three analogous diacrylate monomers it was observed that the polymerization is characterized by a synergistic effect. [29] This was observed in a wide temperature range and occurring for the polymerization rate both in air and in Ar and for final conversions in air. The presence of a heteroatom (sulfur or oxygen) in the ester group of the reactive diluent is beneficial for the polymerization course, especially in air atmosphere The best results were obtained for the sulfur-containing. monomers. [29]

The basic components of light curable coating formulations are presented in Table 3.2.

Table 3.2. Basic Components of Free Radical Photocurable Coatings

Component	Percent	Function
Photoinitiator	1-3	Free-radical initiation
Reactive diluent	15-60	Film formation and viscosity control
Prepolymer or Oligomer	25-85	Film formation and basic properties
Additives and pigments	1-50	Pigmentation, stabilization, flow agents and surfactants to improve flow and surface wetting , etc.

3.2.1.3. The Mechanism of Free-Radical Photopolymerization

Common photopolymerization reactions by free radical mechanism follow the same laws of chemistry as do the thermal polymerizations. The differences are primarily in the formations of the initiating radicals. Typical chain growth polymerization reactions are initiated by free radicals that come from thermal decomposition of the initiators. The initiating free radicals in photo polymerizations, on the other hand, come from photolyses of the photoinitiators.

3.2.1.3.a. The Initiation Reaction

The initiation reaction consists of attacks by the free radicals formed from the photoreactions of the initiators upon the molecules of the monomers and prepolymers. In simple terms the initiation reaction can be described as:

$$\text{Initiator} + h\nu \longrightarrow R\bullet$$

$$R\bullet + M \text{ (monomer)} \longrightarrow RM\bullet$$

where, **R•** represents the radical that forms through the photoreaction of the initiator and **M** is the monomer. The rate of initiation can be described by the equation:

$$R_i = dRM\bullet / dt$$

Many aspects of the polymerization reactions of light curable coatings, however, are more complex. A number of attempts were made to describe the kinetic pictures of photocuring. Some of these are presented below.

Based on simulation experiments Stephenson, Kirks, and Scranton [30] concluded that the initiator concentration has an optimum value. The ability to have both high and low initiator absorptivities is a key advantage when using a polychromatic light source. High initiator absorptivity at one or more wavelengths allows for a higher maximum rate of initiation near the surface, while a lower absorptivity at a different wavelength allows initiation to occur deeper in the sample. The combined effects of all the incident wavelengths in the polychromatic light leads to the more rapid and efficient photoinitiation wave front propagation through the depth of the sample than that observed using monochromatic light. In any system where an element is absorbing over the same wavelengths as the initiator, initiation will be hindered. The wave front of the perfectly nonbleaching system does not propagate through the depth of the film and remains stagnant on the surface. The wave front of the perfectly bleaching and partially bleaching systems start at the surface and eventually reached the bottom of the material. [30]

Recently Fouassier and coworkers, [31] presented a general photo-thermal methods for studying both kinetic and thermodinamic properties of the photopolymerization processes. Photoacoustic and thermal lensing spectroscopies allow the determination. of triplet quantum yields and energy levels of photoinitiators. Beyond the possibility of determining easily and accurately bond dissociation energies of coinitiators, the methods provide important information on their reactivity. The application of photoacoustics was extended by Fouassier and coworkers, [31] to the study of the initiation step. A specific data treatment was developed to determine the rate constants and the enthalpy of the reaction of addition of a radical to a monomer unit. [31]

Kurdikar and Pappas developed a kinetic model for photopolymerization of diethylene glycol diacrylate [32] In this reaction the rate of initiation, $\mathbf{R_i}$ is given by

$$\mathbf{R_i = 2\Phi I_a}$$

where I_a is the intensity of absorbed light and Φ is the number of propagating chains produced per number of light photons absorbed (referred to as the quantum yield for initiation). The term Φ was expressed as $\Phi = f\Phi'$ where f is the initiator efficiency and Φ' is the number of initiating molecules dissociated per photon absorbed. The factor 2 accounts for two radicals that are produced per initiator molecule (this is only true of some photoinitiators). The absorbed light intensity is often expressed as

$$I_a = \varepsilon I_0 [A]$$

where ε is the extinction coefficient of the initiator, [A] is the initiator concentration and I_o is the incident light intensity per unit area. Assuming bimolecular reaction, the rate of termination, R_I could be calculated by

$$R_I = 2k_I[M\bullet]$$

Where[M•] is the concentration of all polymer chain radicals in the system, and the factor 2 accounts for the destruction of two radicals per dead polymer chain that forms.

When the initiator does not absorb the light, however, and all the energy is absorbed instead by a sensitizer, I_a then

$$I_a = (\Phi I_0 L^{-1})(1 - \exp(-\in[S]L))$$

where I_a represents the moles of photons absorbed per liter of the system , [S] is the concentration of the sensitizer, and $\in$ is the molar absorption coefficient of the sensitizer at the wavelength of light that enters the system, and L is the thickness of the film that is being irradiated. [32]

If the monomer absorbs the activating light in a photoinitiated polymerization, then the initiation rate and the sensitizer destruction rate are given by [33] :

$$R_i = 2\Phi\left(\frac{I_0}{L}\right)\left\{\frac{\epsilon[S]}{\epsilon_1[M] + \epsilon[S]}\left(1 - e^{-(\epsilon_1[M]+\epsilon[S])L}\right)\right\}$$

For the initiation process, Terrones and Pearlstein [34] developed an unsteady state one-dimensional model to account for initiator consumption and optical attenuation and to derive relationships for the spatial and temporal variation of the local initiator concentration and initiation rate in a free-radical photopolymerization. With increasing absorbance, the local initiation rate becomes increasingly nonuniform and assumes the form of a highly localized traveling wave propagating with speed $\Phi I_0/C_{A,0}$ (where Φ, I_0, and $C_{A,0}$ are the quantum yield of photoinitiator consumption, incident intensity, and initial photoinitiator concentration, respectively), independent of photoinitiator absorption coefficient. At high attenuation, the maximum photoinitiation rate (as a function of position) is asymptotically one-fourth the initial rate at the front of the layer. Without diffusion, however, the time-integrated production of primary radicals is uniform if the initial initiator concentration is uniform and all initiator is consumed, since each initiator molecule is photolyzed in place. Further, Terrones and Pearlstein, [34] consider the effects of diffusion of a photobleaching initiator for finite values of the ratio of the diffusive time scale l^2/D to the reaction time scale $1/(\phi I_0 \alpha_A)$, where l and I_0 are the layer thickness and incident light intensity at the optical entrance, and D and α_A are the diffusion coefficient and molar absorption coefficient of the photoinitiator, respectively, whose quantum yield of consumption is ϕ. Compared to the limiting case in which diffusion is negligible, diffusion has the effect of shifting the instantaneous initiation rate profiles forward in the layer, where initiator is relatively depleted. On the other hand, for any nonzero initial absorbance, the overall (i.e., time-integrated) consumption of initiator becomes more *non-uniform* as the ratio of the rates of diffusion and reaction, expressed in the dimensionless ratio $\delta = D / (l^2 \phi I_0 \alpha_A)$, increases. When diffusion is fast (large δ), the front-to-back difference in the time-integrated primary radical production varies quadratically. Terrones and Pearlstein [34] developed the following equation for front-to-back variation in the dimensional time-integrated primary radical production:

$$\int_0^\infty [R_i(0,t) - R_i(L,t)]\, dt =$$
$$C_{A,0} \int_0^\infty [\tilde{R}_i(0,\tau) - \tilde{R}_i(1,\tau)]\, d\tau = C_{A,0} \Delta\Theta$$

where R_i is the initiation rate, t is time, L is layer depth, C is concentration of species, τ is dimentionless time, and $\Delta\Theta$ is front-to-back difference in dimensionless time-integrated radical production. [34]

The initiation process can be affected by quenching. Thus, a relationship was demonstrated between the quenching rate constant of benzophenone triplet state and the ionization potential of a series of monomers and amines. [31] If amines are present in the reaction medium, as in the case of photoreduction (this is described in Chapter 2), quenching by the amines can then be a factor. It was shown that the ability of the charge transfer complex to undergo efficiently a proton transfer can be affected by the monomer matrix and the basicity and chemical structure of the amine. In addition, it can also be affected by the solvent, if one is present. [31] Ruhlmann and Fouassier [31] concluded, however, that cleavable photoinitiators that possess short-lived triplet states are not affected by monomer quenching. Such photoinitiators are benzyldimethyl ketal, benzoin ethers, hydroxy alkyls, acetophenones, benzoyl phosphine oxides, benzoyl oxime esters, and dialkoxy acetophenones. In the case of the α-morpholino ketones and α-amino para morpholino benzoyl ketones, the slower α-cleavage process is balanced by a weaker monomer quenching. They also concluded that monomer quenching is reasonably high in benzophenones but quite low in thioxanthones. In the case of efficient cleavable photoinitiators, amine quenching usually does not compete with the photo-scission process. [31] Also, in the highly viscous bulk monomer media, the bimolecular quenching rate constant, k_q decreases due to sharp reduction in the diffusion process. [31] The efficiency of quenching by monomers depends largely upon the monomer's chemical structure. Following is an arrangement of some of the monomers in terms of their quenching efficiency [31]:

Turro [35] shows the quenching rate constant for diffusion controlled reactions as,

$$k_q = 8RT/3{,}000 \, \eta \text{ liters mole}^{-1}\text{sec}^{-1}$$

where R is the gas constant, T is the temperature, and η is the solvent viscosity in poise. Monomer quenching, however, is not a serious problem, even in a medium of high concentrations when highly reactive triplet state are available. [35]

Stephenson, Kirks, and Scranton tried to develop a working model of photoinitiation of thick polymeric systems. [257] They used the generalized governing equations for photochromic illumination. The equations they derived are as follows [30]:

Dependence of initiator concentration on time:

$$\frac{\partial C_i(z,t)}{\partial t} = \frac{C_i(z,t)}{N_A h} \sum_j \frac{\varepsilon_{i,j} \, \varphi_i \, l_j(z,t)}{v_j} + D_i \frac{\partial^2 C_i(z,t)}{\partial z^2}$$

Dependence of photolysis product concentration on time:

$$\frac{\partial C_p(z,t)}{\partial t} = \frac{C_i(z,t)}{N_A h} \sum_j \frac{\varepsilon_{i,j} \, \varphi_i \, l_j(z,t)}{v_j} + D_p \frac{\partial^2 C_p(z,t)}{\partial z^2}$$

Light intensity dependence on sample thickness:

$$\frac{\partial l_j(z,t)}{\partial z} = -\left[\varepsilon_{i,j} C_i(z,t) + A_m + \varepsilon_{p,j} C_p(z,t) \right] l_j(z,t)$$

where, $C_i(Z,t)$ is the initiator molar concentration at depth z and time t, $C_p(Z,t)$ is the photolysis product molar concentration at depth z and time t, $I_j(z,t)$ represent the incident light intensity at depth z and time t, with dimensions of energy/(area•mole); ε_i is the initiator molar absorptivity with dimensions of volume/(length•mole). The photolysis product molar absorptivity is ε_p, with dimensions of volume/(length•mole). It accounts for the photon absorption by all fragmentation species. N_A if the Avogadro number, H is Planck's constant, v is the frequency of light, with dimensions of inverse time, φ is the quantum yield of the initiator, defined as the fraction of absorbed photons that caused the fragmentation of the initiator. D_i, is the diffusion coefficient of the initiator with dimensions of length^2/ time, D_p is the diffusion coefficient for the photolysis product, and A_m is the absorption coefficient of the monomer and polymer repeat unit. [30]

The influence of photoinitiator mobility and solubility on the free-radical polymerization behavior in lyotropic liquid crystalline systems was also investigated. [31] In this study common photoinitiators were used. They were two commercial materials:

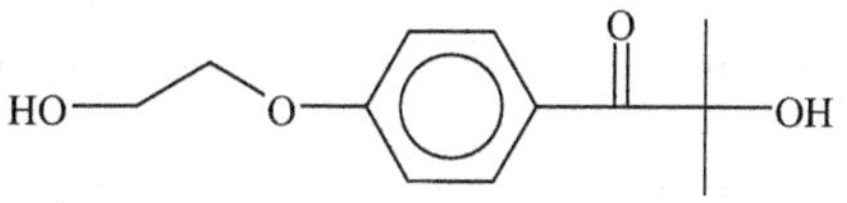

Irgacure 2959

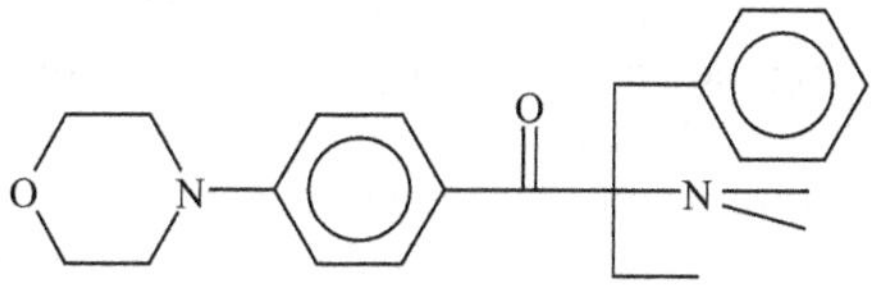

Irgacure 369

The conclusion of this study is that the kinetics of the free-radical polymerization initiated with highly soluble and mobile photoinitiators are governed primarily by monomer ordering. In contrast, initiator efficiency largely controls the polymerization rate for initiations with bulkier initiators. These were found to display lower mobility and solubility in liquid crystalline environment. [31]

3.2.1.3.b. The Propagation Reaction

Free-radical polymerization of multifunctional monomers, such as diacrylates and dimethacrylates forms highly crosslinked, rigid, and glassy polymer networks. One problem with such crosslinking systems is that the conversion of reactive functional groups usually remains incomplete because of dramatic drops in the mobility of the radicals and functional groups during the polymerization reaction. Such dramatic drops in mobility also lead to characteristic kinetic features, such as autoacceleration, and an early onset of trapping of radicals and severe reaction-diffusion behavior. The crosslinking polymerization kinetics is linked to the development of the overall polymer network. Due to the limitation of diffusion, the growing polymer chains tend to cyclize quickly and crosslink into clusters. These clusters are sometimes referred to as microgels. The clusters become linked up into networks. The highly crosslinked clusters can trap the free-radicals and shield them from further polymerization. Free-radicals can also become trapped at high conversion due to vitrification as the microgels connect into a network of the whole material. At some point in the reaction, ordinary diffusion becomes less important to the overall kinetics than the motion of radicals through propagating along their kinetic chains. In addition, due to the fact that some radicals become trapped in highly reacted regions, a spatial unevenness develops. There are variations in conversions from one local to another. Thus, more highly reacted regions coexist with less highly reacted ones.

The diffusional processes become important also during the course of a linear polymerization, causing the controlling step of the reaction to shift from chemical reaction-controlled to diffusion-controlled. For a bulk crosslinking homopolymerization, diffusional limitations exist from the onset of the reaction, causing the reaction tendencies to be even more pronounced. One of the most widely accepted models for the calculation of the diffusion-limited propagation and termination rate constants during linear polymerization was developed by Marten and Hamielec.[40] Using the Doolittle equation of the diffusivity of small molecules in a polymer solution, they derived expressions that could be used to calculate the rate constants throughout the range of conversion. Other models were developed since, but all have limited applicability for the prediction of the kinetic behavior of the crosslinking reaction.[40]

The analysis of the diffusion-controlled features might be simplified by identifying the two types of free radicals: the active and the trapped ones. Electron spin resonance spectroscopy shows that active (mobile) radicals give a 13-line spectrum and trapped (static) radicals give a nine-line spectrum.[8] Also, photopolymerization of a number of neat acrylate monomers used in polymer coatings for optical fiber was studied with photo DSC and with a cure monitor using a fluorescent probe.[36] The acrylates had a functionality of one to six. It was found that conversion of monomers ranges from 40% to 100%. This, however, is depended upon functionality and structure of particular monomers. It can also be a function of the type and amount of the photoinitiator used.

The propagation reactions consist of successive additions of the monomer molecules to the growing polymeric chain radicals. It also involves reactions of the chain radicals with the reactive double bonds of the prepolymers or oligomers:

$$\mathbf{RM\bullet \ + \ M_n \ \longrightarrow \ RM_{n+1}\bullet}$$

Each propagation reaction can result in addition of a thousand or more molecules. The rate of propagation can be expressed :

$$\mathbf{R_p = d[M] \ /dt \ = kp \ [M][\Sigma \ M\bullet]}$$

In these reactions, the quantum yield of polymerization ϕ_m is defined as the number of monomer units polymerized per absorbed photon:

$$\boldsymbol{\phi_m \ = \ R_p \ / \ I_{abs}} = \text{Rate of polymerization (in M s}^{-1}) \ / \ \text{Light intensity (in Einstein l}^{-1}\text{s}^{-1})$$

If monochromatic light is used and a steady state hypothesis is assumed, then the following relationships apply for the propagation rate, R_p, and the degree of polymerization, DP_n is [36,37]:

$$R_p = k_p / k_t^{0.5} \text{ [M] } R_i^{0.5} \qquad DP_n = Kk_p/k_t^{0.5}[M]/R_i$$

Where $R_i = (2.3 I_0 \times \text{optical density} \times \Phi_1)$ and k_p and k_p are rate constants for propagation and termination, respectively. There is, however, an inherent spatial no uniformity of the reaction rate, due to the dependence of local rates on local light intensity. Because the light intensity $I(x,t)$ decreases along the beam path according to the equation [4]:

$$\delta I (x,t) / \delta x = [\Sigma \alpha_i C (x,I)_i] I (x,t)$$

where α and C are the absorption coefficient and concentration of the i-th absorbing species. There are $N \geq 1$ absorbing species. [17] Decker and coworkers [38] investigated the kinetics of free-radical photopolymerization of a mixture of acrylic monomers and oligomers, using triphenylphosphine combined with bisacylphosphine oxide as the initiators. Their conclusion was that from the individual values of k_p (rate constant for propagation) and k_t (rate constant for termination) and Φ_i, the quantum yield of initiation, the rate of propagation, R_p, can be determined from the following rate equation [38]:

$$R_p = (k_p / k_t^{0.5})(\Phi_i \times I_0)^{0.5} \text{ [M]}$$

The ratio of $k_p / k_t^{0.5}$ was calculated by them from rate measurements under steady state conditions. [38]

Wen and McCormick [39] developed a kinetic model for radical trapping in photopolymerization of multifunctional monomers. Their model assumes that the trapping of radicals behaves as a unimolecular first-order reaction. Also, the trapping rate constant is presumed to increase exponentially with the inverse of the free volume. The model predicts the experimental trends in trapped and active radical concentrations: (1) the active radical concentration passes through a maximum while the trapped radical concentration increases monotonically; (2) a higher light intensity leads to a lower fraction of trapped radicals at a given conversion of functional groups but to a higher trapped radical concentration at the end of the reaction. It also predicts that the polymerization rate depends more on the light intensity, showing that higher light intensity can lead to a higher final conversion. [39] In this work, Wen and McCormick [39] define trapping of radicals as the conversion of active radicals into inactive ones as the mobility becomes lost. Qualitatively, radical trapping has been considered as a pseudo-first-order termination process. Such an approach, combined with a pseudo-steady-state approximation, has been used to rationalize the experimental finding that the rate of photopolymerization can depend more on light intensity than might be expected without trapping at high conversions. [38]

The functional-group reaction scheme for free-radical polymerizations of multifunctional monomers with the formation of trapped radicals is shown below [39]:

$$\text{initiation: I} \longrightarrow 2R\bullet$$

$$R\bullet + M \rightarrow R\bullet_{1a}$$

$$\text{propagation: } R\bullet_{na} + M \xrightarrow{k_P} R\bullet_{(n+1)a} \qquad (n > 0)$$

$$\text{trapping: } R\bullet_{na} \xrightarrow{k_b} R\bullet_{nb} \qquad (n > 0)$$

$$\text{termination: } R\bullet_{na} + R\bullet_{ma} \longrightarrow P_{n+m} \qquad (n,m > 0)$$

In this scheme, I is an initiator, $R\bullet$ is a primary radical, $R\bullet_{1a}$ is a chain initiating radical, $R\bullet_{na}$ and $R\bullet_{nb}$ are active and trapped (buried) radicals with n functional groups reacted, M is a functional group, P_{n+m} is a dead polymer with $(n + m)$ functional groups reacted, k_p is the rate constant for propagation, k_b is the rate constant for radical trapping (burying), and k_t is the rate constant for termination. [39]

In the initiation step, the initiator splits into two primary radicals $R\bullet$, each of which reacts with a functional group to form a chain initiating radical $R\bullet_{1a}$. Because reactions by trapped radicals, as defined in this model by Wen and McCormick [39] is very much slower than the reactions by active radicals, propagation and termination can be taken to involve only active radicals. Propagation consists of the growth of active radicals by the successive addition of functional groups. Because radical trapping is presumed to be permanent, it affects the rate in a manner similar to unimolecular termination. [39]

To develop their model, Wen and McCormick[39] adopted a number of simplifying assumptions. These are: (1) initiation produces two equally reactive radicals, (2) chain transfer reactions are neglected, (3) the rate constants for radicals of different sizes are assumed identical, (4) the propagation rate constant k_p, termination rate constant k_t, and the rate constant for radical trapping k_b are all simple functions of free volume as shown below, and (5) there is no excess free volume. The material balance equations for the initiator, the functional group, the active radical, and the trapped radical concentrations are [39]:

$$d[A]/dt = -I_0 b \in [A] \qquad (1)$$

$$d[M]/dt = -k_P [MI][R_a] \qquad (2)$$

$$d[R_a]/dt = 2fI_0 b \in [A] - k_b[R_a] - 2k_t [R_a]^2 \qquad (3)$$

$$d[R_b]/dt = k_b[R_a] \qquad (4)$$

where [A] is the current (not initial) photoinitiator concentration, [M] is the functional group concentration, $[R_a]$ is the active radical concentration, $[R_b]$ is the trapped radical concentration, I_0 is the incident light intensity in the units of Einstein / (L s), b is the film thickness which will be canceled out when it is multiplied by I_0, ϵ is the extinction coefficient of initiator at the applied wavelength, and f is the initiator efficiency. The initiation term in equation 3 assumes only weak absorption (e.g., $\epsilon[A]b \ll 1$). In principle, this model can be applied to either thermal initiated or photoinitiated systems, but the development here was focused on photoinitiation. [39]

The propagation rate constant k_p and the termination rate constant k_t are calculated as functions of free volume, following a model developed by Anseth and Bowman.[41] In that model, the two constants are calculated by summing different resistances for diffusion, intrinsic reaction, and reaction—diffusion (for termination).

$$k_P = \frac{k_{P0}}{1 + e^{A_P(1/u_f - 1/u_{fcp})}}$$

$$k_t = k_{t0}\left(1 + \frac{1}{\dfrac{Rk_p[M]}{k_{t0}} + e^{-A_t(1/u_f - 1/u_{fct})}}\right)$$

where A_p is the dimensionless activation volume which governs the rate at which propagation rate constant decreases in the diffusion-controlled period, u_{fcp} is the characteristic fractional free volume at which the diffusional resistance equals the intrinsic reaction resistance for propagation, k_{p0} is a preexponential factor, A_t is the dimensionless activation volume which governs the rate at which termination rate constant decreases in the diffusion-controlled period, f_{ct} is the characteristic fractional free volume at which the diffusional resistance equals the intrinsic reaction resistance for termination, k_{t0} is a preexponential factor, and R *is* the reaction—diffusion parameter. [39]

Radical trapping becomes more and more severe as polymerization proceeds. This means that the rate constant for radical trapping k_b, should increase with the extent of reaction. Because no model is available to predict how k_b changes during the course of polymerization by a mean-field method, it is presumed [3] that it is inversely proportional to the diffusion coefficient of active radicals. This simple assumption appears to be reasonable because active radicals become less and less mobile as their diffusion coefficient drops and leads to radical trapping. Here, it was postulated by Wen and McCormick [39] that segmental diffusion controls the trapping process. Consequently, the diffusivity is not a function of polymer radical molecular weight but changes with the fractional free volume exponentially:

$$k_b = k_{b0}e^{A_b / u_f}$$

where A_b is the dimensionless activation volume which governs the rate at which radical trapping increases as a function of fractional free volume and k_{b0} is the preexponential factor. This relationship implies that the rate of radical trapping increases dramatically with the loss of free volume. [39]

There is a question about what takes place in a photopolymerization reaction in the event that the photoinitiator is depleted before all the monomer or monomers are converted to polymers in these reactions. The kinetics of such polymerizations were studied by several investigators. [42-47] Tobolsky termed it "dead end polymerization" [47] and assumed that the decomposition of the photoinitiator is a first-order reaction. Subsequently, Hill and O'Donnell modified the Tobolsky analysis [48] that simplified data analysis and made it possible to estimate the kinetic parameters without measuring the monomer conversion at infinite time. Later Joshi and Rodriguez [33] derived expressions for a photoinitiated polymerization that does not go to completion. At very high conversions the propagation rate constant can decrease with increase in conversion but Joshi and Rodriguez point out that is reasonable to assume that the rate of propagation and termination remain constant to simplify the derivation. The limiting conversion at infinite time is then expressed by them as follows [33]:

$$1 - \frac{[M]_\infty}{[M]_0} = \frac{1}{[M]_0} \int_0^\infty k_p \left(\frac{R_i}{2k_t} \right)^{1/2} [M]\, dt$$

3.2.1.3.c. The Termination Reaction

The termination reaction can be by combination or by disproportionation, depending upon the monomer used [49] :

$$RM_n\bullet + \bullet M_mR \longrightarrow RM_{nm}MR$$
$$RM_n\bullet + \bullet M_mR \longrightarrow RM_n + M_mR$$

The rate of termination is:

$$R_t = k_t \left[\sum M\bullet \right]^2$$
$$(RM_n\bullet + R\bullet \longrightarrow RM_nR)$$

Nie, Lovell, and Bowman [50] studied dependence of termination on the chain length in crosslinked methacrylate systems that takes place before the onset of diffusion controlled termination.. Research has shown that. in linear systems, the termination rate constant is dependent mainly on the length of the macroradical undergoing termination; Their results do suggest that chain length dependent termination is important in cross-linked methacrylate systems before the onset of reaction diffusion controls termination. [50] In addition, Bowman *et al.*, [50] found that shifting the kinetic chain lengths toward shorter chains has little visible effect on multiacrylate systems. In contrast, similar changes in the corresponding multimethacrylate polymerizations changed the kinetics significantly. Shorter kinetic chains led to the delayed onset of reaction-diffusion-controlled termination behavior, as well as an increase in the ratio of k_t /k_p[M] at all conversions prior to the onset of reaction-diffusion control. Additionally, the magnitude of the kinetic constant ratio in the reaction-diffusion-controlled regime was affected by the kinetics at low conversion in the polymerization of a rubbery system,, based on polyethylene glycol dimethacrylate. This behavior was independent of the method used to alter the kinetic chain length distribution and thus implies that chain length dependent termination may potentially impact the network formation in polymerizations occurring above the T_g of the system. Thus, their results illustrate that chain length termination is an important factor in cross-linking free radical polymerizations.

When the initiation rate constant is increased, the polymerization rate also rises because of the higher population or active radicals. The final conversion, however, drops. This is due to an increase in active radicals and results in lower kinetic chain length. The cumulative chain length, v, is the number of propagation steps that are carried out be a free radical. It can be calculated from the following equation [3]:

$$v = \frac{xn_{m0}}{n_{rg}}$$

where x is the conversion of functional groups, n_{m0} is the initial number of functional groups, and n_{rg} is the total number of radicals generated.

Chang *et al.*, [51] studied the relationships between photocuring, shrinkage, and chemical structure of the monomers. They used six silicon containing methacrylates. The following monomers were prepared and used in their study: Si(2-hydroxyethyl methacrylate)$_4$, CH_3Si-(2-hydroxyethyl methacrylate)$_3$, HSi-(2-hydroxyethyil methacrylate)$_3$, $(CH_3)_2Si$-(2-hydroxyethyl methacrylate)$_2$, n-C_5H_{11}-Si-(2-hydroxyethyl methacrylate)$_3$, and (2-hydroxyethyl methacrylate)-$[Si(CH_3)_2$-$O]_4$-(2-hydroxyethyl methacrylate). Volumetric curing shrinkages of the monomers were measured by a water displacement technique using a molar monomer/initiator ratio of 100:1. They found that the polymerization shrinkage of the methacrylates decreased with increasing molecular weight of the monomer This demonstrates that in formulating light curable compositions it preferable to use bulky monomers when possible.

Stolov et al., [52] studied stress buildup in light cured coatings and its relationship to sample thickness. They observed that stress buildup in ultraviolet radiation-induced coatings can vary significantly as a function of processing conditions. Depending on radiation intensity and photoinitiator concentration, the average stress in the film may either increase or decrease with film thickness. The conclusion is that two main factors affect stress in radiation curable coatings. These are (1) degree and distribution of crosslinking and (2) temperature at the solidification point. In certain experimental conditions, an uneven distribution of the conversion within the film may occur, leading to a decrease of the average film stress with increasing film thickness. In contrast, the temperature jump in the sample when irradiated may lead to an increase of the ultimate stress with film thickness. [52]

Stolov et al., [52] concluded that the quality of polymer coating is strongly affected by the magnitude of internal stress that develops as the liquid (solution or dispersion) transforms into an elastic or viscoelastic solid. Since the coating by definition adheres to a substrate, shrinkage can occur only in the thickness direction. The constrained or frustrated volume change in the direction parallel to the substrate leads to an in plane stress which can ultimately lead to defects such as buckling, cracking, curling, and delamination, all of which degrade final coating quality. The quality of polymer coatings, however, depends not on the average stress but on stress distribution with the distance from the coating/substrate interface. [52] Generally the conversion of vinyl groups, P, as a function of film depth, y, is described by Stolov et al., as follows [52]:

$$P(y) = P_0 \exp(-\alpha y) \qquad (1)$$

where P_0 is the conversion at the outer surface and α is a coefficient that depends on the absorptivity and diffusion constant. The average conversion in the system was determined by them from equation 1:

$$P_{AV} = P_0/t_F\alpha \, [1 - \exp(-\alpha t_F)] \qquad (2)$$

Average stress in the coating is proportional to crosslink density that is a function of conversion. In case of a hexafunctional monomer the following was obtained previously [52]:

$$\sigma_F \sim P^3(4 - P) + 4P_C^{\,3}P(P_C - 3) + P_C^{\,3}(8 - 3P_C) \qquad (3)$$

Stolov et al.[52] explain that in this expression P_c is the vinyl group conversion at the gel point which for this monomer equals 0.23 ± 0.02. In the vicinity of P_0, equation 3 can be represented as power series of $(P - P_0)$. As determined from ATR spectra, 0.7 is a reasonable estimate for P_0. Substituting P_C and P_0 in the power series and assuming that inside the film $(P_0 - P)$ does not exceed 0.15, the linear term in the power series accounts for more than 80% of the changes of σ_F with P. Thus, the average stress in the film, (σ_F), can be derived as[52]

$$(\sigma_F) = (\sigma_0 / T_F \alpha) [1 - \exp(-\alpha t_F)] \qquad (4)$$

3.2.1.3d. Effect of Oxygen

Cao et al., [53] studied the effect of oxygen on the photopolymerization processes in acrylate coatings and the effects of O_2 inhibition on a polymer network structure and its mechanical properties. The effect of O_2 inhibition in polymer coatings is limited to a thin surface layer. By using slow beam positron annihilation spectroscopy and nano-indentation they observed strong indications that the O_2 effect is maximal at surface and diminishes gradually with depth. [53] One might actually expect that. The effect of oxygen on free-radical polymerization is well known. It is also known that high concentration of free-radicals on the surface of the film offsets the diffusion of oxygen into the coating during the curing process.

Oxygen inhibits free radical polymerization through two pathways. First, an oxygen molecule can quench the active triplet state of the photoinitiator and be excited to singlet state itself. The second path is scavenging of the photoinitiator and polymer radicals through oxidation with the yield of peroxy radicals. The overall chain reactions of photopolymerization with the presence of O_2 is schematically shown as follows [55]:

$$
\begin{array}{ll}
PI + h\nu \longrightarrow (PI^*)_T \longrightarrow R_1\bullet & \textbf{initiation} \\
R_1\bullet + R \longrightarrow R\bullet & \textbf{initiation} \\
(PI^*)_T + (O_2)_T \longrightarrow PI + (O_2^*)_S & \textbf{quenching} \\
R\bullet + R \longrightarrow RR\bullet \ldots \longrightarrow R_n\bullet & \textbf{propagation}
\end{array}
$$

The termination is by recombination of radicals.

$$
\begin{array}{ll}
R_1\bullet \text{ or } R_n\bullet + O_2 \longrightarrow R_1OO\bullet \text{ or } R_nOO\bullet & \textbf{scavenging} \\
R_1OO\bullet \text{ or } R_nOO\bullet + DH \longrightarrow R_1OOH \text{ or } R_nOOH + D\bullet & \textbf{chain transfer}
\end{array}
$$

The conclusion is that as a result of oxygen inhibition there is a decrease in hardness and toughness of the surface layer. [53]

Several techniques were proposed, and some are used commercially, to reduce the effect of oxygen. One is inert gas atmosphere over the material cured, while it is being irradiated. Another is addition of paraffin waxes to the formulation. The waxes bloom to the surface and form a protective shield against oxygen. A third us making use of oxygen scavengers, like amines, phosphines, and thiols.

Gou *et al.,* [54] reported that presence of dimethylanthracene in the polymerization mixture effectively scavenges singlet oxygen:

The singlet oxygen trapper is claimed to effectively reduce the incubation period and improve the reaction rate even without nitrogen purging. [54]

Fouassier and Ruhlman [55] presented evidence, however, that the effect of oxygen is not of major importance in photocurable reactions that proceed very rapidly. When efficient photoinitiators are used that react within a very short time scale, with triplet states less than 100 ns., oxygen quenching is not significant. This is true in organic solvents or media, where the rate of quenching is typically 2×10^9 M^{-1}s^{-1}. In viscous media the effect of oxygen is still less important even for long-lived triplet states. [55]

3.2.1.3e. Depth of Cure

Gatechair and Tiefenthaler [56] studied the depth of cure profiling of UV light cured coatings. They utilized Beer's law to determine the fraction of incident radiation that was absorbed at various points of the coating and tried to correlate that with the cure. This interesting work provides a potential explanation for effects of the concentration of photoinitiator ratios on surface cures and on total cures. They also showed how changing lamps can affect the choice of the initiator and that some initiators perform consistently better than others with a given lamp and a given thickness of the film. In addition they point out that screening, the internal filter effect, can result in loss of adhesion or poor through-cure. Unfortunately, their calculation are not currently capable of yielding quantitative guides for formulating best possible compositions for light curing. [56]

3.2.1.3.f. Effect of Pigments

Presence of pigments in the coating, like titanium dioxide or colored pigments, affects the rate of cure, because pigments act as light filters and are detrimental to light absorption by the initiators. This often restricts pigmentation to thin coatings. Because the absorbtion maxima for pigments like titanium dioxide is between 280 and 300 nm, use of visible light photoinitiators have an advantage. Combinations of morpholino ketones with sensitizers like thioxanthone was reported to enhance cures of pigmented systems [2]:

Not all pigments, however, are detrimental to the rate of cure. Photocuring kinetics of UV-initiated free-radical photopolymerizations of acrylate systems with and without silica nanoparticles were investigated. [57] The kinetic analysis revealed that the presence of 10-15 % of silica nanoparticles actually accelerates the cure reaction and the cure rate of the UV-curable acrylate system. [57] This was attributed to a synergistic effect of silica nanoparticles during the photopolymerization process. A slight decrease in the polymerization reactivity, however, was observed when the silica content increased beyond 15 wt %. This was attributed to aggregation between silica nanoparticles. [57] It was also reported that the addition of silica nanoparticles lowers the activation energy for the UV-curable acrylate system, and that the collision factor for the system with silica nanoparticles is higher than that obtained for the system without silica nanoparticles. [57]

3.2.2. Compositions that Are Based on Thiol-Ene Reactions

The photolyses of mixtures of difunctional thiols and difunctional ene compounds that produced polymers was originally reported in the 1950s. [58] The crosslinking mechanism of the thiol-ene reactions is based on step-growth anti-Markovnikov condensations of thiols with multiolefin compounds. [58] These are photoinitiated polymerizations. Morgan et at., [59] showed that irradiation with ultraviolet light of a mixture of multifunctional thiols and multifunctional enes results in crosslinking. They proposed the use of benzophenone to absorb the light and to initiate the polymerization through a radical chain process. This involves hydrogen transfer reactions involving the excited triplet state of benzophenone and the ground state of the thiol. Compounds like peroxides, azobisobutyronitrile, and ketones can initiate these reactions in the presence of ultraviolet light. [58]

A mechanism based on chain transfer was described recently by Bowman et al., [60] :

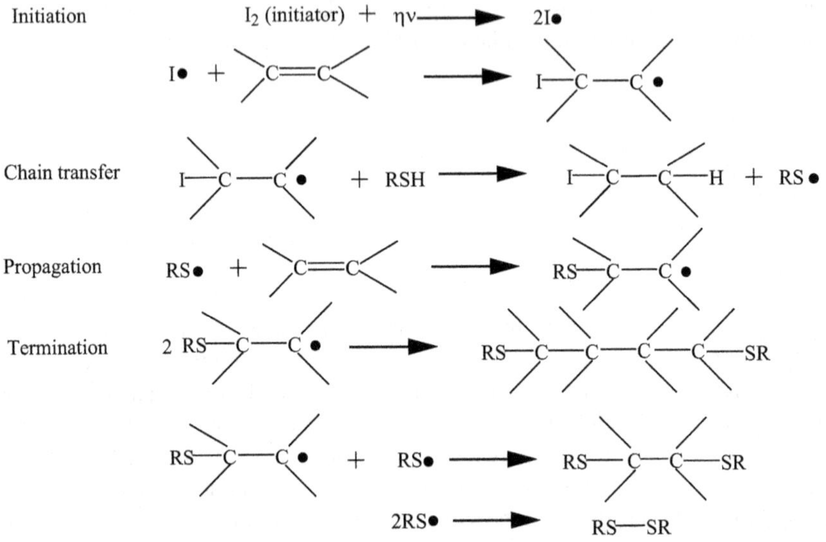

Initiation I_2 (initiator) $+$ $\eta\nu \longrightarrow$ $2I\bullet$

Chain transfer

Propagation

Termination

The addition of the thyil radical, that forms from chain transferring, to an olefin results in formation of a thioether carbon radical. The newly formed radical generates a new thyil radical, either by a hydrogen abstraction reaction or by an oxidation reaction (if oxygen is present). This is then followed by hydrogen abstraction and the process repeats itself. Coupling or radicals decreases the number of free radicals available for the condensation reaction. The degree of polymerization depends upon the chemical structures of both co-reactants, the olefin and the mercaptan. It was found that when the point of gelation is reached depends upon the olefin. Thus, in a thio-norbornene condensation it occurs at a higher degree of conversion (~60%) [58] than it does in acrylate resins (~ ≤ 15%). [61]

Based on the above scheme, Bowman et al., [60] predicted that the rate might be proportional to the square root of the product of monomer concentration and rate of initiation. Cook and Patisson, on the other hand, claim that their investigation shows that influence of irradiation intensity and initiator concentration on the cure rate does not follow the classical square root relationship. [61]

Bowman et al., [62] wrote a kinetic model for this step growth mechanism. The governing equations for this model are [62]:

$$d[SH] / dt = - k_{CT}[SH][C\bullet]$$

$$d[C{=}C] / dt = - k_{pSC} [C{=}C][S\bullet]$$

$$d[S\bullet] = R_i - R_t (S\bullet) + k_{CT}[SH][C\bullet] - k_{pSC}[C{=}C][S\bullet]$$

$$d[C\bullet] / dt = R_i - R_t (C\bullet) - k_{CT}[SH] [C\bullet] + k_{pSC}[C=C] [S\bullet]$$

Chain transfer is modeled as a rate-limiting step with the rate parameter (k_p) for the propagation reaction being a factor of 10 greater than for the chain transfer process $(k_{CT})(k_p/k_{CT} = 10)$. The reaction is first order overall and independent of the concentration of ene functional groups,[62]

$$R_p = \alpha[SH]^1 [C=C]^0$$

Polymerization rate behavior vs. functional group concentration change is shown to be only a function of the ration of propagation to chain transfer kinetic parameters.[62]

The initiation rate, R_i was calculated by Cramer, Bowman and coworkers[227] from the following equation

$$R_i = d[I]/dt = 2.303\, f\epsilon[I]\, I_0\lambda\, /\, N_{av}hc$$

where f is efficiency; ϵ is molar absorptivity 150 L/mol/cm for 365 nm light; [I] is the initiator concentration; I_0 is the light intensity; λ is the wave length; N_{AV} is the Avogadro number; h, is Plank's constant; c is the speed of light. The termination rates for thiyl and carbon radicals are given by[60] the following equations:

$$R_t(S^\bullet) = 2k^{t1}[S^\bullet]^2 = k_{t2}[S^\bullet][C^\bullet]$$

$$R_t(C^\bullet) = k_{t2}[S^\bullet][C^\bullet] + 2k_{t3}[C^\bullet]^2$$

All the kinetic parameters are assumed to remain constant throughout the reaction. In addition, radical-radical recombinations for termination are assumed to be equally likely and thus, the termination kinetic parameters are assumed to be equal. This assumption for equal termination may not be correct. No information, however, is currently available and until it does become available this assumption can not be contradicted. The following olefins were utilized in that particular study[60]:

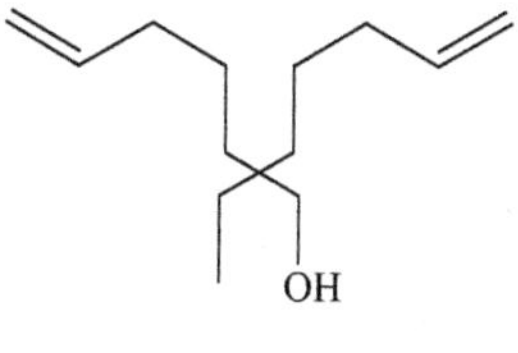

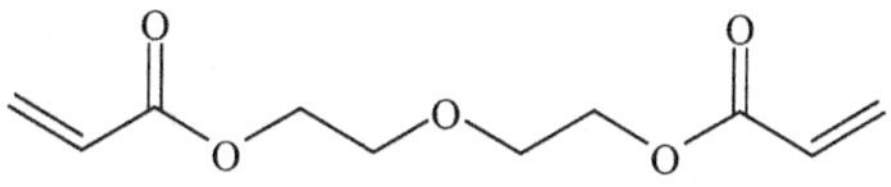

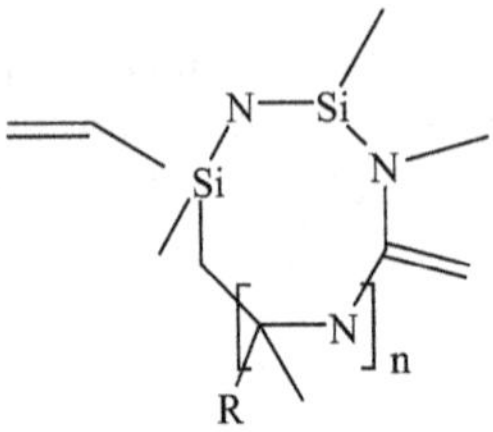

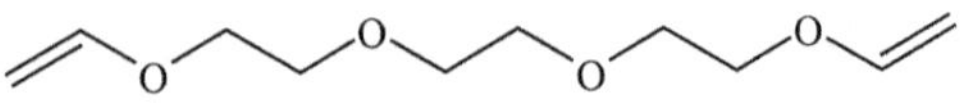

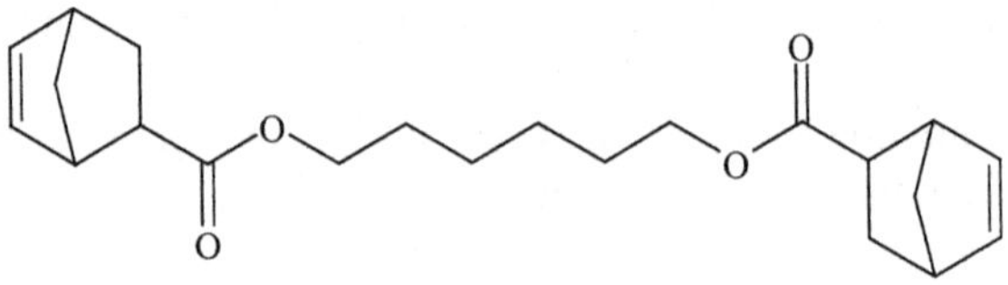

These polymerization processes, as mentioned above, are repeated anti-Markovnikov additions of thiols across the double bonds. Because the thiols also act as chain transferring agents, they limit the chain propagations to single steps. High molecular weight materials form, because the monomers are polyfunctional. The rate of conversion of this step growth mechanism can be expressed by the following equation [58]:

$$\alpha = [1/r \, (f_{thiol}^{-1})(f_{ene}^{-1})]^{0.5}$$

where α represents the fractional conversion required to attain an infinite gel network, r is the stoichiometric thiol/ene functionality ratio, and f_{thiol} and f_{ene} are the functionalities of the thiol and the ene respectively. Production of a high molecular weigh polymer requires a functionality of the monomers to be three or greater. [63] The thiol-ene reaction was illustrated by Jacobine, Glazer, and Nakos as follows [58]:

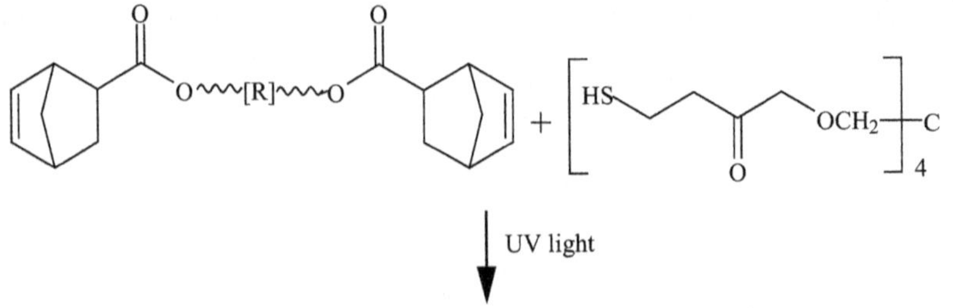

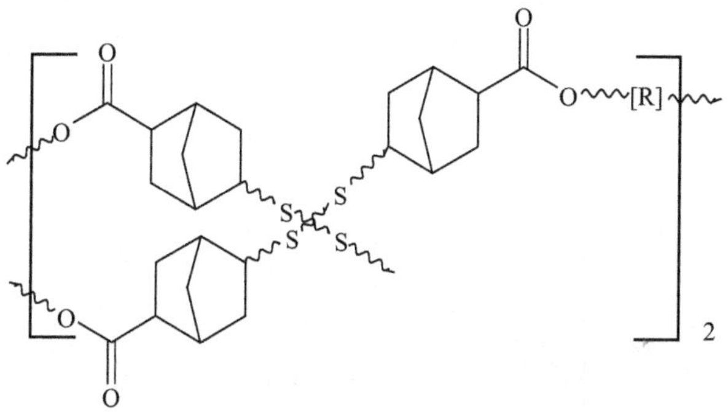

It has been stated that the propagation processes in the thiol-ene reactions are less sensitive to oxygen than are typical free radical chain-growth polymerizations. [7] In the presence of excess oxygen, however, peroxy radicals and hydroperoxides are generated. [64] In addition, it was shown that photoinitiation can take place through formation of charge transfer complexes. [65] Some examples of compounds that are used in thiol-ene polymerizations are shown in Table 3.3.

The reactivity of alkenes towards thiyl radicals decreases in the following order [66]:

vinyl ether > allyl ether > n-alkene > acrylates

In addition it was also reported that compounds like bicyclo[2,2,1]-heptene derivatives react up to 30 times as fast as allyl ethers. [70] Also, it was reported early that internal olefins exhibit lower reactivity than do the terminal ones. [58] Cyclic olefins are claimed to be less reactive than terminal ones, however, Jacobine et al., [58] reported that resins functionalized with norbornenes react rapidly and exothermally in the thiol-ene reaction with multifunctional thiols due to ring strain:

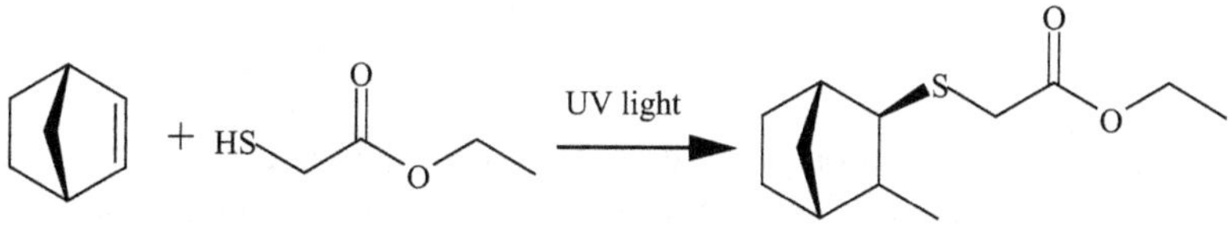

Jacobine et al., [58] also reported preparation of norbornene functionalized siloxanes:

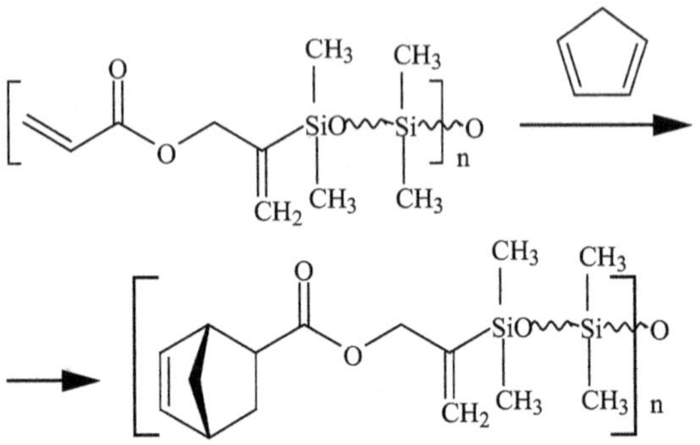

Resins with varying properties from toughened plastics to high elongation elastomers can be formed. [58] Many acrylate thiol hybrid polymerizations that result in modifications of mechanical properties of the polymeric network are possible.[67-70]

Okay, Reddy, and Bowman investigated the molecular weight development during the thiol-ene photopolymerizations [71] As a result they developed a kinetic model for the pregelation period of thiol-ene polymerizations of multifunctional thiol and ene monomers. The model involves the moment equations for thiyl radicals, carbon radicals, and polymers, and it predicts the chain length averages as a function of the reaction time or functional group conversion. They assign dependent variables Sxyc •, Rxyc •, and Pxyc • to represent thiyl radicals, carbon radicals, and dead polymer or monomer molecules, respectively. The total concentration of free radicals, vinyl, and thiol groups is defined by them as follows:

$$[R^{\bullet}] = \sum_x \sum_y \sum_c [R_{x,y,c}^{\bullet}]$$

$$[S^{\bullet}] = \sum_x \sum_y \sum_c [S_{x,y,c}^{\bullet}]$$

$$[M] = \sum_x \sum_y \sum_c M_{x,y,c}[P_{x,y,c}]$$

$$[SH] = \sum_x \sum_y \sum_c SH_{x,y,c}[P_{x,y,c}]$$

Cyclization reactions are expected with the assumption that the probability of cyclization is proportional to the number of local pendant vinyl groups connected to the thiyl radical centers. The cyclization reaction is arbitrarily truncated at a critical size, N, in that only cycles with sizes less than N are allowed during gelation. The relative rate constant of cyclization with respect to propagation was calculated using the kinetic model and the experimental gelation data obtained for a thiol-ene system consisting of divinyl and trithiol monomers. The rate equation for polymer molecules, P_{xyc} was derived by them to be [71]:

$$\frac{d[P_{x,y,c}]}{dt} = k_{tr}[R^{\bullet}][SH]\left\{\phi'_{x,y,c} - P_{x,y,c}\left(\frac{M_{x,y,c}}{[M]} + \frac{SH_{x,y,c}}{[SH]}\right)\right\};$$

$$[P_{xyc}](0) = 0$$

Where $x \geq 1$, $y \geq 1$

The average chain length of polymer molecules X_n was calculated by them[71] as

$$X_n = (1 + r_0) / (1 + r_0 + x_M f_{ave}(1 - \xi))$$

Where f_{ave} is the average functionality of the monomers:

$$f_{ave} = ([M]_0 + [SH]_0) / ([P_{1,0,0}]_0 + [P_{0,1,0}]_0)$$

and ξ is the fraction of the reacted functional groups in cycles:

$$\xi = \Psi_{0,0,1} / [M]_0 x_M$$

The functional group conversion is taken as the independent variable, the reaction kinetics of the ring-free thiol-ene system is expected to be identical to that of usual step-growth reactions. In the rate expressions for intermolecular reactions in thiol-enes are first-order reactions overall. The cyclization rate in thiol-ene reactions is slower compared to that in step-growth reactions. This result arises from intramolecular chain transfer reactions reducing the probability of favorable intramolecular collisions between the functional groups.

Some examples of thiols and olefins are listed in Table 3.3.

Table 3.3. Examples of Thiols and Olefins that Cure by the Thiol-Ene Reaction

Thiol	Ene	Ref.
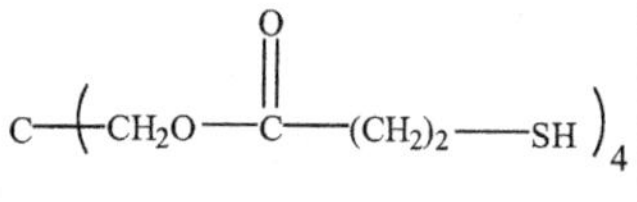	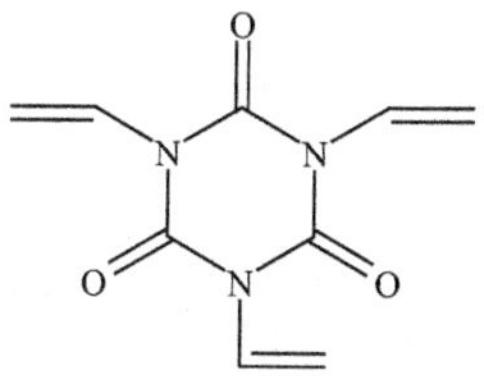	72
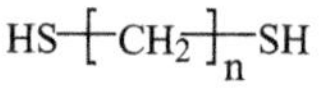	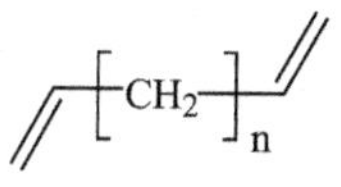	73

Table 3.3. (Continued)

Thiol	Ene	Ref.
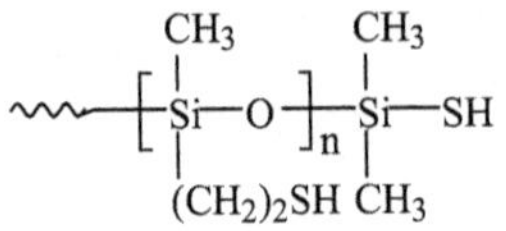	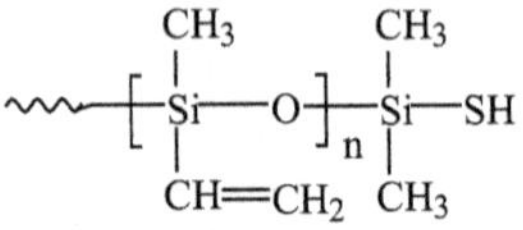	74
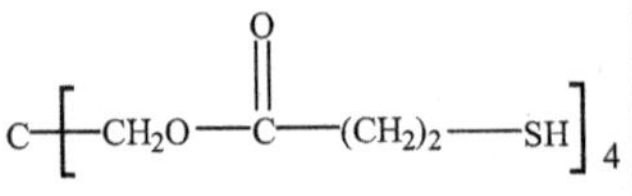	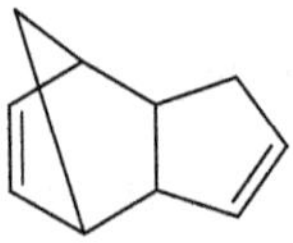	75
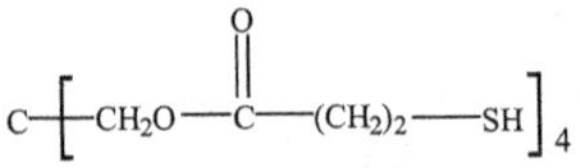	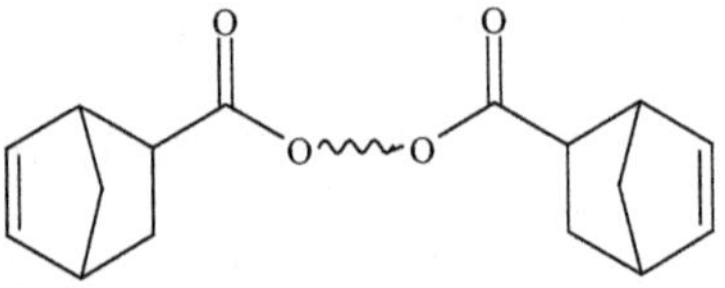	76

Additional examples of thiol-ene compounds can be found in work by Hoyle et al.,[77] who reported using a trifunctional thiol

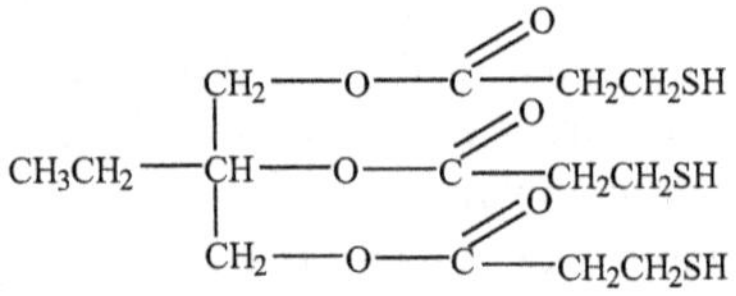

and four enes:

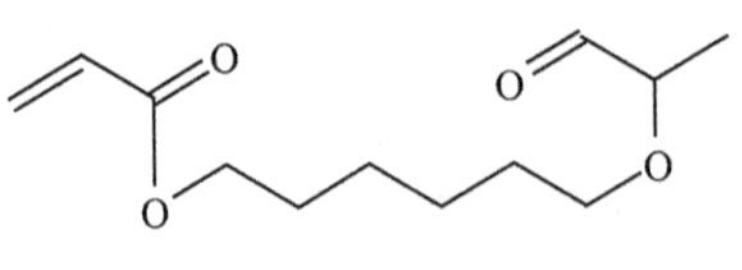

hexane diacrylate 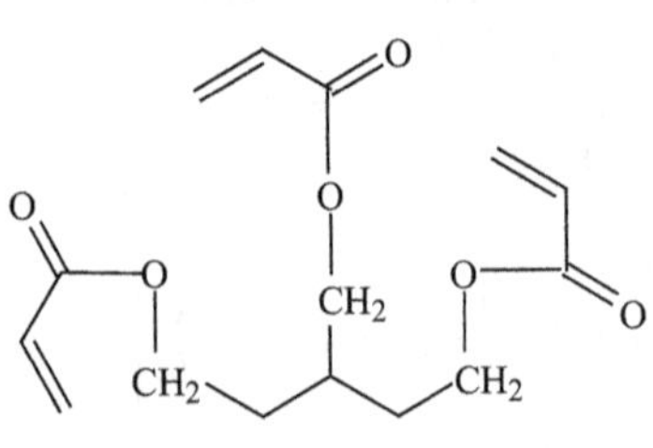　　　　　　　trimethylolpropane triacrylate

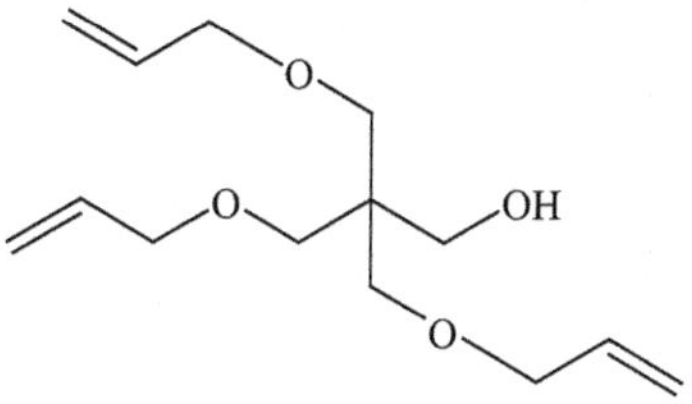

triallyl ether 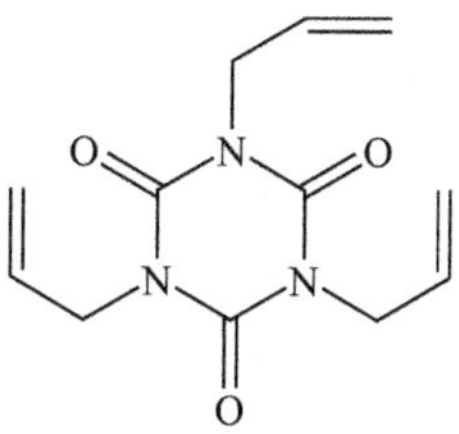   triallyl triazine

Also, Rehnberg et al.,[78] reported preparation of two allyl ethers and two thiols. These materials phtocured at a speed, that they felt, should be adequate for many applications:

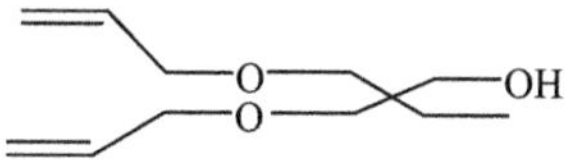

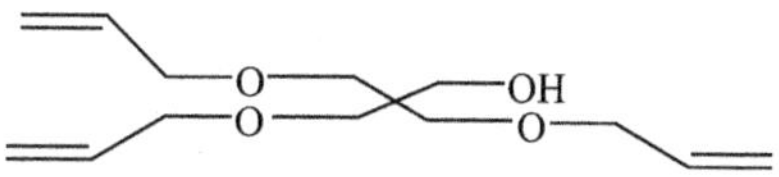

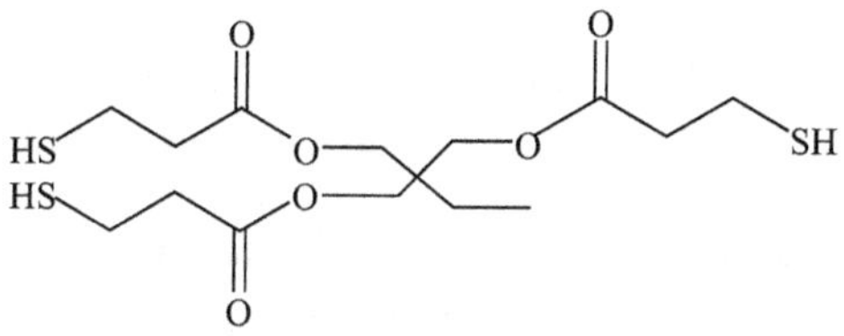

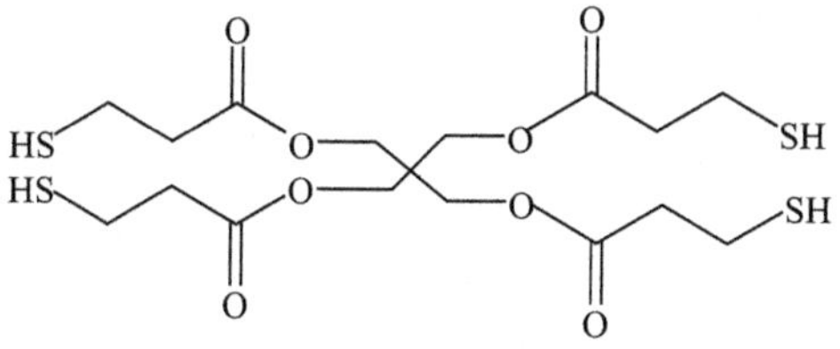

Lee et al., [79] reported preparation of several novel mono- and multifunctional vinyl esters for thiol-ene reaction of photocuring with mercaptans:

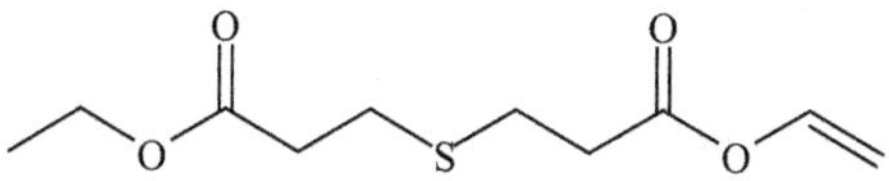

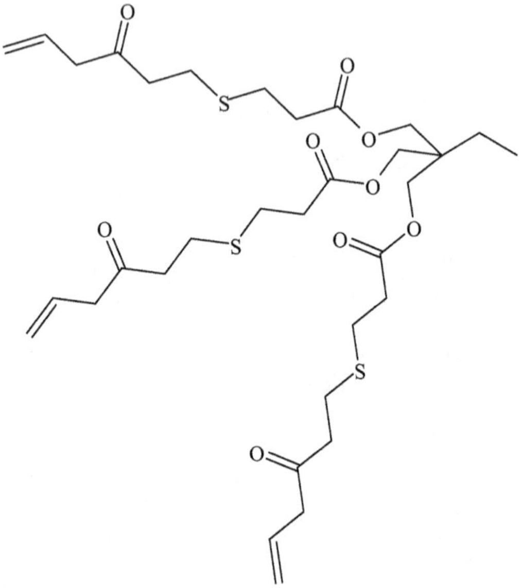

These esters, however, show lower polymerization rates under a nitrogen atmosphere than similar acrylics. But in an open air environment they were found to be faster than the acrylates. [79]

An investigation of the effect of alkene structure on the reaction kinetics was carried out on seven different alkenes and two thiols [80]:

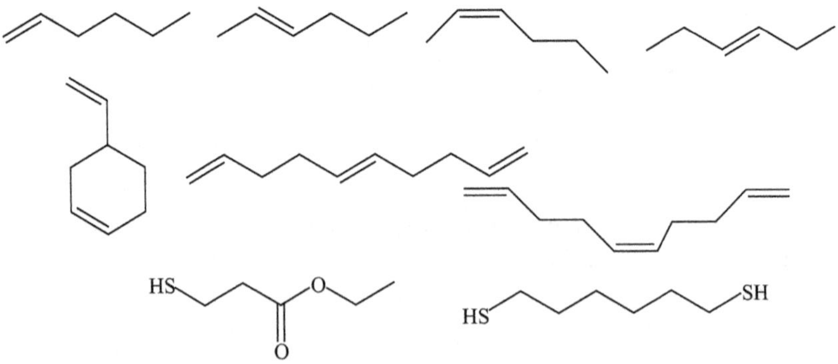

It was found that the terminal ene, 1-hexene, exhibits reaction rate that is one order of magnitude greater than the conversion rate of the internal enes, 2-hexene and 3-hexene.

Clark, Hoyle, and Jonsson reported preparation of highly functional thiols that were formed by an amine catalyzed thiol-ene reaction [81]:

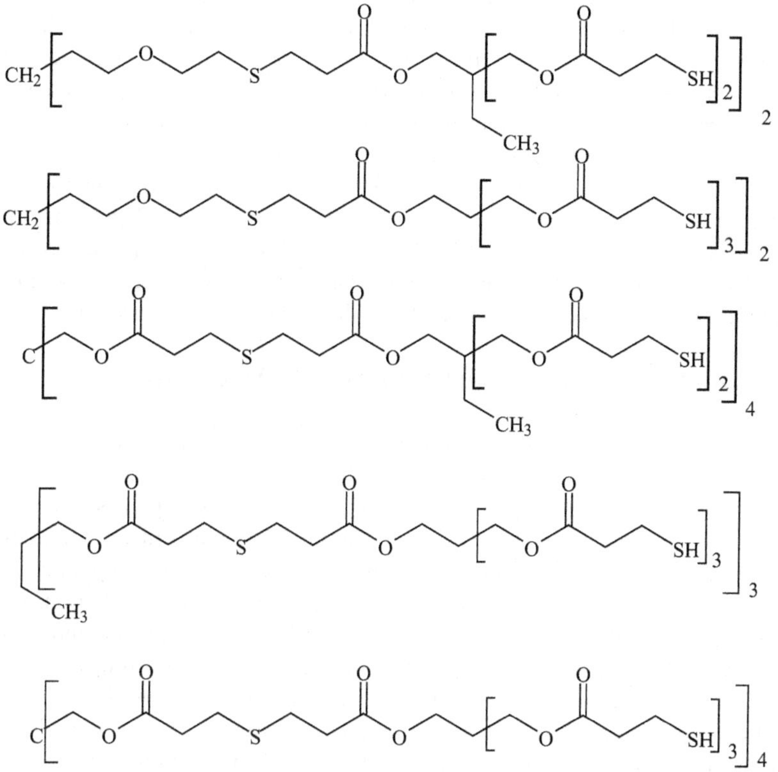

Films formed with these thiols were reported to possess good physical and mechanical properties and good adhesion and abrasion resistance. [81]

Carter *et al.*, [82] reported using pentaerythritol tetrakis(2-mercaptoacetate) and two equivalents of 1,4-cyclohexane dimethanol divinyl-ether to form fast thiol-ene photopolymers:

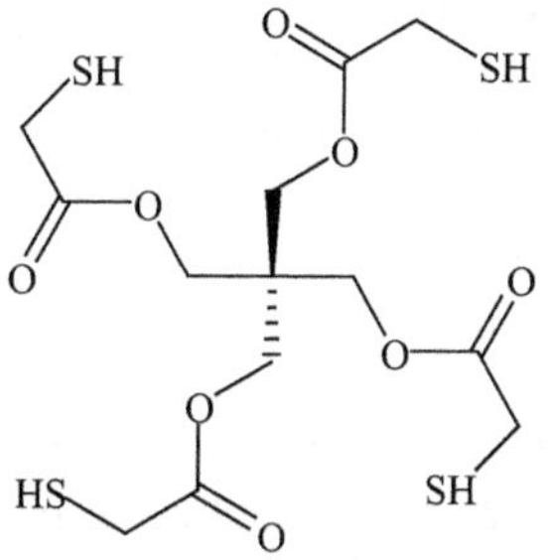

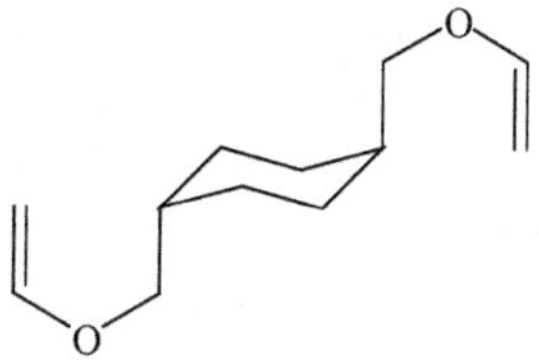

These were used in nano-contact molding and the polymerizations were carried out in the absence of a photoinitiator. [82]

In addition, Cramer and Bowman studied the kinetics of the thiol-ene reaction by means of real-time Fourier transform infrared.[63] They found that the conversion of the ene functional groups was up to 15.5 greater than the conversion of the thiol functional groups . For stoichiometric thiol-acrylate polymerizations, the conversion of the acrylate functional groups was roughly twice that of the thiol functional groups. In addition they reported that with the kinetic expressions for the thiol-acrylate polymerizations, the acrylate propagation kinetic constant was 1.5 times greater Thus, depending upon the monomer used, the reaction may proceed with equal consumption of both, the thiol and the ene functional groups, or with the ene functional groups exhibiting higher conversion than the thiol, due to homopolymerization.[84]

3.3. Compositions that Cure by Cationic Mechanism

There are three types of compositions that cure by cationic mechanism. One of them uses aryldiazonium salts to initiate the reaction. The second one utilizes onium salts and the third one organometallic complexes. The most prominent ones are those that cure with the aid of onium salt photoinitiators. Many cationic curable compositions consist of mixtures of compounds with oxirane rings. They may also be mixtures of vinyl ether. In addition, some compositions contain both, epoxides and vinyl ethers. More recent compositions might also include silicone based monomers with epoxide groups.[26] Thus, Crivello and Lee described a synthesis of a series of silicon-epoxy monomers that undergo rapid and efficient photoinitiated cationic polymerizations.[85] Such compounds can be prepared by direct hydrosilylation of olefinic epoxides.[85]

The structures of the oligomers or prepolymers are just as important in determining the final properties of the coatings that are cured by cationic mechanisms as are those that are cured free-radical mechanism. As a result, considerable effort has gone into preparations and studies of prepolymers and oligomers.

3.3.1. Compositions that Utilize Diazonium Salts

The same type of materials are used in formulations that are light cured with the aid of diazonium salts as are those that are cured with the aid of onium salts. Such materials are discussed in subsequent sections. There are some serious disadvantages to using of diazonium salts in photocuring compositions. As stated in Chapter 2 these are a tendency to gel, unless kept in the dark., a strong tendency of the cured films to yellow with time and nitrogen evolution from the photoinitiators during the cure process that causes bubbles or pinholes in thicker films.[26] As a result, there does not appear to be a large commercial acceptance of the system for use among the light curable coatings to date.

3.3.2. Compositions that Utilize Onium Salts

Photocationic polymerization based on photogeneration of acid from onium salts induced by UV light and consecutive cationic polymerization initiated by the generated acid was first proposed in 1970s. Typical oligomers that cure by cationic mechanism are epoxy resins, like a diglicidyl ethers of 4.4'-isopropylidine-diphenol or its higher analogs:

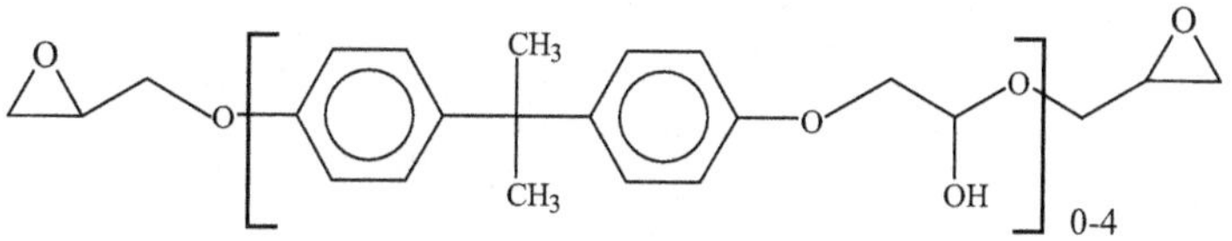

They can also be diglycidyl ethers of a silicon oligomer:

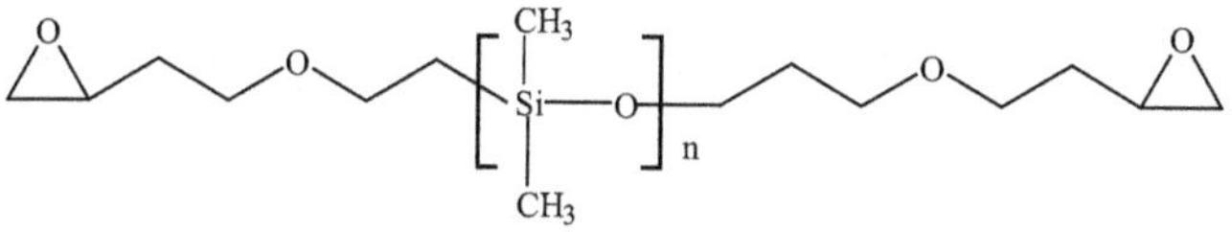

Epoxides are known to give cured coatings with high thermal capability, excellent adhesion and good chemical resistance. The curing speed of commercially available epoxides, such as glycidyl ether derivatives, however, is rather slow, when compared to the rapid free-radical cures of acrylic esters. This slower reactivity of the epoxides is a drawback to their use in some industrial applications.

The viscosity of the epoxide prepolymers in many instances is reduced by additions of thin liquid epoxy compounds that may include cycloaliphatic epoxides or lactones. Various epoxidized vegetable oils, like epoxidized linseed, sunflower, soybean, and others are also used in some formulations. Table 3.4. lists come cycloaliphatic epoxy monomers used for such purposes. Vinyl ethers polymerize at a faster rate than do cycloaliphatic epoxides. They are, therefore, included in some formulations. Other materials that are included for such purposes or to control viscosity can be epoxy novolacs, glycidyl ethers of long chain alcohols, To control viscosity, polyols might also be used, because it is assumed that hydroxy group bearing compounds crosslink with the epoxy resins. [121-126]

Crivello and Lee reported that epoxy-functional silicone monomers are particularly attractive in their application to UV curing. [85] Some of these monomers can be shown as follows [85]:

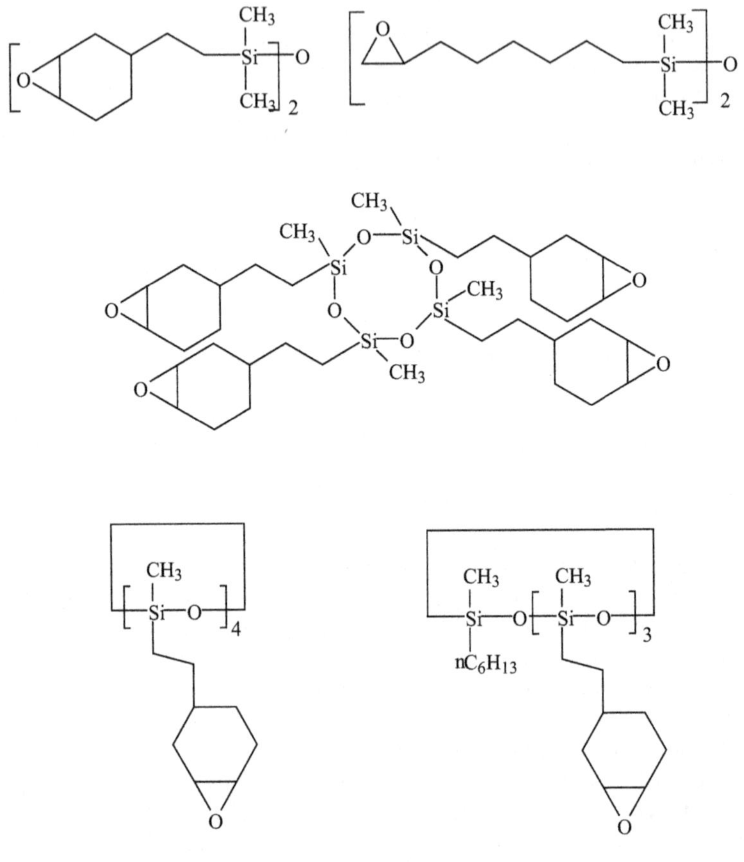

Crivello and Lee [85] also reported that the silicone monomers, some of which are shown above, can be used in all three applications in coatings, inks, and adhesives. These monomer are freely miscible with other epoxy monomers and when added in modest amounts increase the cure rates. [85]

While alkyl vinyl ethers polymerize readily by cationic mechanism, [15] the films that they form often lack good physical properties. Aromatic analogs of the aliphatic vinyl ethers, on the other hand, were reported by Crivello and Ramdas [86] to yield thermally stable materials with improved properties. In addition, they photopolymerize readily in the presence of diaryliodonium salts. The syntheses of the monomers was carried out according to the following scheme [86]:

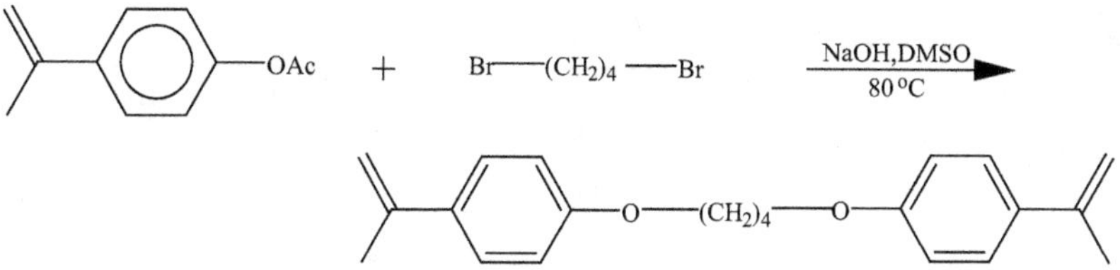

Lapin reported forming vinyl ether terminated urethane oligomers by reacting a number of different hydroxyvinyl ethers with isocyanates. [87] This can be illustrated as follows:

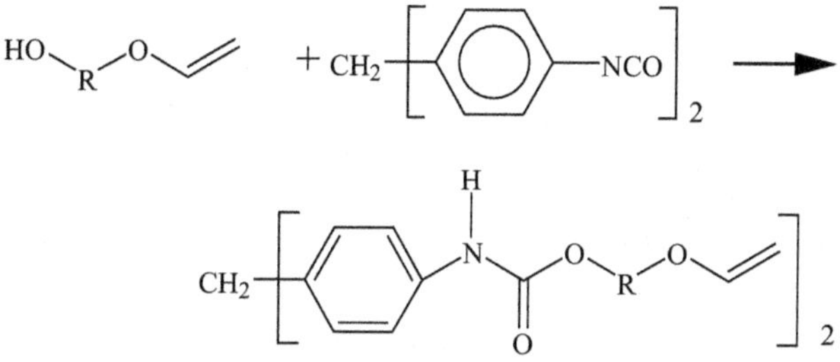

These prepolymers are claimed to cure rapidly in mixtures with vinyl ether monomers when exposed to ultraviolet irradiation in the presence of onium salt photoinitiators. The light curing is not inhibited by oxygen, but apparently is affected by high atmospheric humidity. [87] Table 3.4. lists some common diluents used in formulations for cationic polymerizations.

Table 3.4. Some Examples of Diluents Used in Cationic Polymerization Systems [a]

diluent	chemical structure
Cyclohexene oxide	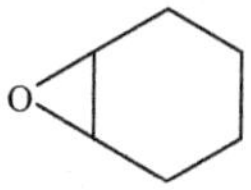
Vinyl cyclohexene dioxide	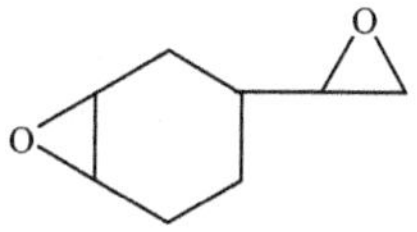
3,4-Epoxycylohexylmethyl-3.4-epoxycyclohexene carboxylate	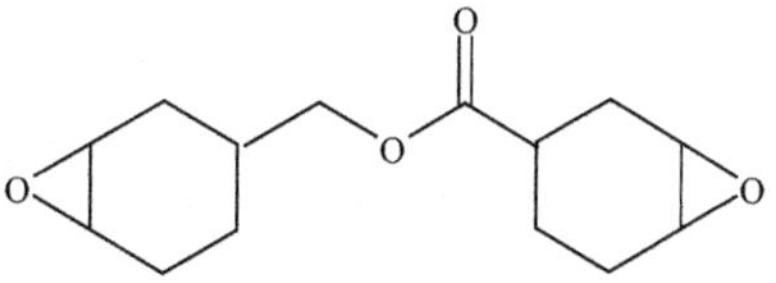

Table 3.4. (Continued)

diluent	chemical structure
Dicyclopentadiene dioxide	
Lactones	
Cyclic formals and acetals	
Cyclic ethers	

[a] from various literature sources

.   Novel epoxide monomers and oligomers were reported by Crivello and
Song that are based on dicyclopentadiene [88] The different reactivity of the two
double bonds in dicyclopentadiene,

makes it possible to conduct the addition of alcohols and silanes preferentially at
one of the double bond. By epoxidation of the residual unsaturation, they formed
mono- and difunctional monomers bearing epoxycyclopentyl groups. [88] They
also found that these epoxide monomers display excellent reactivity in the
photoinitiated cationic ring-opening polymerizations using diaryliodonium salt
and other onium salt photoinitiators. The high reactivity of these monomers was
attributed to the high ring strain present in the epoxycyclopentyl ring system. [88]
Crivello and Song also prepared two inorganic / organic hybrid systems in which
epoxy dicyclopentadiene moieties are placed as pendant groups onto a siloxane
backbone.[88] Photopolymerization of these oligomers was also found by them to

take place very rapidly under UV irradiation conditions and yield glassy resin matrixes with very good thermal oxidative resistance. [88]

A different approaches to the synthesis of siloxane-based hybrid organic-inorganic resins were also explored by Crivello *et al.* [89] A new, generally applicable, synthetic approach was reported that can be employed to prepare a family of silicone-epoxy resins using simple, readily available precursors. These resins undergo facile cationic photopolymerization to yield organic-inorganic hybrid resins with a wide range of properties. The epoxy resins were formed by a three-step process.

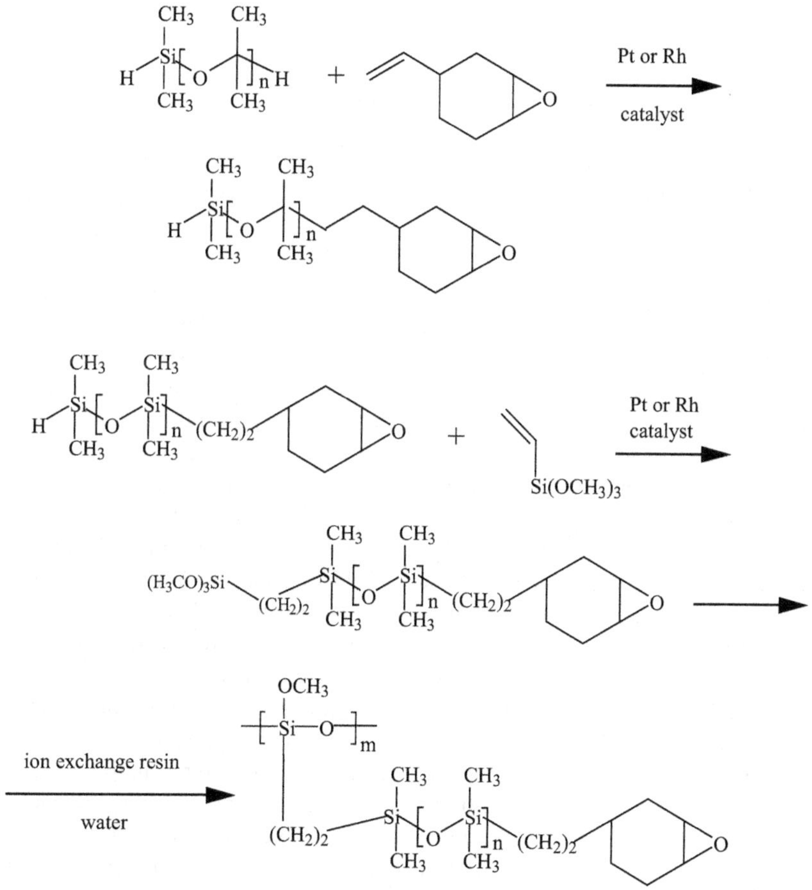

Subsequent technology [89] consists of preparation of difunctional silicon-containing monomers by monohydrosylilation of an α,ω-difunctional Si-H-terminated siloxane with a vinyl-functional epoxide or oxetane .This is followed by the dehydrodimerization of the resulting Si—H—functional intermediate. The method uses simple, readily available starting materials and it can be conducted as a streamlined one-pot, two-step synthesis. It was reported that the monomers containing epoxycyclohexyl groups display excellent reactivity in cationic ring-opening polymerizations in the presence of lipophilic onium salt photoinitiators. [89]

Nikolic and Schultz, [90] reported preparation of radiation curable epoxy resins based on styrene oxide structures:

The authors claim that these materials exhibit higher rates of cure and higher enthalpies of photopolymerization than do the corresponding cycloaliphatic epoxides.

Crivello and Ortiz [91] also studied a series of novel cationically photopolymerizable epoxide monomers bearing benzyl, allyl, and propargyl acetal and ether groups that can stabilize free radicals. These monomers were found to display enhanced reactivity in cationic photopolymerization in the presence of certain onium salt photoinitiators. [91] The explanation by Crivello and Ortiz is that during irradiation with ultraviolet light the aryl radicals that are generated from the onium slats abstract labile protons present in the above monomers to form the corresponding carbon-centered radicals. The newly formed radicals in turn interact with the onium salts by a redox mechanism to induce further decomposition of the salts. The overall result is that additional cationic species are generated by this mechanism with an accompanying increase in the rate and extent of the cationic ring-opening polymerization of the epoxide monomers. [91] Kato and Sasala [92] pointed out that three major factors, basicity, ring strain and steric hindrance contribute to the reactivity of cyclic ethers in the cationic ring-opening polymerization. Oxetane, four member cyclic ether, possesses rather high ring strain energy (107kJ/mol) and basicity (pKa = 2.02). These properties of oxetane ring would suggest that the oxetanes are more reactive than epoxides. [92] Previously, however, it was shown that they possess quite different polymerization characteristics than oxiranes. [92] The initiation of oxetane polymerization can be rather slow, but the polymerization proceeds smoothly until high conversion of monomers takes place. The products possess high molecular weights. The slow initiation of oxetane was explained by the high energy barrier for the ring opening of protonated oxetane and high polymerizability was elucidated by the high nucleophilicity of oxygen on the ring. [92]

Kato and Sasala [92] also investigated the reactivity of 2-phenyl-3, 3-dimethyl-oxetane and 2-(4-methoxyphenyl)-3, 3-dimethyl-oxetane in photo-cationic polymerizations and copolymerization. These compounds can be illustrated as follows:

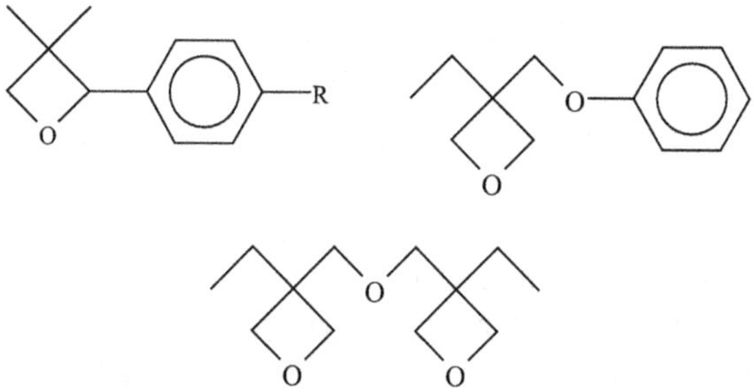

With 1% of diphenyl-4- thiophenoxyphenyl sulfonium hexafluoroantimonate as the photoinitiator. these monomers give the highest yield of polymers and also the highest number average molecular weights. [92]

Preparation of silicon-containing oxetane monomers and their polymerization was also investigated by Sangermano et al. [93] Following monomers were reported:

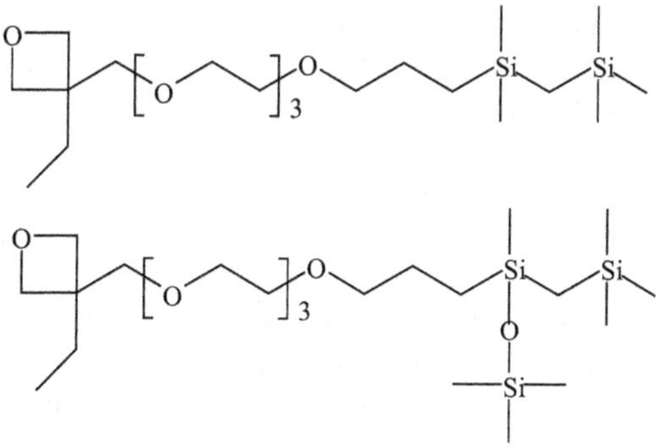

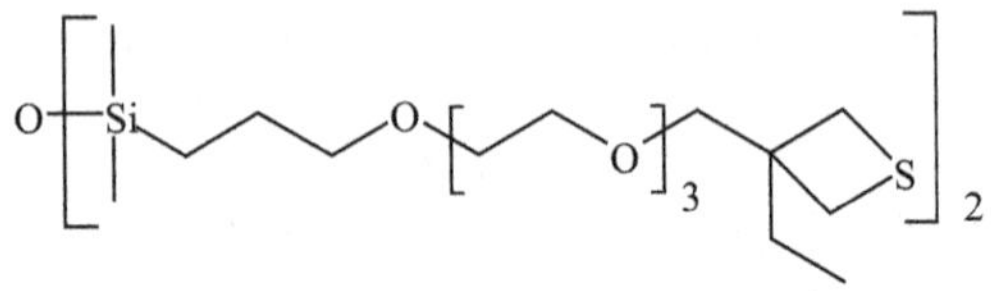

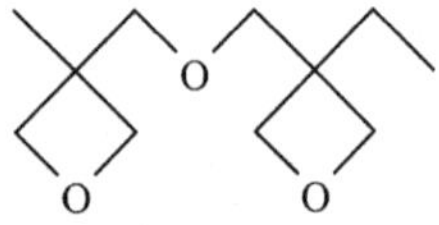

Shim *et al.*, [94] reported preparation of several novel mono- and di-propenyl ethers by condensation of perfluoroalkyl alcohol with allyl bromide followed by the ruthenium-catalyzed isomerization of the corresponding allyl ethers. These fluorinated propenyl ethers undergo rapid photoinitiated cationic polymerization with the aid of triaryl-sulfonium salt bearing a long alkoxy group. [94]

The monomers were photo cured in various combinations and all showed a high gel-fraction upon curing (greater than 95%). [93]

Bongiovanni, *et al.*, [97] reported syntheses of fluorinated vinyl ethers $(H_2C=CHOCH_2CH_2C_nF_{2n+1}$, (n-6 or 8) and their copolymerization with bis(4-vinyloxybutyl) isophthalate. When added in low concentrations (0.1-3.0 wt %) to formulations containing bis(4-vinyloxybutyl) isophthalate, they did not affect the kinetics of the cationic photopolymerization. The cured films were transparent and showed interesting properties in terms of wettability, hardness, cross-cut adhesion, and chemical inertness. The fluoro-monomers increased the hydrophobicity of the film surface, whereas the adhesion on various substrates such as glass and wood was unchanged. An increase in the methyl ethyl ketone resistance was also observed. [97]

Photoinitiated cationic ring-opening polymerization of a six-membered monothiocarbonate, 5,5-dimethyl-1,3-dioxane-2-thione by using onium salts was investigated by Yagci, et al. [98] Diphenyliodonium, triphenylsulfonium, and N-ethoxy-2-methylpyridinium salts were employed as photoinitiators in CH_2Cl_2 under a dry nitrogen atmosphere. It was found that oxygen had no effect on the reaction as should be expected. The reaction can be illustrated as follows [98]:

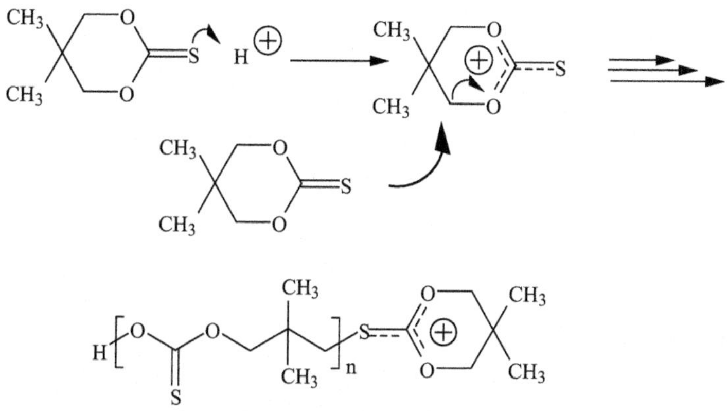

In this study was also included an investigation of the use of photosensitizers, perylene and anthracene. [98]

3.3.3. Hybrid Systems

Hybrid systems are understood to be photocurable compositions that cure by both ionic and free radical mechanisms. As can be seen in Chapter 2, section 2.3.4. photoinitiators were actually designed for use with systems that contain both type of monomers, or prepolymers, and both to cure by both mechanisms. One such example are combinations of phosphonium and sulfonium salts described in that chapter:

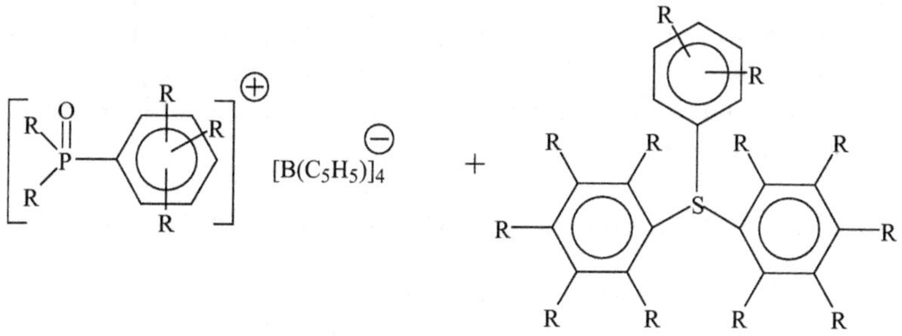

or

$[B(C_5H_5]]_4$

where R is hydrogen or methyl.

Also, as can be seen in Chapter 2, the photodecompositions of onium salts yield free radicals as by products. These free radicals are capable of initiating polymerizations of acrylic and methacrylic monomers.

Oxman et al., [95] studied controlled, sequentially curable cationic/free radical hybrid photopolymerization of diepoxide/acrylate hybrid material with the aid of photodifferential scanning calorimetry. The polymerizations were carried out in the presence of various concentrations of either ethyl-4-dimethylamino benzoate or 4-tert-butyl-N,N,-dimethylaniline as electron donors and camphoquinone/diphenyliodonium hexafluoroantimonate as the sensitizing system. The results showed that the free-radical acrylate reactions always precede the cationic epoxy polymerizations. [95]

Maruyama et al., [96] studied photopolymerization of addition products of multifunctional acrylates and alkoxysilyl alkyl thiol. Three monomers with multiple (trimethoxysilyl)propylthio groups were prepared by them via Michael addition of the multifunctional acrylates. These were neopentyl glycol diacrylate, trimethylol propane triacrylate , and pentaerythritol tetraacrylate that were reacted with (γ-mercaptopropyl)trimethoxysilane. These photocurable oligomers were then UV-irradiated using photo cationic and photo radical initiators. They found that polymerization with photo cationic initiator provided films with a high hardness and mar resistance. [96]

3.3.4. Mechanism and Kinetic Aspects of Cationic Photocuring

Decker   and   Moussa [98]   investigated   the   kinetics   of   typical photoinitiation of cationic polymerization. They concluded that the viscosity of the starting material has a strong effect on the rate of propagation, R_p, and the residual monomer content. The quantity of residual monomer increases with the thickness of the coating. On the other hand, the crosslink concentration of the three-dimensional network that forms can reach values as high as 4 mol/l. The quantum yield of the polymerization is estimated at 200 and the kinetic chain length at approximately 400. The postpolymerization is often large and reach 80% in some cases. Once the polymerization is initiated, it can probably be expected to follow the path of typical cationic polymerization.

The initiation process in cationic photocuring with the aid of onium salts consists of two steps:

(1) photodecomposition of the salt. This is described in Chapter 2.

(2) reaction of the initiating cation with the resins and monomers of the curable compositions. This is described in various textbooks of polymer chemistry. [49] It need not, therefore, be belabored here and is only be illustrated briefly:

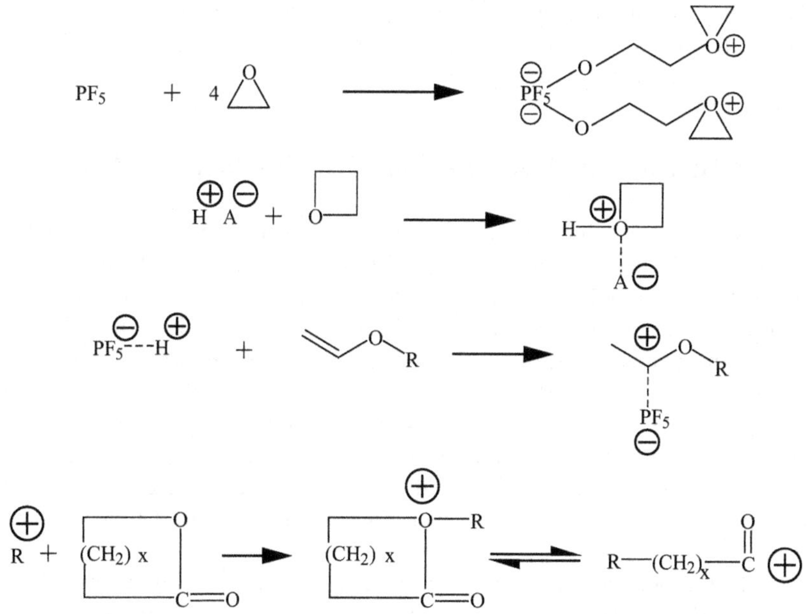

The propagation process too is typical of cationic polymerizations. It can be illustrated on the polymerization of epoxides as follows:

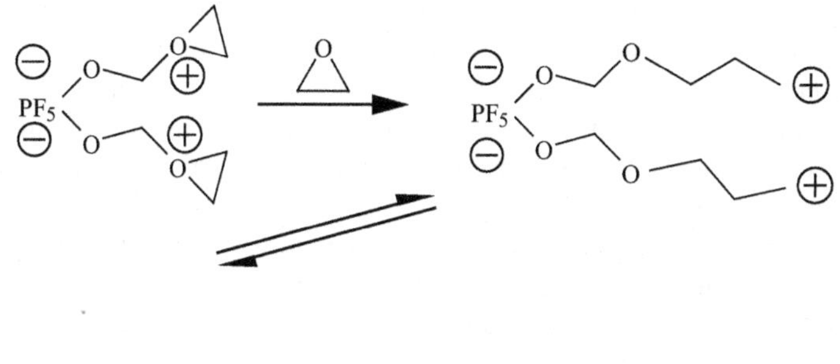

The termination process is usually by a reaction with any available anion, including some hydroxyl groups that are available. It can also be by chain transferring. These compositions are usually mixtures of resins and diluents. The composition of the products, therefore, will depend upon reaction conditions, the counter ions, and the temperature. The initiator system can be very important when cyclic monomers with different functional groups are present in the reaction mixture. Also, the propagation rates of different propagating centers can differ considerably.

3.4. Investigations and Monitoring of the Photocuring Process

The photocuring process, as described above, result in complex microstructures. [2] In addition to crosslinking, there are pendant double bonds, cycles, unreacted prepolymers, and microgels. Various methods were devised to monitor the progress of photoinitiated polymerization. Also, it is important to determine the optimum exposure time and cure. Some of the methods used, however, are quite simple, yet, to a certain extent, effective. In UV commercial curing of coatings, for instance, some simply rely on rubbing a light exposed coating surface with a rag soaked in methyl ethyl ketone and base the degree of cure upon the number of rubs required to penetrate the cured coating. Much more sophisticated approaches, however, are desirable and are being explored by many. [99]

It was pointed out, however, by Gasper et al., [100] that many of the published studies, like some described below, were carried out on simple systems. These systems would consist of one or two mono- or a difunctional materials and a photoinitiator. And they claim correctly that the link between such systems and actual commercial coating compositions, that are usually complex mixtures of compounds, is tenuous.

As a result, O'Brien and Bowman [101] developed a comprehensive photopolymerization model. It incorporates heat and mass transfer effects, diffusion-controlled propagation and termination, and temporal and spatial variation of species concentration, temperature, and light intensity. This model is applied to systems with varying thermal and optical properties. The absorbance of the polymerizing system is varied by altering either the initiator concentration, sample thickness, or molar absorption coefficient of the initiator. Based on simulations they concluded that the choice of initiator and sample thickness limits the initiator concentration usable to achieve complete monomer conversion in a sample. Similarly, the initiator and its concentration should independently be chosen since each impact the polymerization differently. Three different thermal boundary conditions and their effects on polymerization were also considered by them. These boundary conditions include isothermal, perfectly insulating, and perfectly conducting. Simulations show that a higher absorbance sample polymerizes completely when perfectly insulating boundary conditions are assumed. Thus, it was found that the choice of initiator and its concentration should be determined not only for the desired film thickness but also by considering the thermal conditions that affect the sample during photopolymerization. [101]

Gornez et al., [192] investigated the photopolymerization of acrylamide initiated by the safranine-T in the presence of triethanolamine in aqueous solutions. The addition of diphenyliodonium chloride to the system resulted in marked acceleration of polymerization. In the absence of diphenyliodonium chloride an inhibiting effect of the amine was observed by them when included at high concentrations. This effect is suppressed by the presence of the onium salt. At the same time the photobleaching of the dye caused by the amine is suppressed by the presence of acrylamide or the diphenyliodonium chloride salt. Laser flash photolysis experiments show that the presence of the onium salt increases the yield of triplet state of the dye. Also the yield of radicals in the quenching of the dye triplet by the amine increases by the presence of the iodonium salt. The semireduced form of the dye decays faster in the presence of the iodonium salt, suggesting a possible way of generating extra active radicals at the same time that the dye is regenerated. [102]

Terrones and Pearlstein, [34] used a photo-bleaching initiator to study the kinetics of free-radical photopolymerization with a photo-bleaching initiator to determine chain propagation and termination rate constants k_p and k_t. By assuming that termination occurs only by recombination, they felt that they were able to account for Beer—Lambert attenuation and initiator consumption. They also claimed to be able to predict how spatial variation of the final chain length distribution depends on k_p, k_t incident light intensity I_0, layer thickness L, photoinitiator absorption coefficient α_A and initial concentration. $C_{A,0}$, and quantum yield of photoinitiator consumption φ, for a typical value (10^{-4}) of the initial ratio of initiator and monomer concentrations $C_{A,0}/C_{M,0}$. They show how spatial variation of the final chain length distribution depends on initial absorbance $\gamma = \alpha_A C_{A,0} L$ and a parameter $\beta = k_p[fC_{A,0}/(\varphi \alpha_A I_0 k_t)]^{0.5}$ where f primary radicals are produced per photoinitiator molecules consumed. For small γ, the number averaged mean chain length increases with depth at each β and with β at each depth. The chain length at which the chain length distribution achieves its maximum value, along with a measure of polydispersity (half the chain length distribution width at half-maximum, divided by the number averaged mean), increase with depth at small β and decrease with depth at large β, with the chain length distribution having its minimum nonuniformity at intermediate β. Front-to-rear chain length distribution variation increases as γ increases. At small β, nonuniformity is confined to a progressively smaller portion of the front of the layer as γ increases, while for large β, spatial variation is more evenly distributed. The chain length distribution can be related to the spatiotemporal variation of initiation rate, with "down beam" conversion of monomer at very low radical concentration playing an important role. Based on their results a rheological method for monitoring laser-initiated photocuring reactions with a quartz microbalance was developed. This flow cell experimental setup is of limited utility, however, in coatings or thin films.

3.4.1. Use of Infrared Spectroscopies

Qiao *et al.*, [103] reported studying the photopolymerization of four ethoxylated bisphenol A dimethacrylates with real-time FTIR monitoring. They were able to observe that increased photoinitiator concentration leads to maximum conversion of double bonds for each methacrylate. They also observed that it increased the rate of polymerization.

Cho, *et al.*, [104] investigated the effects of photosensitizers, 2,4-diethylthioxanthone and anthracene and dual curing on the kinetic and physical properties of the product from cationic photocuring of clear and pigmented coatings. These coatings were based on cycloaliphatic diepoxide / ϵ-caprolactone polyol. The cationic photopolymerizations. were investigated by Fourier-transform infrared spectroscopy with attenuated total internal reflection, real-time infrared spectroscopy, and differential scanning photo-calorimetry. The coatings continued to polymerize in the dark, after the light source was removed, due to the living character of cationic polymerization. Measurements of the yellowness index revealed that the yellowing of the coatings containing. a blend of the diethoxythioxanthone with anthracene was less than that of the coatings containing the thioxanthone derivative only. Depth-profile analyses of FTIR-ATR spectroscopy data showed that such dual curing significantly increased both surface and interior curing of the coating films. The real time infrared spectroscopy and photodifferential scanning chromatography analyses showed that the polymerization reactivity and the ultimate percentage conversion for the cationic clear and pigmented formulations containing photosensitizers were higher than the corresponding values for the formulation without the photosensitizers. [104]

While the high resolution nuclear magnetic resonance spectroscopy is very useful in characterizing microstructures in polymers, it cannot be used to study details of cross-linked UV cured materials due to their insolubility. Solid-state NMR however, was utilized to study structure heterogeneity in polymerization of multifunctional acrylates. [105-107] Other techniques can range from photo differential scanning calorimetry to real time Fourier transform infrared spectroscopy. [108,109] Decker reported, therefore that the crosslinking reactions that takes place in the UV curing process can be monitored in situ by real time infrared spectroscopy. [110] This technique allows recording conversion vs. time. For acrylate-based resins UV-curing proceeds with long kinetic chains in spite of high initiation rates. The infrared spectroscopy, is claimed to be very valuable in assessing the influence of various parameters. These can be initiation efficiency, chemical structure of the telechelic oligomer, light intensity, and the inhibiting effect of oxygen on the polymerization kinetics. [110]

Scott et al., [111] carried out Fourier transform infrared and electron spin resonance spectroscopic investigations of the photopolymerization of vinyl ester resins These consisted of studying reaction kinetics during photopolymerization of bis glycol methacrylate/styrene blends using commercial photoinitiators, Irgacure 819 and Lucirin TPO:

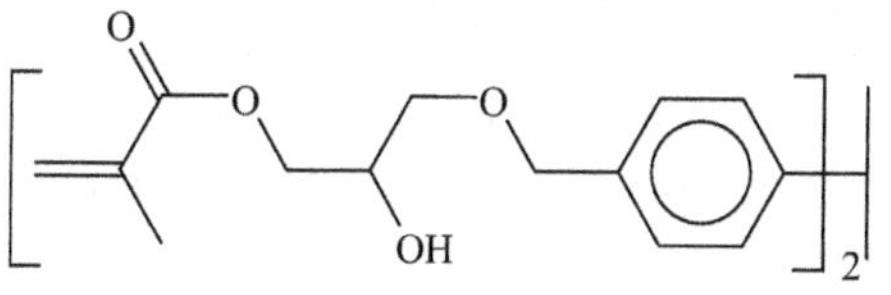

bisGMA 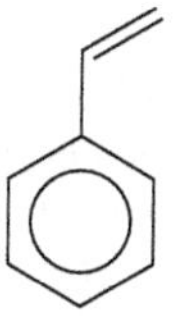   styrene 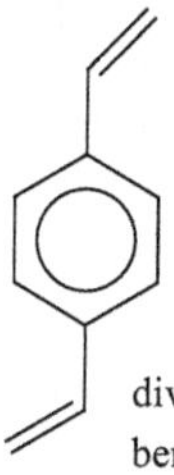   divinyl
                                            benzene

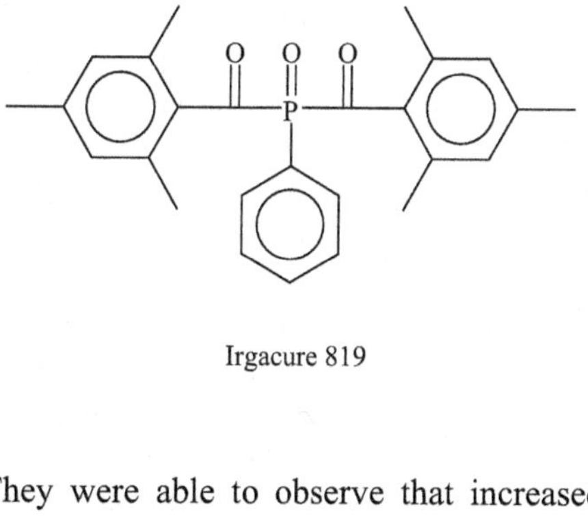

Irgacure 819 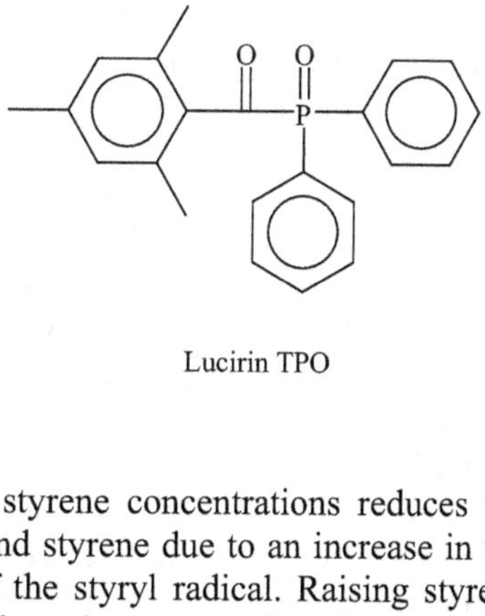   Lucirin TPO

They were able to observe that increased styrene concentrations reduces the polymerization rates of both methacrylate and styrene due to an increase in the termination rate and due to the stability of the styryl radical. Raising styrene concentrations also increases the final methacrylate conversion, but the final styrene conversion decreases because styrene plasticizes the network, allowing methacrylate conversion to rise at higher styrene concentrations. The final concentration of radicals is reduced at higher styrene concentrations, because of an increase in the bimolecular termination rate for networks with low cross-link densities. The proportion of styryl radicals trapped in the vitrified matrix was found to be markedly higher than the proportion predicted from the ratio of styrene monomer in the feed resin or from the copolymerization rate constants due to the higher mobility of styrene relative to the methacrylate groups. Increasing isothermal cure temperature results in a raised polymerization rate, increased conversion, and a decrease in the final concentration of radicals. This was attributed to the effects of an increase in the propagation rate and the bimolecular termination rate. The proportion of styryl radicals trapped in the vitrified matrix increases with raised isothermal cure temperature because the fraction of styrene reacting increases relative to the methacrylate groups. Scott *et al.*, also found that the concentration of radicals released by photo cleavage of the photoinitiator is sufficient to account for the increase in the concentration of radicals even after the sample enters the vitrification region. [111]

Thus, the applicability of real time infrared spectroscopy for monitoring high-speed photopolymerizations allows determining the rate of polymerization and the final monomer conversion. This can be used with either thin or thick samples exposed to intense polychromatic radiations. [112]

Dias *et al.*, [113] used, what they called, a hyphenated rapid real-time dynamic mechanical analysis (RT DMA) and time resolved near-infrared spectroscopy to simultaneously monitor photopolymerization of acrylate coating compositions. This allowed them to determine the rate of conversion and the mechanical properties of the finished films. It is claimed that up to 374 near infrared spectra and to 50 dynamic analysis points can be accumulated within a second. They observed that modulus buildup does not linearly follow chemical conversion of acrylate bonds. The gel point is detected after passing a certain critical acrylate conversion. Their experimental data revealed a critical dependence of the mechanical property development during the later stage of acrylate conversion.

Dias *et al.*, [113] point out that a common practice in understanding development of mechanical properties during UV cure is done by plotting of so-called dose-modulus curves, i.e., the modulus versus the applied UV dose. This approach is useful to monitor development of properties with the amount of UV exposure. It is of less utility, however, in understanding the relationship between chemical conversion and mechanical properties of a particular formulation in a time-resolved manner. In contrast, the RT-DMA/NIRS results, they claim, allow plotting of chemical conversion versus modulus curves during the course of the reaction. Such curves yield valuable information for designing and optimizing photocurable materials. [113]

Dias et al. [113] draw attention to the fact that a significant fraction of the final network density is developed at conversions above 98%, which is at the limits of accuracy with infrared spectroscopy. This emphasizes the importance in getting as complete a chemical conversion as possible so as to achieve the desired properties; i.e., the systems are very sensitive to incomplete acrylate conversion.

3.4.2. Use of Optical Pyrometry and Pyrolysis Gas Chromatography

Matsubara et al., [115] applied reactive pyrolysis gas chromatography in the presence of organic alkali to study the network structure of a UV.-cured acrylic ester prepared from polyethylene glycol diacrylate. An α-amino-alkylphenone type photoinitiator was used . The network formation can be illustrated as follows [115]:

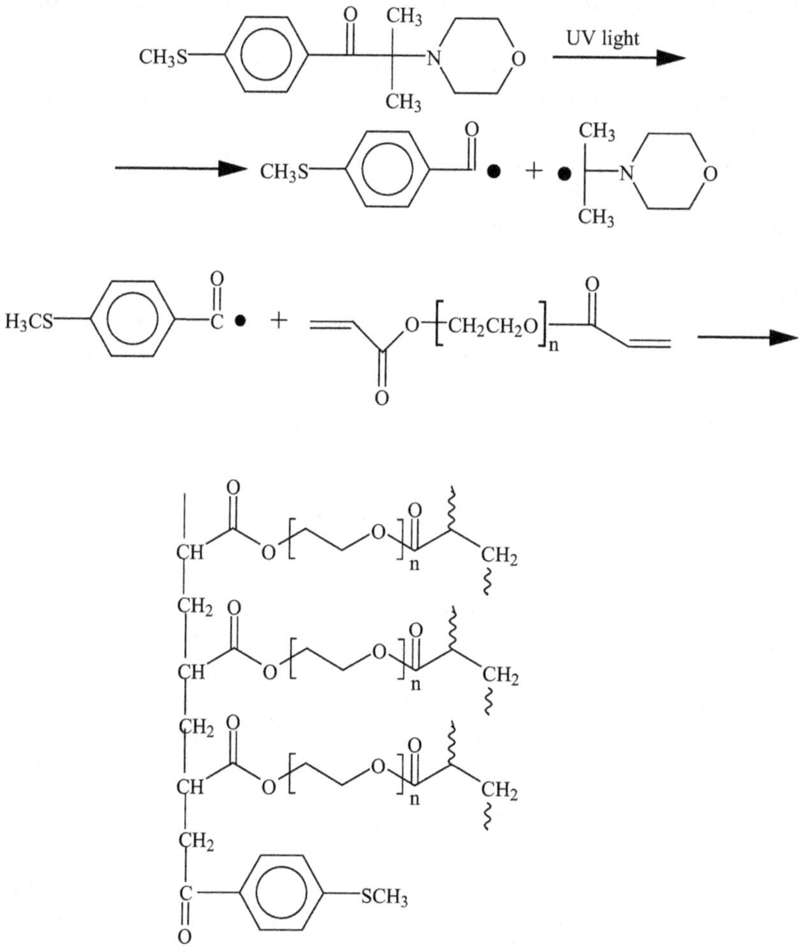

The above shown networks that formed in the polymerization were then submitted to pyrolysis in the presence of tetramethyl ammonium hydroxide. The decomposition products were separated and identified by gas chromatography / mass spectra. [115]

Crivello et al., [114] monitored photopolymerizations with aid of optical pyrometry. It was used to study the progress of both free radical and cationic ring-opening photopolymerizations. The monomers that were polymerized were diethylene glycol diacrylate, 1.6-hexanediol diacrylate, and 4-vinylcyclohexene dioxide. It was possible to record the temperature profiles of the reactions and the effects of critical parameters, such as light intensity and colored pigments on the course of the polymerization. [114]

3.4.3. Electron Proton Resonance with Near Infrared Spectroscopy

Bowman et al., [116] combined electron proton resonance spectroscopy with near infrared spectroscopy to study crosslinking photopolymerization kinetics of poly(ethylene glycol) dimethacrylate. The system was monitored in situ. Also, the environment of the propagating radicals throughout the majority of the reaction, propagation kinetic parameters, unsteady state determination of termination kinetics, and the quantification of persistent radical population could be studied. [116] This allowed to develop experimental protocols to provide a more complete picture of the mechanistic and kinetic foundation of the free radical curing polymerization. It also made it possible to compare the kinetics of two reactions, one yielding a rubbery and the other one a glassy polymer network. Revealing differences in their attributes were thus observed. Steady and non-steady states of radical concentration profiles, propagating radical environments during the polymerization and persistent radical populations were explored. [84]

3.4.4. Microwave Dielectric Measurements

Rolla et al., [117] used microwave dielectric measurements to monitor the polymerization process of monofunctional n-butyl acrylate as well as 50/50 w/w blends with a difunctional hexane-diol diacrylate that gave highly cross-linked networks. In these real time cure experiments the decreasing acrylate monomer concentration was studied via a linear correlation with the dielectric loss index at microwave frequencies. This correlation is a result of the largely different time scales for dipolar polarization in the monomer on one hand and in the polymerized reaction product on the other hand.

Direct time-resolved measurement of mechanical properties was initially developed using an oscillating plate rheometer fitted with a quartz parallel plate system. [118] Upon UV irradiation, the dynamic viscosity of these photocurable compositions increased and this was observed via the changes in the phase angle and amplitude of oscillation. The measurements, however, were limited to films or coatings of approximately 10 μm thickness. Also, care has to be taken to ensure that in spite of high shrinkage, the two plates remained truly parallel. Guthrie et al., [119] compared cone and plate versus the parallel plate geometry in a rheometer modified with a quartz plate. The parallel plate geometry was chosen due to practical considerations to monitor cationic and free radical photopolymerization including pigmented photocurable systems. [254]

3.4.5. Fourier Transform Mechanical Spectroscopy

Khan and co-workers [120] applied real-time Fourier transform mechanical spectroscopy to study cure, by fitting a rheometer with quartz glass plates. This allows monitoring the evolution of rheological mechanical properties in a temperature-controlled cell. The technique was used to monitor thiol-ene step growth photocross-linking polymerization. It enabled Khan et al., [120] to apply the Winter-Chambon criterion and show that the loss tangent (tan δ) becomes independent of frequency at the gel point. RT-FTMS was used to demonstrate for thiol-ene photopolymerization (tri- and tetrafunctional thiols copolymerized with triallyl isocyanurate) that the crosslinking rate increases by increasing the monomer (thiol) functionality or the temperature. [120]

3.4.6. Differential Photocalorimetry

Dickens, et al., [121] studied photopolymerization kinetics of typical dental photocurable materials by differential photocalorimetry. The naming of the components, shown below, is done according to Dickens et al., [121] publication

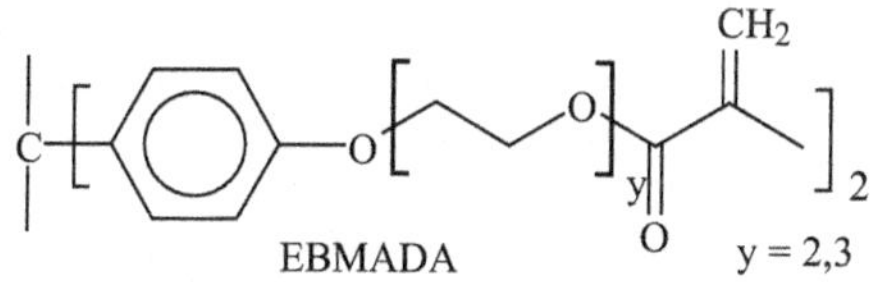

EBMADA $y = 2,3$

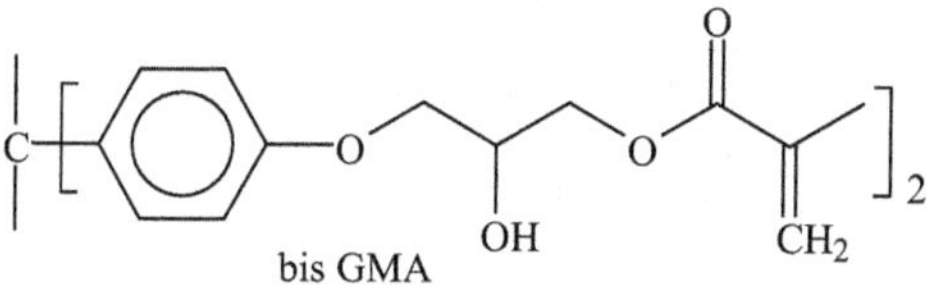

bis GMA

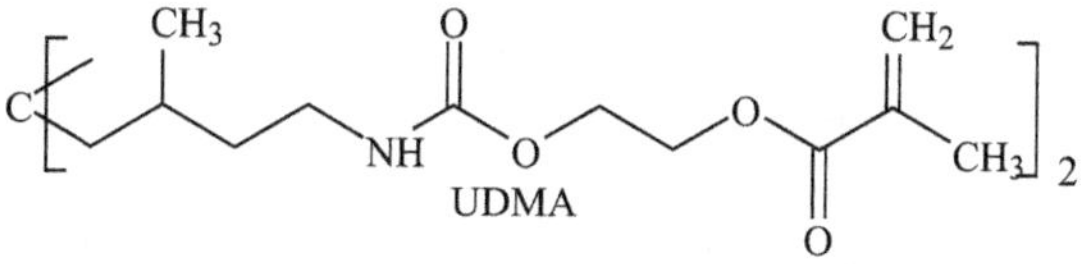

UDMA

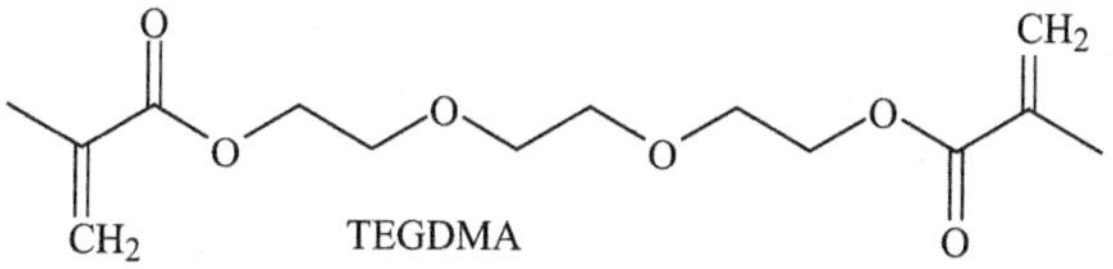

TEGDMA

The basic monomers were combined with the diluent in various proportions with the fractions increasing from 0 to 0.875. The compositions also contained camphoquinone, 0.6% mole faction, and ethyl-4-(dimethylamino)-benzoate, 2.0% mole fraction for photoinitiation. This yielded three base resins that differed in their hydrogen-bonding potential and, therefore, resulted in compositions covering a broad range of viscosities. When compared at similar diluent concentrations, UDMA resins were reported [121] as being significantly more reactive than Bis-GMA and EBADMA resins. At higher diluent concentrations, EBADMA resins provided the lowest photopolymerization reactivities. Optimum reactivities in the UDMA and EBADMA resin systems were obtained with the addition of relatively small amounts of TEGDMA, whereas the Bis-GMA/TEGDMA resin system required near equivalent mole ratios for highest reactivity. Synergistic effects of base and diluent monomer on the polymerization rate and the final conversion were observed [121] for the two base resins having hydrogen-bonding interactions. The structures of the individual monomers and, consequently, the resin viscosities of the comonomer mixtures strongly influence both the rate and the extent of conversion of the photopolymerization process.[121]

Decker *et al.*, [112] carried out an investigation of photoinitiated polymerization of diacrylate and vinyl ether resins. These diacrylates were described in the publication as follows:

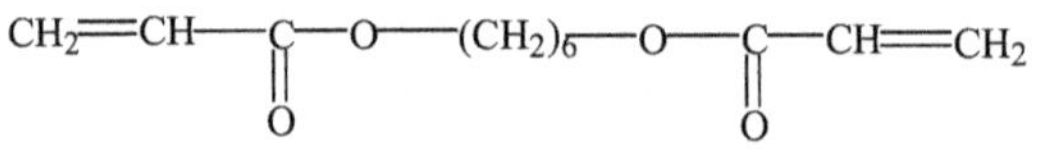

hexanediol diacrylate

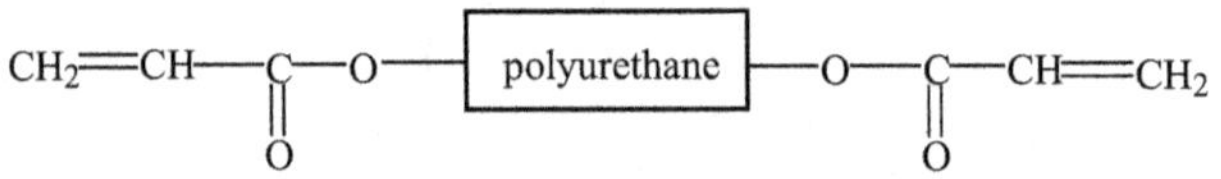

polyurethane diacrylate

The material included a bis-acylphosphine oxide photoinitiator:

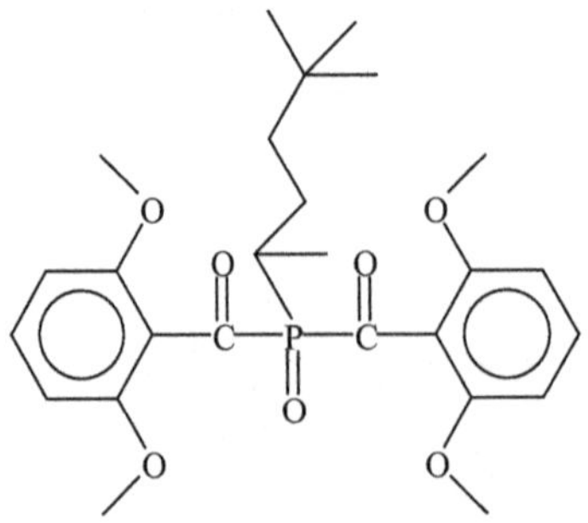

phosphine oxide photoinitiator

The cationic type photopolymerizable resins, vinyl ethers were illustrated as follows [112]:

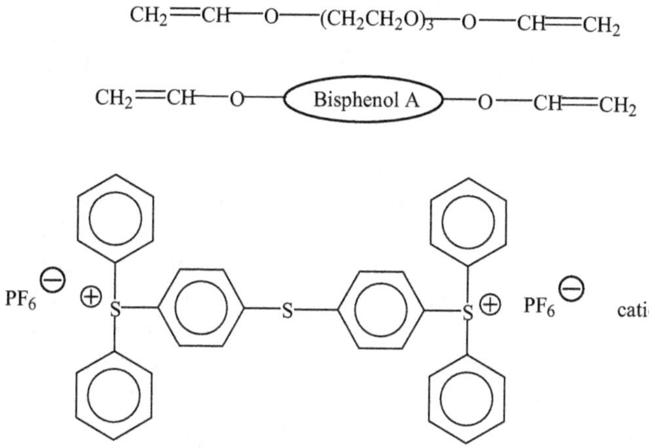

$$CH_2{=}CH{-}O{-}(CH_2CH_2O)_3{-}O{-}CH{=}CH_2$$

$$CH_2{=}CH{-}O{-}\boxed{Bisphenol\ A}{-}O{-}CH{=}CH_2$$

cationic photoinitiator

The photoinitiated polymerizations were followed by real-time infrared spectroscopy on thin films. radiation. The rates of polymerization were reported by them to increase with the light intensity according to a nearly square root law, up to an upper limit. The upper limit or the saturation effect was attributed by them to a fast consumption of the photoinitiator under intense illumination. A strong correlation was found to exist between the rate at which the temperature increases and the rate of polymerization. The temperature shows the same light intensity dependence as the reaction rate, and levels off to a maximum value under intense illumination. Photopolymerization experiments carried out at a constant temperature of 25°C show that thermal runaway is not responsible for the increase of the polymerization rate observed at the beginning of the UV exposure. [112]

The conclusion reached by Decker *et al.*, [112] is that the temperature reached by a monomer undergoing photopolymerization plays a key role on the reaction kinetics, in particular on the ultimate degree of conversion and, therefore, on the physico-chemical properties of the UV-cured polymer. It is strongly dependent on the formulation reactivity, the film thickness, as well as on the light intensity. A sharp rise of sample temperatures upon UV exposure of diacrylate monomers was shown to be directly related to the polymerization process and to depend on chemical and physical parameters, such as the photoinitiator concentration and the light intensity. On the other hand, they also concluded that for both the acrylate and the vinyl ether resins studied, thermal runaway is not responsible for the large increase in the polymerization rate occurring soon after the beginning of the UV exposure. [112]

3.5. Some Examples of Light Curable Coating Compositions

In this section a few typical photocurable compositions were assembled to illustration the formulations that are used commercially as photocurable coatings. There is no attempt, however, to present here descriptions of various coating formulations. The descriptions of the compositions listed below appeared in chemical and patent literature. How well these materials actually perform as coatings is not known to this writer.

A photocurable coating composition intended as a hard protective coat was disclosed by Yoshikawa and Yamaya.[122]The photocurable composition comprises (A) 100 parts of a silicone compound that has a ≥ 3 epoxycyclohexyl-bearing groups each directly attached to a Si atom, but free of alkoxy groups. The molecular weigh of the compound is 500-2100 and an epoxycyclohexyl-bearing groups are equivalent to 180—230. (B) 0.1-5 parts of a photoacid generator, and optionally, (C) 30-400 parts of a pigment, inorganic oxide .

Matsunami and Suzuki disclosed [123] a UV-curable composition that consist of (A) urethane-modified alkyd resins, (B) compounds that have ≥ 2 methacryloyl groups, and (C) photoinitiators. The coating is intended for application on fiber-reinforced plastic molded products. Thus, soybean oil 564 part, glycerol 110 parts, and pentaerythritol 22 parts are combined with HCO_2Na and PPh_3 and then mixed with phthalic anhydride 262 parts, tall rosin 43 parts, and xylene 40 parts, are heated to 210° C for 4 hours. Xylene is removed at 150° under reduced pressure to give an oil—modified alkyd resin. 100 parts of this resin is treated with 0.07 part of isophorone diisocyanate in xylene to give a 60% urethane—modified alkyd resin solution. A UV curable composition is then formed from this resin 100 parts, dipentaerythritol hexaacrylate 20 parts, trimethylolpropane triacrylate 20 parts, benzophenone 2 parts, isoamyl dimethylaminobenzoate 2 parts, ethylacrylate 113 parts. A leveling agent 0.02 parts, and methylhydroquinone 0.2 parts were included. [122]

Nakajima patented a UV-curable coating compositions for uses in metal cans and bottles. [130] The coating material consists of (A) pigments, (B) ≥ 2 alicyclic epoxy group-containing cationic polymerizable compounds., (C) ≥ 2 oxetane group, or oxetane and OH group-containing compounds., (D) iodonium salt cationic polymerization photoinitiators that are described as, $R^1C_6H_4I^{\oplus}C_6H_4R^2\text{-}PF_6$ $\ominus$ (where; R^1, R^2 - alkyl), and (E) photosensitizer, a thioxanthone derivative that is shown below, (where R_3-R_6 are H, alkyl, alkoxy, halogen). The formulation contains a total of B+C 100 parts, D of 1-20 parts, and E of 0.5—10 parts. As an illustration, a composition, containing aluminum paste, 35 parts, 3,4-epoxycyclohexylmethyl-3,4-epoxycyclohexane carboxylate, 40 parts, bis(3-ethyl-3-oxetaneylmethyl) ether, 20 parts, photoinitiator ($R^1 = 2$-methylpropyl at p-position, $R^2 =$ methyl at p-position) 0.5 parts, and isopropyl-thioxanthone 3 parts was sprayed on a poly(ethylene terephthalate)/steel laminate and cured with 240 W/cm lamps for 30 seconds to form a film showing pencil hardness of H.

The photosensitizer is shown as follows:

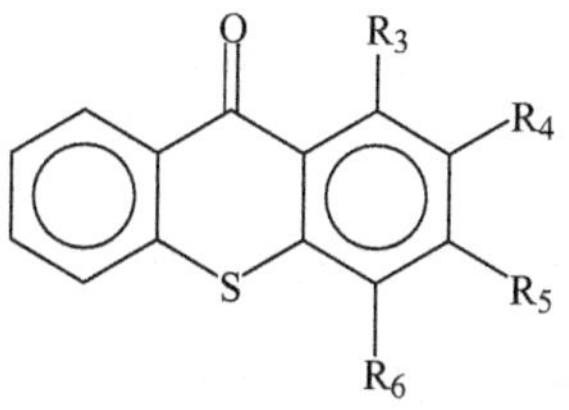

Yasuda and Tano [128] describe a Japanese patent for light curable compositions that are useful as coating materials. These consist of (A) 15-70% of a polyol, polyglycidyl ether,

$$(HO)_l R^1 [O(CH_2 CHR^2 O)_m R^3]_n$$

where R^1 is the polyol residue; R^2 is methyl of hydrogen; R^3 is glycidyl or hydrogen; n = 2-6; m = 0-2; (B) 30-85% of epoxidized polybutadiene. In addition there are 0.1—10 % of photoinitiators. Based on this patent, a composition of glycerol polyglycidyl ether (60 parts), epoxidized polybutadiene (40 parts), and photoinitiator (2 parts) was applied on poly(ethylene terephthalate) film and irradiated with UV to form a coating showing good cross-cut adhesion and no change by rubbing with solvents. [128]

Wang and Cheng reported [129] preparation of ultraviolet light-curable resin for coating optical fibers. It consists of polydimethylsiloxane epoxy acrylate and polyethylene glycol urethane acrylate. To this were added in different proportions trimethylolpropane triacrylate to modify the properties of blends. [129] After addition of a photoinitiator the choice formulation photo cured in 5 seconds.

References

1. J.F.G.A. Jansen, A.A. Dias, M. Dorschu, and B. Coussens, *Macromolecules*, **2003**, *36*, 3861
2. J.-P. Fouassier, *Photoinitiation, Photopolymerization and Photocuring,* Hanser/Gardner Publications, inc., Cincinnati, 1995; S.P. Pappas, ed. , *UV-curing Science and Technology. 2.* Technology Marketing Corp., Stanfort, CT (1985);
3. *Radiation Curing of Polymeric Materials*, Am.Chem.Soc. Symposium series # 417, C.E. Hoyle and J.F. Kinstle, eds., Am, Chem. Soc., Washington, D.C., 1900

4. L.R. Gatechair and A.M. Tiefenthaler, in *Radiation Curing of Polymeric Materials*, Am. Chem. Soc. Symposium series #417, C.E. Hoyle and J.F. Kinstle, eds., Am. Chem. Soc., Washington, D.C., 1900

5. C. Decker, in *Material Science and Technology*, Vol. 18, H. Meijer, ed., Wiley-VCH , Weinheim, 1997

6. E. Andrzejewska and M. Andrzejewska, *J. Polymer Sci., Part A., Polym. Chem.*, 1998, *36*, 665

7. C. Decker and K. Moussa, *Makromol. Chem. Rapid Commun*, **1990**, *11*,159; *Makromol Chem.*, **1990**, *192*, 507; *Eur. Polym. J.* **1991**, *27*, 403, 881

8. J.F. Jansen, A.A. Dias, M. Dorschu, and B. Coussens, *Macromolecules*, **2002**, *35*, 7329

9. E.R. Beckel, K.A. Berchtold, N. Nie, H. Lu, J.W. Stansbury, C.N. Bowman, *Am. Chem. Soc. Polymer Preprints*, **2003**, *44* (1), 31

10. N.B. Cramer, S.K. Reddy, A.K. O'Brien, and C.N. Bowman, *Macromolecules*, **2003**, *36*, 7964; N.B. Cramer, T. Davis, A.K. O'Brien, and C.N. Bowman, *Macromolecules*, **2003**, *36*, 4631

11. J.F.G.A. Jansen, A.A. Dias, M. Dorschu, and B. Coussens, *Macromolecules*, **2003**, *36*, 3861

12. H. Kilambi, E.R. Beckel, J.W. Stansbur, and C.N. Bowman, *Am. Chem. Soc. Polymer Prelprints*, **2004**, *45* (2), 73

13. T.Y. Lee, T.M. Roper, C.A. Guymon, E.S. Jonnson, *Am.Chem.Soc. Polymer Prelprints*, **2004**, *45* (2), 49

14. C. Decker and K. Moussa, *Die Makromoleculare Chemie*, **1991**, *192*, 507

15. B. Elzaouk and C. Decker, J. Macromol. Sci., Pure Appl. Chem. **1996**, Aee, 173

16. T.Y. Lee, C.A. Guymon, E.S. Jonsson, and C.E. Hoyle, *Am. Chem. Soc. Polymer Preprints*, **2004**, *45* (2), 25; T.Y. Lee, C.A. Guymon, E.S. Jonsson, S. Hait, and C.E. Hoyle, Macromolecules, **2005**, *38*, 7530

17. R. Pretot, A. Muehiebach, W. Peter, K. Dietliker, T. Jung, C. Auschra, K. Clemens, J. Hans, and P.A. Van Der Schaaf, PCT Int. Appl. WO 03 74, 466 (Cl. C07C69/54), (12 Sep 2003)

18. J. McConnell and F. K. Willard in *Radiation Curing of Polymeric Materials*, Am. Chem . Soc. Symposioum, Series # 417, C.E. Hoyle and J.F. Kinstle, eds., p. 258, Am. Chem. Soc., Washington, D.C., 1900

19. I. Osagawa, Japan Kokai Tokkyo Koho JP # 137,964 (2003); T. Endo, Japan Kokai Tokkyo Koho JP # 137,990 (2003)

20. K. Yukyasu and K. Hamada, JP 2000 273,128, (Oct. 3, 2000)

21. Y. Heschkel, K. Menzel, W. Paulus, C. Decker, and R. Schwan, *Am. Chem. Soc. Polymer Preprints*, **2003**, *44*(1), 9

22. C. Huimin, E. Cume, M. Tilley, and Y.-C. Jean, *Am. Chem. Soc. Polymer Preprints*, **2001**, *42*(2), 735

23. M. Jacobi and A. Henne, *J. Radiation Curing*, **1983**, *1*, 16

24. Nashikkubo, Tadaatomi, Kameyaina, and Atsushi, *Am. Chem. Soc. Polymer Preprints*, **2001**, *42*(2), 722

25. B. Yang and B. Schaeffer, *Am. Chem. Soc. Polymer Preprints*, **2001**, *42*(2), 795

26. J.J. Gummeson, in *Radiation Curing of Polymeric Materials*, Am. Chem. Soc. Symposium series #417, C.E. Hoyle and J.F. Kinstle, eds., Am. Chem. Soc., Washington, D.C., 1900

27. H.-G. Kou, A. Asif, and W.-F. Shi, *Chinese Journal of Chemistry*, **2003**, *21*(2), 91

28. V.V. Krongauz and C.P. Chawla, *Polymer* **2003**, 44(14), 3871-3876

29. S. Shimizu, S. Fumya, and T. Urano, Japan. Kokai # 6,344,651 (1988)

30. N. Stephenson, D. Kirks, and A. Scranton, *Am. Chem. Soc. Polymer Prelprints*, **2004**, *45* (2), 27

31. P. Fouassier, D. Ruhlman, Y. Takimoto, M. Harada, and M. Kawabata, *J. Imaging Sci. and Tech.*, **1993**, *37*, 208

32. D.L. Kurdikar and N.A. Peppas, *Macromolecules*, **1994**, *27*, 4084

33. M.D. Joshi and F. Rodriguez, *J. Polymer Sci., Polymer Chem. Ed*, **1988**, *26*, 819

34. G. Terrones and A.J. Pearlstein, *Macromolecules* **2003**, 36(17), 6346

35. N. Turro, *Molecular Photochemistry*, W.A. Benjamin, New York, 1967

36. M. Jacobi and A. Henne, *J. Radiation Curing*, **1983**, *1*, 16

37. J.G. Klusterboer, *Adv.Polymer Sci.*, **1988**, *84*, 1-66; I.R. Bellobono, E. Selli, and L. Righetto, *Makromol. Chem.*, **1989**, *190*, 1945

38. C. Decker, F. Masson, and L. Keller, *Polymeric Materials Science and Engineering*, **2003**, *88*. 215-216.

39. M. Wen and A. V. McCormick, *Macromolecules* **2000**, *33*, 9247-9254; M. Wen, L.E. Scriven, and A.V. McCormick, *Macromolecules*, **2003**, *36*, 4151

40. G. Odian,*Principles of Polymerization*, III ed., J. Wiley, New York, 1983

41. K.S. Aneseth and C.N. Bowman, *Polym. React. Eng.* **1993**, *1*,499

42. W. Wang and K. Cheng, *Eur. Polymer J.*, **2003**, *39*, 1891

43. A.V. Tobolsky, *J. Am. Chem. Soc.*, **1958**, *80,* 5727

44. R.D. Bohme and A.V. Tobolsky, "Dead-End Polymerization," *Encyclopedia of Polymer Science and Technology*, Wiley-Interscience, New York, 1966, Vol. 4, p. 599

45. A.V. Tobolsky, C.E. Rogers, and R.D. Brickman, *J. Am. Chem. Soc.*, **1961**, *82*, 1227

46. R.H. Gobran, M.B. Berenbaum, and A.V. Tobolsky, *J. Polym. Sci.*, **1961**, *46*, 431

47. E. Senogles and L.A. Woolf, *J. Chem. Educ.*, **1967**, *44*, 157

48. D.J.T. Hill and J.H. O'Donnell, *J. Polym. Sci. Polymer Chem. Ed.*, **1982**, *20*, 241

49. A. Ravve, *Principles of Polymer Chemistry*, I or II editions, Springer, New York, 1995; 2000

50. J. Nie, L. Lovel, and C.N. Bowman, *Am. Chem. Soc. Polymer Preprints*, **1999**, *40*(2), 1332

51. C.M. Chang, M.S. Kim, Y.-S. Roh, J.-P. Kim, *J. Mater. Sci. Letters*, **2002**, *21*(4), 1093

52. A.A. Stolov, T. Xie, J. Panelle, and S.L. Hsu, *Macromolecules*, **2001**, *34*, 2865

53. H. Cao, E. Hume, M. Tilley, and Y.-C. Jean, *Am. Chem. Soc. Polymer Preprints*, **2001**, *42*(2), 735

54. L. Gou, C.N. Chorestopoulus, and A.B. Scranton, *Am. Chem. Soc., Polymr Preprints,* **2004**, *45*(2), 39

55. J.P. Fouassier and D. Ruhlman, *J. Photochem.,* **1991**, *61*, 47

56. L.R. Gatechair and A.M. Tiefenthaler in *Radiation Curing of Polymeric Materials*, Am. Chem. Soc. Symposium Series #417, C.E. Hoyle and J.F. Kinstle, eds., Am. Chem. Soc., Washington, D.C., 1900

57. J.-D. Cho, H.-T. Ju, and L.-W. Hong, *Journal of Polymer Science, Part A: Polymer Chem.* **2005**, *43*(3), 658-670

58. A.F. Jacobine, D.M. Glaser, and S.T. Nakos, *Radiation Curing of Polymeric Materials,* C.E. Hoyle and J.F. Kinstle, (eds.), *Am. Chem. Soc., Symposium,* Series #417, 1990

59. C.R. Morgan, F. Magnotta, and A.D. Kettley, *J. Polym. Sci., Chem. Ed.* **1977**, *15*, 627

60. T.F. Scott, W.D. Cook, J.S. Forsythe, C.N. Bowman, and K.A. Berchtold, *Macromolecules* **2003**, *36*(16), 6066-6074

61. W.D. Cook and D.W. Paterson, *Am. Chem. Soc. Polymer Preprints*, **2004**, *45* (2), 62

62. N.W. Cramer, T. Davis, A.K. O'Brien, and C.N. Bowman, *Am. Chem. Soc. Polymer Preprints*, **2003**, *44* (1), 23

63. N. Cramer and C.N. Bowmna, *J. Polyme. Sci., Polym. Chem.,* **2001**, *39*, 3311

64. E. Klemm and S. Stenfuss, *Makromol. Chem.* **1991**, 192, 159

65. V.T. Desouza, V.K. Iyer, and H.H. Szmant, *J. Organic. Chem.,* **1987**, *52*, 1725

66. S. Davidson, *Exploring the Science, Technology, and Application of UV and EB Curing*, Sita Technology Ltd., London, 1999

67. G.K. Noren and E.J. Murphy, in *Radiation Curing of Polymeric Materials*, Am. Chem. Soc. Symposium Series #417, C.E. Hoyle and J.F. Kinstle, eds., Am. Chem. Soc., Washington, D.C., 1900

68. L.R. Gatechair and A.M. Tiefenthaler, in *Radiation Curing of Polymeric Materials*, Am. Chem. Soc. Symposium series #417, C.E. Hoyle and J.F. Kinstle, eds., Am. Chem. Soc., Washington, D.C., 1900

69. J.C. Graham and D.J. G.osky, in *Radiation Curing of Polymeric Materials*, Am. Chem. Soc., Symposium series #417, C.E. Hoyle and J.F. Kinstle, eds., Am. Chem. Soc., Washington, D.C., 1900

70. J.G. Kloosterboer, *Adv. Polym Sci.*, **1988**, 84, 1; J.R. Bellobono, E. Selli, and L. Righetto, *Makromol Chem.* **1989**, *190*, 1945

71. O. Okay, S.K. Reddy, and C.N. Bowman, *Macromolecules,* **2005**, *88*, 4501–4511

72. A.F. Jacobine in *Radiation Curing in Polymer Science and Technology*, J.P. Fouassier and J.f. Rabek (eds.), Elsevier, London, 1993,

73. C.S. Marvel and A.H. Markhart, *J. Polymer Sci.*, **1951**, *6*, 71

74. C.L. Lee and M.A. Lutz, U.S. Patent #4,780,486 (1988)

75. S.K. Dirlikov, U.S. Patent # 4,663,416 (1987)

76. K.D. Ahn U.Y. Kim, C.H. Kim, *J. Macromol. Sci.,-Chem.***1986**, *A23*, 169

77. C.E. Hoyle, M. Cole, M. Bachemin, W. Kuang, B. Yoder, N. Nguyen, and S. Jonsson, *Am. Chem. Soc. Polymer Preprints*, **2001**, *42* (2), 697

78. N. Rehnberg, A. Harden. S. Landmark, A. Manea, and L. Svenson, *Am. Chem. Soc. Polymer Preprints*, **2001**, *42* (2), 701

79. T.Y. Lee, K.A. Lowery, C.A. Gyuman, E.S. Johnsson, and C.E. Hoyle, *Am. Chem. Soc. Polymer Preprints*, **2003**, *44* (1), 21

80. T.M. Roper, C.A. Guymon, E.S. Jonsson, and C.E. Boyle, *Am. Chem. Soc. Polymer Prelprints*, **2004**, *45* (2), 67

81. T.S. Clark, C.H. Hoyle, and S. Jonsson, *Am. Chem. Soc. Polymer Prelprints*, **2004**, *45* (2), 57

82. M. Malkoch, E.C. Hagberg, C.J. Hawker, and K.R. Carter, *Am. Chem. Soc. Polymer Prelprints*, **2004**, *45* (2), 59

83. A.F. Jacobine, D.M. Glaser, and S.T. Nakos, in *Radiation Curing of Polymeric Materials*, Am. Chem. Soc. Symposium Series #417, C.E. Hoyle and J.F. Kinstle, eds., Am. Chem. Soc., Washington, D.C., 1900

84. N.B. Cramer, S.K. Reddy, A.K. O'Brien, and C.N. Bowman, *Macromolecules*, **2003**, *36*, 7964; *ibid*, 4631

85. J.V. Crivello and J.L. Lee in *Radiation Curing of Polymeric Materials*, Am. Chem. Soc. Symposium Series #417, C.E. Hoyle and J.F. Kinstle, eds., Am. Chem. Soc., Washington, D.C., 1900

86. J.V. Crivello and A. Ramdas, *Am. Chem. Soc. Polymer Preprints*, **1991**, *32*(2), 174

87. C.S. Lapin in *Radiation Curing of Polymeric Materials*, Am. Chem. Soc. Symposium, Series # 417, C.E. Hoyle and J.F. Kinstle, eds., Am. Chem. Soc., Washington, 1900

88. J.V. Crivello, and S. Song, *Chem. Mat.* **2000**, *12*, 3674

89. K.Y. Song, J.V. Crivello, and G. Kamakrishna, *Am. Chem. Soc. Polymer Preprints*, **2001**, *42* (1), 785; b. M. Jang and J.V. Crivello, *Journal of Polymer Science, Part A: Polymer Chemistry* **2003**, 41(19), 3056-3073

90. N.A. Nukolic and R.V. Schultz, *Am. Chem. Soc. Polymer Preprints*, **2001**, *42*(2), 743

91. J.V. Crivello and R.A. Ortiz, *J. Polymer Sc;i. Polym Chem.* **2001**, *39*, 2385

92. H. Kato and h. Sasala, *Am. Chem. Soc. Polymer Preprints*, **2001**, *42* (2), 729

93. M. Sangermano, R. Bongiovanni, G. Malucelli, A. Priola, A. Harden, J. Olbrych, and N. Rehnberg, *Am. Chem. Soc. Polymer Prelprints*, **2004**, *45* (2), 7

94. S.-Y.Shim and D.H. Suh, *ACS Symposium Series* **2003**, *847*, 277-284;

95. J.D. Oxman, W. Dwight, and M.C. Trom, *Polymeric Materials, Science and Engineering*, **2003**, *88*, 233

96. T. Maruyama, N. Kusumoto, and K. Seko, *J. Photopolym. Sci. Technol.* **2001**, 14(2), 165-170

97. Y. Yagci, B. Ochiai, and T. Endo, *Macromolecules*, **2003**, *36*, 9257

98. C. Decker and K. Moussa, *J. Polymer Sci., Polymer Chem Ed.*, **1990**, <u>*28*</u>, 3429

99. G.R. Noren, J.M. Zimmerman, J.J. Krajewski, and T.E. Bishop, in *Radiation Curing of Polymeric Materials*, Am. Chem. Soc. Symposium series #417, C.E. Hoyle and J.F. Kinstle, eds., Am. Chem. Soc., Washington, D.C., 1900

100. S.M. Gasper, D.N. Schissel, L.S. Baker, D.L Smith, R.E. Youngman, L.-M. Wu, S.S. Sonner, S.R. Givens, and R.R. Hancock, *Am. Chem. Soc. Polymer Preprints*, **2003**, *44* (1), 27

101. A.K. 0'Brien and C.N. Bowman, *Macromolecules* **2003**, 36, 7777-7782

102. M.L. Gornez, V. Avila, H.A. Montejano, and C.M. Previtali, *Polymer* **2003**, 44(10), 2875-2881

103. J. Qiao, V. Stone, G. Francois, A. Herssens, and H.E. Tweedy, *Polymeric Materials, Science and Engineering*, **2001**, *84*, 597

104. J.-D. Cho, H.-K. Kim, Y.-S. Kim, and J.-W. Hong, *Polymer Testing* **2003**, 22(6), 633-645

105. P.E.M. Allen, D.J. Bennett, H. Hagias, G.S. Ross, G.P. Simon, D.R.G. Williams, and E.H. Williams, *Eur. Polymer J.*, **1989**, *25*, 785

106. P.E.M. Allen, G.P. Simon, D.R.G. Williams, and E.H. Williams, *Macromolecules*, **1989**, *22*, 809

107. G.P. Simon, P.E.M. Allen, D.J. Bennett, D.R.G. Williams, and E.H. Williams, *Macromolecules*, **1989**, *22*, 3555

108. C.D. Hoyle in *Radiation Curing Science and Technology*, S.P. Pappas, ed., Plenum Press, New York, 1992

109. C. Decker and K. Moussa, *Makromol. Chem.*, **1988**, *189*, 2381

110. C. Decker, *Macromolecular Rapid Communications*, **2002**, *23* (18), 1067

111. T.F. Scott, W.D. Cook, J.C. Forsyth, C.N. Bowman, and K.A. Berchtold, *Macromolecules,* **2003**, *36*, 6066

112. C. Decker, D. Decker, and S. Morel, *ACS. Symposium, Series,* **1996**, *673*, 63

113. P.A.M. Steeman, A.A. Dias, D. Wienke, and T. Zwartkruis, *Macromolecules*, **2004**, *37*, 7001

114. B. Falk, S.M. Vallinas, J.V. Crivello, *Polymeric Materials, Science and Engineering*, **2003**, *88* 209

115. H. Matsubara, A. Yoshida, Y. Konodo, S. Truge, and H. Ohtani, *Macromolecules*, **2003**, *36*, 4750

116. K.A. Berchtold, T.W. Randolph, and C.N. Bowman, *Polymeric Materials, Science and Engineering*, **2003**, *88*; K.A. Berchtold, T.W. Randolph, and C.N. Bowman, *Macromolecules*, **2005**, *38*, 6954; K.A. Berchtold, B. Hacioglu, L. Lovell, J. Nile and C.N. Bowman, *Macromolecules*, **2001**, *34*, 5103

117. C. Carlini, P.A. Rolla, and E. Tambari, *J. Polymer Sci., Polym Phys.,* **1989**, *27*, 189; C. Carlini, P.A. Rolla, and E. Tambari, *J. Appl Polymer Sci.,* **1990**, *41,*

118. T. Nakamuchi, *Prog. Org. Coat.* **1986**, *14,* 23; K. Watanabe, T. Amari, and Y. Otsubo, *J. AppL Polym. Sci.* **1984**,*29,*57

119. J. Davison, and J.T. Guthrie, *Surf. Coat. Int., JOCCA* **1992**, 75, 316.

120. B.-S. Chiou, R. English, and S.A. Khan, *Macromolecules* **1996,** *29,* 5368

121. S.H. Dickens, J.W. Stansbury, K.M. Choi, and C.J.E. Floyd, *Macromolecules* **2003,** *36,* 6043-6053

122. Y. Yoshikawa, and M. Yamaya, U.S. Patent # 2003 187,088 (April 1, 2002)

123. H. Matsunami and H. Suzuki, J. Patent #2003 221,408, 5 Aug 2003

124. J.V. Koleske, *Proc. RadTech North America*, **1988**, 352, New Orleans

125. J.V. Sinka and D. Mazzoni, *Proc. RadTech North America*, **1988**, 378, New Orleans

126. H.G. Gaube and J. Oblemacher, *Proc. RadTech Europe*, **1989**, 699, Florence

127. R.F. Eaton, B.D. Hanrahan, and J.K. Braddok, *Proc. RadTech North America*, **1988**, 384, Chicago
128. R. Yasuda and T. Tano, J. Patent # 2003 301,027 (Oct 21, 2003)
129. W. Wang and K. Cheng, *Europ. Polymer J.* **2003,** 39(9), 1891-1897
130. Y. Nakajima, JP # 2003 253,196 (10 Sep 2003)

Chapter 4

Photocrosslinkable Polymers

Some photocrosslinking of polymers can be traced back to ancient days, when pitch was photocrosslinked for decorative purposes. [1] In modern times, wide varieties of photocrosslinkable polymers were developed. The early practice of photo imaging relied mainly upon the photodimerization reactions. These reactions are common, photoinduced, reactions of organic chemistry, namely intermolecular cyclization. This reaction of cyclization that takes place between two reactive species, with one of them electronically excited, is actually predicted by the Woodward-Hoffmann rules. [2]. In contrast, the reactions of thermally excited ground states of molecules proceed by different pathways. Many polymers were synthesized that possess pendant groups capable of photo-cyclization intermolecularly to be used in the photoimaging technology. Photo-crosslinking technology today, however, also uses coupling reactions of radicals, chain growth polymerizations that result in photocrosslinking and some ionic reactions. The light-induced polymerizations of multifunctional monomers that transform liquid resins into solid polymers almost instantly and selectively in the illuminated areas are now versatile processes. They can, however, be achieved in a variety of ways. Thus, for instance, Decker discussed recently work in the less explored areas of photocuring, namely laser-induced ultra-fast polymerization and UV curing of binary polymer systems. [3] By using highly sensitive acrylate photoresists, relief images of micronic size can be obtained by fast scanning with a focused laser beam.. Also, polymer networks of different architectures can be obtained by UV irradiation of various monomer blends, e.g., acrylate-epoxide, acrylate-vinyl ethers, acrylate-polyene, vinyl ether-maleate and thiol-polyene. [3] This does not mean, however, that photocross-linking of polymers is now unimportant technologically or scientifically. The fact that considerable research still continues in the field is a direct indication of that.

The full kinetic picture of photocrosslinking of polymers is still evolving. Wen *et al.*,[4] studied the kinetics of models for gelation that simulate free-radical polymerization on fixed lattices, such as photocrosslinking, where propagation and termination reactions are restricted to occur only between nearest neighbors. Such a model was used with bifunctional sites and with kinetics recast as a Markov process through a stochastic approach. They concluded that as the polymerization proceeds, the evolution of structure is characterized by pair correlation functions of three types—of reacted sites, of doubly reacted sites, and of monomers. As the polymerization proceeds, reacted sites and doubly reacted sites come to be distributed more uniformly in space; monomers come to be distributed less uniformly. A higher initiation rate constant and concentration, and lower propagation rate constant can result in a uniform distribution of reacted sites. These factors, however, can also lead to lower average amount of reactions between reacted sites. These trends are strongest at low conversions. In contrast, an enhanced primary cyclization leads to less uniform distribution of reacted sites but to more uniform distribution of monomers. It also results in higher connectivity between reacted sites that are close together but to lower connectivity between reacted sites that are far apart. Finally, at high conversions it leads to a more uniform distribution of doubly reacted sites. [4]

The light crosslinkable reaction, like all crosslinking reactions results in *gelation* and the extent of gelation is important in this process. This extent is tied to the quantity of the functional groups in the reaction mixture. Carothers equations relate the critical extent of the reaction, p_c, at the gel point to the functionality of the reactants:

$$p_c = 2 / f_{ave}$$

where f_{ave} is the functionality. This equation, however, was written for reaction mixtures that contain two different functional groups in stochiometric proportions to each other. All the functional groups, however, for various reasons, might not participate in photocrosslinking reactions. Even though these reactions may involve identical groups, this equation probably would not apply.

Also, a kinetic gelation model that simulates free-radical network polymerization on a lattice with a stochastic kinetic approach to enable real time calculation was used to assess how initiation rate and primary cyclization affect the overall kinetics of polymerization of difunctional monomers. Changes that cause a more uniform distribution of reacted sites-higher initiation rate or less primary cyclization increase the accessibility of free radicals to functional groups, lower the fraction of trapped radicals, and consequently raise the apparent propagation rate constant. On the other hand, the final conversion, determined by kinetic chain length at a given initiator concentration, drops when termination becomes more severe such as under higher initiation rate or when radical trapping worsens such as under enhanced cyclization. In addition, the model simulates the contribution of pendant functional groups to the formation of different structures. The higher the radical concentration brought by higher initiation rate or by less preferred primary cyclization, the lower the fraction of pendant functional groups to form primary cycles and the higher the fractions of pendant functional groups to form crosslinks and secondary cycles.

A statistical approach was developed by Flory [5,6] and by Stockmayer [7] to derive an expression for predicting the extent of reaction at the gel point. It is expressed as

$$p_c = 1 / [r(f_{w,A} - 1)(f_{wB} - 1)]^{0.5}$$

where $f_{w,A}$ and f_{wB} are weight average functionalities of A and B functional groups. Here the functionalities are defined as,

$$f_{w,A} = \Sigma f^2_{Ai} N_{Ai} / \Sigma f_{Ai} N_{Ai} \text{ and } f_{wB} = \Sigma f^2_{Bi} N_{Bi} / \Sigma f_{Bi} N_{Bi}$$

When the crosslinking, however, takes place by cyclization through dimerization of pendant groups, it appears that the gelation should probably be treated instead as a case of multiple dimerizations.

4.1. Polymers that Photocrosslink by Formation of Cyclobutane Rings

Many of the photocrosslinkable polymers for photoimaging in use today react by a $2\pi + 2\pi$ type dimerization with the accompanying formation of the cyclobutane rings. [8,9] The formation of the cyclobutane ring can be simply shown on the photocrosslinking reaction of poly(naphthyl vinyl acrylate), a polymer that also undergoes this type of dimerization [10]:

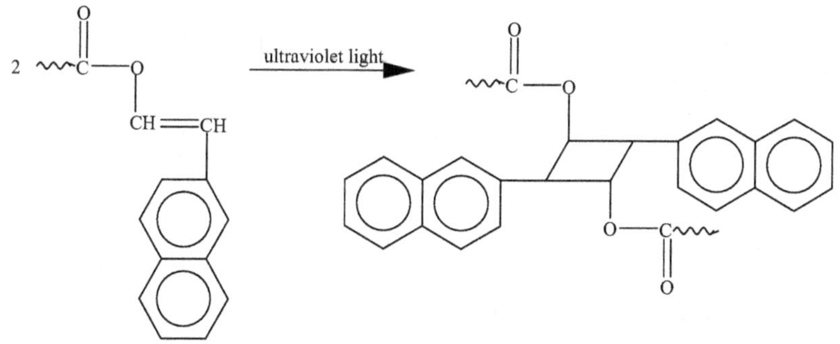

The naphthalenes become bonded to cyclobutane ring in 1,2 and 1,3 positions. Many polymers with other functional groups can also photocrosslink by $2\pi + 2\pi$ addition. Following is an illustration of some of these groups [11]:

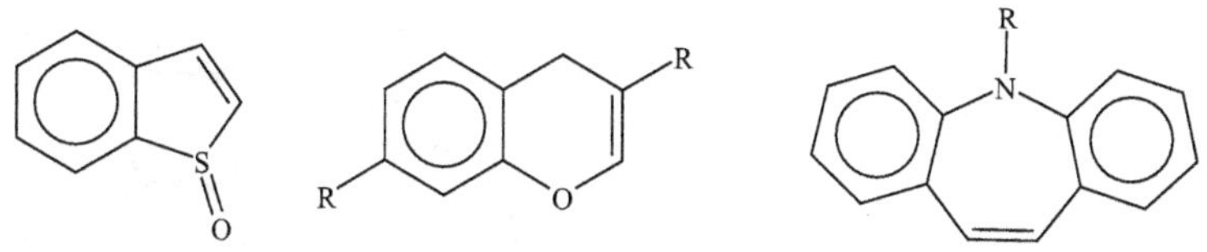

banzothiophene oxide   coumarin

dibenzazepene

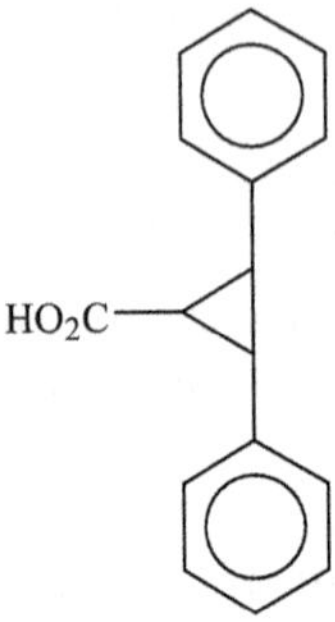

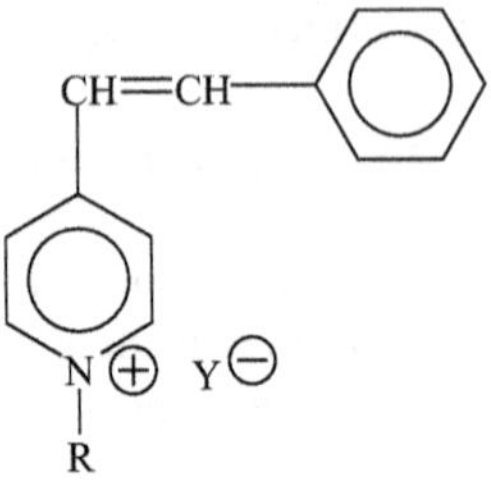

stilbazole 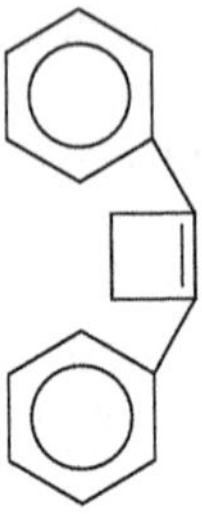   diphenylcyclobatane

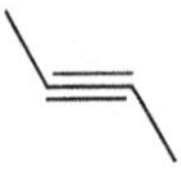

alkyne 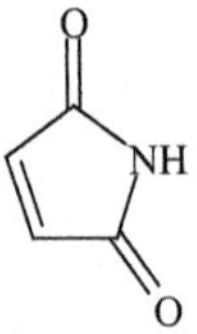   maleimide 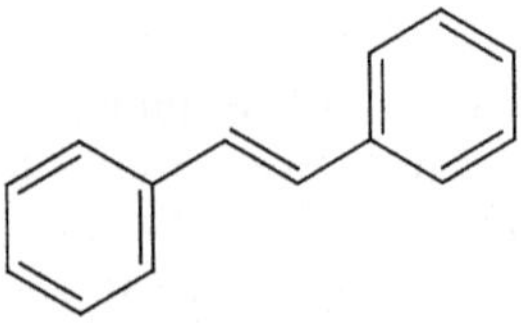   stilbene

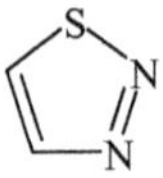

1,2,3-thiadiazole 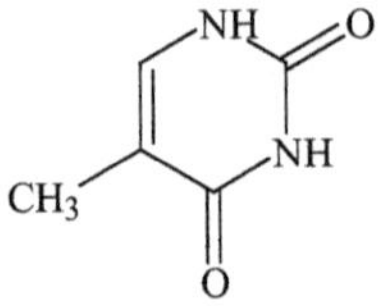   thiamine 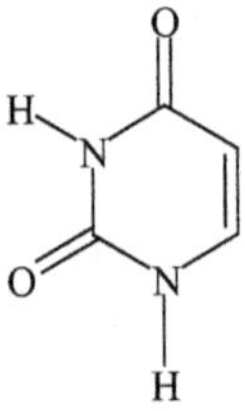   uracil

Pendant groups with anthracene moieties, however, are believed to crosslink by a $4\pi + 4\pi$ type cycloaddition[11]:

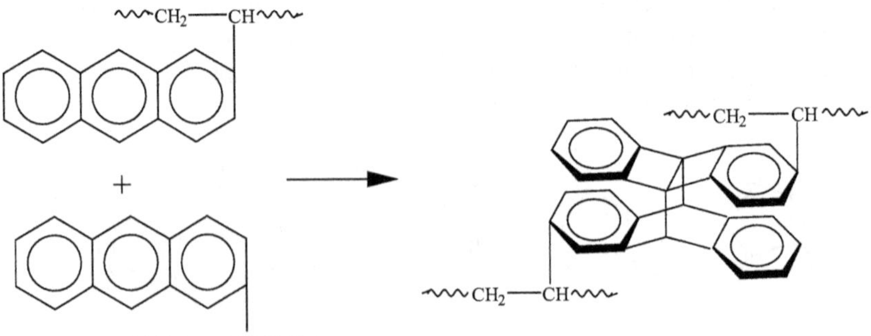

Many photodimerizations of functional groups, like the ones shown above, require the presence of photosensitizers. These compounds may be selectively excited to their triplet state by sensitizers with the right type of energy levels. For efficient energy transfer to occur, the triplet energy of the donor should be approximately 3 kcal/mole greater than that of the acceptor. [12,13] This type of sensitization is believed to be diffusion controlled. [14]

Trecker [14] lists the steps of the sensitized dimerization reaction as follows:

$$
\begin{aligned}
\text{excitation:} &\quad S\,(\text{sensitizer}) + h\nu \longrightarrow S^* \\
\text{energy loss:} &\quad S_3^* \longrightarrow S \\
\text{sensitization:} &\quad S_3^* + D \longrightarrow \bullet S\text{—}D\bullet \\
\text{dimerization:} &\quad \bullet S\text{—}D\bullet + D \longrightarrow S + \bullet D\text{—}D\bullet \\
\text{dimer:} &\quad \bullet D\text{—}D\bullet \longrightarrow \text{dimer(s)}
\end{aligned}
$$

The quantum yield expression for this scheme of dimerization is shown as follows [14]

$$1 / \Phi = 1/\phi\alpha + k_3 / \phi\alpha k_4(D)$$

where Φ represents the quantum yield of dimer formation, ϕ is the efficiency of sensitizer intersystem crossing, D is the initial concentration of the reactive groups and k's are the specific rate constants for the reactions shown in the above scheme.

4.1.1. Polymers with Pendant Cinnamoyl Functional Groups

Minsk [15] may have been the first to synthesize a photocrosslinkable polymer, namely poly(vinyl cinnamate). The photochemistry of this compound is similar to the photocylization of cinnamic acid that is discussed below in this section. It is interesting that the reaction of cyclization of cinnamic acid can take place even in the solid crystalline stage. This illustrates that the reaction requires very little molecular motion. Similar reactions occur in polymeric materials that are functionalized with cinnamate groups. The photocrosslinking of poly(vinyl cinnamate) is illustrated below:

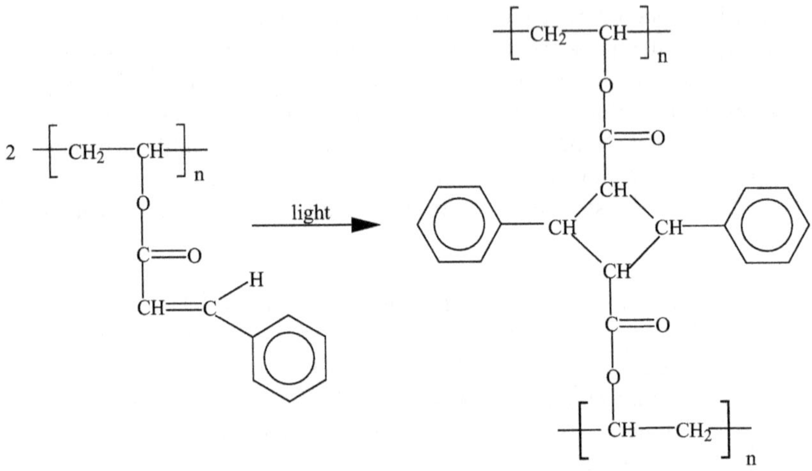

Much earlier, well before Minsk, in 1895, Berham and Kursten [16] recognized that solid cinnamic acid undergoes a chemical change when exposed to light. Following this, Ruber [17] established that the change is a dimerization of the acid to form a cyclobutane derivative. This dimerization results in formation of truxillic and truxinic acids:

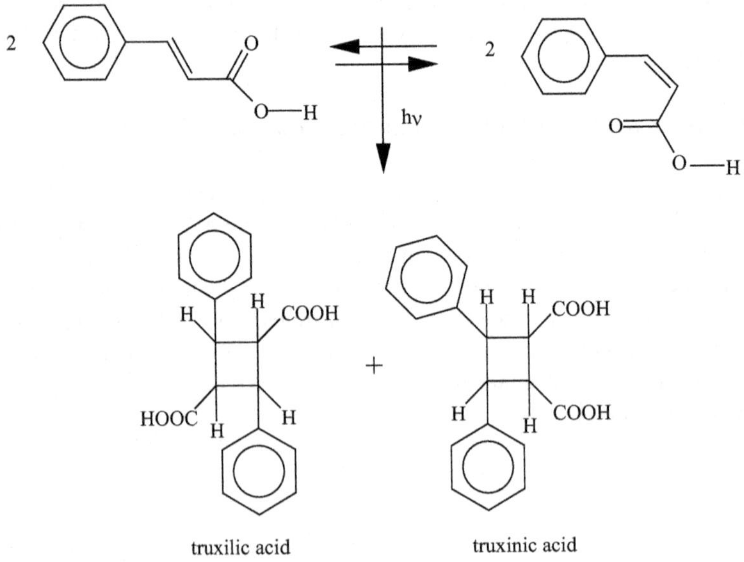

truxilic acid   truxinic acid

Schmidt and coworkers [18-20] studied the reaction mechanism and came to the following conclusions:

1. Photodimerization of cinnamic acid and its esters is controlled by the crystal lattice,

2. Dimerizations are possible when olefinic double bonds of the two neighboring molecules in the crystals are 4.1 Å or less apart.

3. Dimerizations are not possible when the double bonds are 4.7 angstroms or more apart.

The dimerization reaction takes place upon irradiation with light of a wavelength longer than 300 nm. It was demonstrated subsequently, [21] however, on poly(vinyl cinnamate) that adducts dissociate again upon irradiation with light of 254 nm. Photodimerization and formation of cyclobutane groups is enhanced as a result of sensitized irradiation of ethyl and other cinnamate. [22]

When dealing with poly(vinyl cinnamate) it is reasonable to assume that the degree of order in the relationship of one cinnamic group to another is much lower than is found in a crystal lattice of cinnamic acid. On the other hand, it should be higher than in solutions of cinnamic acid, where the groups are far enough apart so that very little photodimerization takes place. Photo-crosslinking of poly(vinyl cinnamate) can include the following reactions [23]:

1. Truxinic acid type dimerization in irradiated poly(vinyl cinnamate) that can occur intramolecularly. It can be shown as follows:

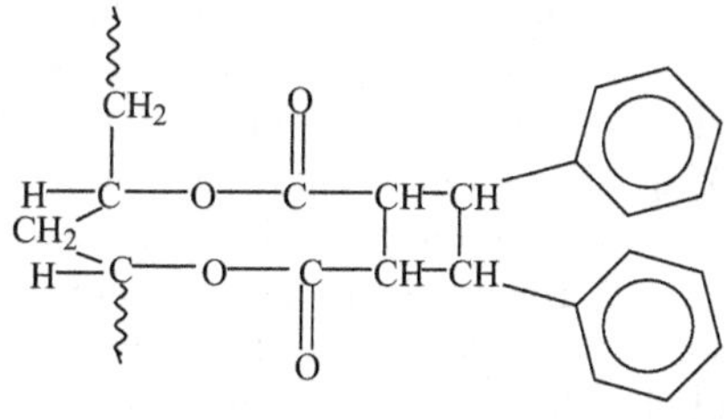

This is accompanied by formation of both folded and parallel chains

2. Truxillic acid type intermolecular dimerization in irradiated poly(vinyl cinnamate), on the other hand, can be illustrated as follows:

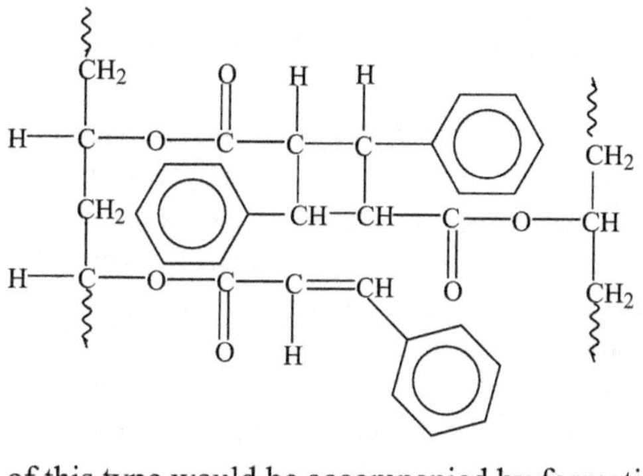

dimerization of this type would be accompanied by formation of folded chains.

Reactions of formations of folded and parallel chains are similar with the exception that the reacting cinnamic groups are further apart in folded chains and come together only by virtue of chain folding. Chains located parallel to each other but at the right distance can also conceivably yield truxinic acid type dimerization. This would be similar to the arrangements in crystal lattices. Formation of a truxillic dimer, like in 2, shown above, requires favorable folding or two different chains. Also, there is accompanying possibility that the double bonds may simply polymerize by a chain propagating reaction. [23] This was observed with some cinnamate esters.[24] Attempts were made to determine the reaction products of photocrosslinked poly(vinyl cinnamate) by first hydrolyzing it, and then by isolating and identifying the acids. The results showed that α-truxillic acid does form. Formation of β-truxillic acid, however, was not demonstrated. In addition, among the reaction products there is also a large quantity of unreacted cinnamic acid. This indicates that only a small portion of the double bonds participate in the reaction. Also, it should not be forgotten, that only small changes in unsaturation in polymer molecules, as a result of crosslinking, can profoundly effect solubility.

Studies [25] of electronic structures in photoisomerization and photo-dimerization of cinnamic acid showed that phosphorescence of cinnamic groups occurs at about 20,000 cm^{-1}. Also, it was demonstrated when photosensitizers are present, the critical distance between donor, sensitizer, and acceptor molecules (cinnamic acid) is about 10 angstroms.[25] Although all the details of incipient photocrosslinking of poly(vinyl cinnamate) have still not been fully worked out, most accept that all three mechanisms take place. These are: dimerizations to truxillic and truxinic acid type structures and polymerizations through the double bonds. The excited states of the molecules can be produced by direct irradiation and also through intersystem crossing from an appropriate photosensitizer.[25]

Several criterions were derived from proper selection of sensitizers. [26] These are:

(1) The triplet state must be at the energy level close to 50-55 K cal/mole for the cinnamate moiety.

(2) The quantum yield of the ratio of phosphorescence to fluorescence should be higher than unity.

(3) The mean lifetime of a triplet state must be greater than 0.01 sec. The photosensitizing activity (characterized by the triplet state) of derivatives of cinnamic acid is beyond the phosphorescence of the cinnamate group (at about 20,000 cm^{-1}, as stated above). An energy transfer diagram for poly(vinyl cinnamate) photosensitization with a sensitizer like 2-nitro-fluorene was published. [27]The rate of dimerization obeys first order kinetics. In addition, polymers consisting of flexible segments exhibit higher rates of photo-dimerization than do those composed of rigid segments. It was also demonstrated that greater photosensitivity can be obtained by separating the

cinnamic   group from the polymer backbone by introducing   -CH$_2$-CH$_2$-O- spacers as follows [28]:

In   addition,   Tsuda   and   Oikawa   carried   out   molecular   orbital calculations of the electronic structures in the excited states of poly(vinyl cinnamate).[29] They based their calculations on the reaction of intermolecular concerted cycloaddition that take place according to the Woodward-Hoffmann's rule. This means that the cyclobutane ring formation takes place if a nodal plane exists at the central double bond in the lowest unoccupied MO(LLUMO) and not in the highest occupied MO(HOMO) of the ground state cinnamoyloxy group. This is within the picture of Huckel MO or Extended Huckel MO theory. The conclusion is that the concerted cycloadditions occur favorably in the lowest triplet state T$_1$ and in the second excited singlet state S$_2$.[29]

The effectiveness of photosensitizers in accelerating the crosslinking reaction of poly(vinyl cinnamate is illustrated in Table 4.1.

Some four decades after the original development of poly(vinyl cinnamate) into a useful photocrosslinkable polymer, a novel optical property of the polymer was observed. When the material is irradiated with linearly polarized light it exhibits polarization holography.[30,31] The exposure of thin films of poly(vinyl cinnamate) to linearly polarized ultraviolet light causes uniaxial reorientation into liquid crystal layers.[32-39] Poly(vinyl cinnamate) and its derivatives have the ability to align in thin films the liquid crystal moieties in the direction that is perpendicular to the polarization axis of the linearly polarized ultraviolet light.[41-43] Schadt and coworkers[92] suggested that the surface-settled homogeneous alignment of nematic liquid crystals results from photodimerizions of the cinnamate moieties and formation of cyclobutane rings (as shown earlier) with an azimuthally oriented order. This, he feels, determines the direction of the liquid crystal alignment.[92] Ichimura and coworkers[41] suggested a different photoalignment process. They claimed that the photoinduced homogeneous liquid crystalline alignment is caused by polarization of photochromophores at the uppermost surfaces of the substrates due to repeated A/Z photoisomerizations, similarly to azobenzenes (see chapter 5).[41-47] This was also shown to take place with stilbenes.[92] In addition it was demonstrated by them[41] that both photoisomerization and photodimerization contribute to liquid crystalline alignment. Photoregulation in a polymethacrylate with o-cinnamate side chains displays preferential formation

Table 4.1. Effectiveness of Sensitizers on Relative Speed of Crosslinking of Poly(vinyl cinnamate)[a]

Sensitizer	Relative speed	Sensitizer	Relative speed
(none)	1	4-Nitroaniline	100
Naphthalene	3	3-Nitrofluorene	113
Benzanthrone	7	4-Nitromethylaniline	137
Phenanthrene	14	4-Nitrobiphenyl	200
Crysene	18	Picramide	400
Benzophenone	20	4-Nitro-2.6-dichlorodimethylaniline	561
Anthrone	31	Michler's ketone	640
5-Nitroacenaphthene	84	N-Acyl-4-nitro-1-nathylamine	1100

a. from reference 28.

of Z-isomer. Dimerization, on the other hand, takes place more favorably in other polymers, including poly(vinyl cinnamate). [41]

The liquid crystals alignment in films prepared from materials with cinnamate group after irradiating the films with linearly polarized UV light is quite uniform. All the aggregate structures, lamellar crystals, produced by the photocrosslinking reaction were found to be square in shape. [48] Because this has a potential application in flat panel liquid crystal displays, Lee *et al.*, [49] synthesized a soluble photoreactive polyimide with cinnamate chromophore side-groups.   The   polymer,   poly(3,3'-bis(cinnamoyloxy)-4,4'-biphenylene hexafluoroisopropylidene diphthalimide), has a reasonably high molecular weight and forms good quality films through conventional solution spin-casting and drying.

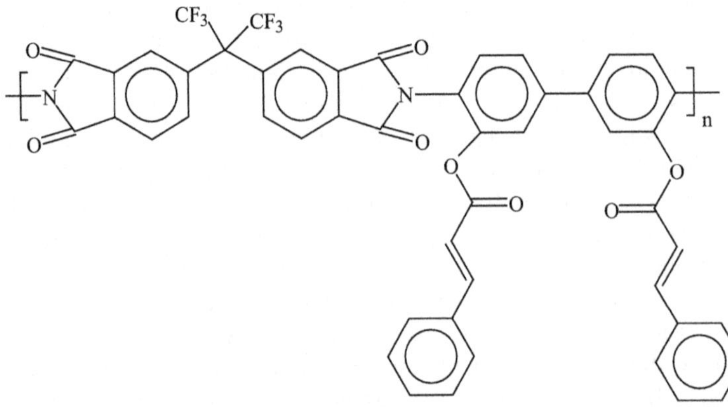

The polymer is thermally stable up to 340 °C and positively birefringent. The photochemical reactions of the polymer in solution and in films, as well as its molecular orientations are induced by exposure to linearly polarized ultraviolet light. As one might expect, the cinnamate chromophores undergo both photo-isomerization and photodimerization. Also, exposure to UV light induces anisotropic orientations of the polymer main chains and of the cinnamate side groups in the films. The irradiated films homogeneously align nematic liquid crystal molecules along a direction at an angle of 107° with respect to the polarization of the linearly polarized ultraviolet light. This coincides with the orientation direction of the polyimide chains. Thus, the liquid crystal alignment process is principally governed in irradiated polyimide films by the polymer main chains and the unreacted cinnamate side groups. [49]

Nagata et al., [50] reported preparation of a series of photocrosslinkable biodegradable polymers by condensation of dichlorides of 4,4'-adipoyldioxy)dicinnamic acid and alkane diols of various methylene lengths.. They also used various poly(ethylene glycols) with molecular weights ranging from 200 to 8300.

Among other interesting polymers with cinnamate functional groups are high polymeric phosphazenes that bear cinnamate groups. [51] A typical polymer synthesis is a follows:

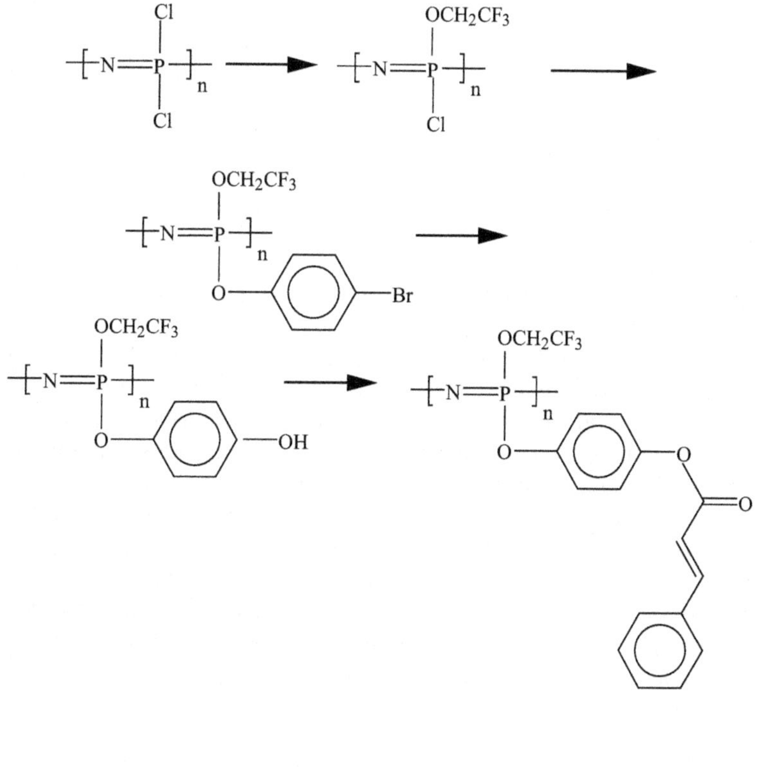

It was found, however, that the T_g-s of these polymers are too low for use as photoresists. Phosphazenes with chalcone functionality, on the other hand, appear to be more promising.[51]

Nitrocyannamates and similar structures can be introduced into poly(glicidyl methacrylate) to form photocrosslinkable macromolecules. Such compounds are 4-nitrophenyl cinnamate and 4-nitro-1-naphthyl cinnamate.[52]

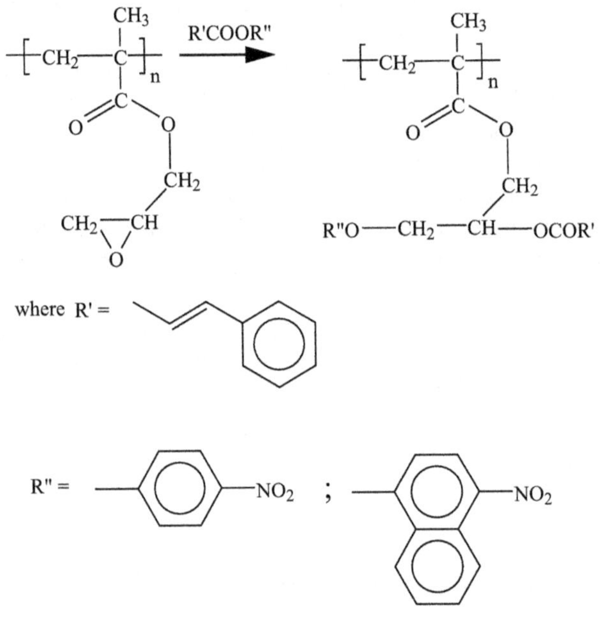

The above reaction was carried out in the presence of quaternary amine salts as catalysts.[52]

Shindo, Sugimura, Horie, and Mita[53] investigated intra- and intermolecular photocrosslinking reactions of poly(2-cynnamoyloxy-ethyl)methacrylate and its copolymer with methyl methacrylate in dilute solutions under ultraviolet light irradiation at 30 °C. The reactions were followed with ultra violet light spectroscopy, by light scattering measurements, gel permeation chromatography, and viscosity measurements. It was observed that the contraction of polymer coils proceeds due to predominant intramolecular photodimerization in very dilute solution. When the polymer concentration increases, competition between intra- and intermolecular photodimerization was observed. The volume of the polymer coils increases gradually due to the increase in molecular weight caused by the intermolecular reactions overcoming the contractions caused by the intramolecular reactions.[53]

Park *et al.*, [54] reported the synthesis and application of terpolymer bearing cyclic carbonate and cinnamoyl groups. The syntheses of photopolymer with pendant cinnamic ester and cyclic carbonate groups was achieved by the addition reaction of poly(glycidyl methacrylate-co-styrene) with CO_2 and then with cinnamoyl chloride. Quaternary ammonium salts showed good catalytic activity for this synthesis. Photochemical reaction experiments revealed that terpolymer with cinnamate and cyclic carbonate groups has good photosensitivity, even in the absence of sensitizer. In order to expand the application of the obtained terpolymer, polymer blends with poly(methyl methacrylate) were also prepared.

Lee *et al.*, [55] reported a synthesis of a photopolymer bearing pendant cinnamic ester and cyclic carbonate groups by the addition. reaction of poly(glycidyl methacrylate-co-acrylonitrile) With carbon dioxide and then with cinnamoyl chloride. Quaternary ammonium salts showed good catalytic activity for the synthesis of poly[3-chloro-2-cinnamoyloxypropyl methacrylate-co-(2-oxo-1,3-dioxolane-4-yl)methyl methacrylate-co-acrylonitrile] The conversion of poly(glycidyl methacrylate-co-acrylonitrile) increased with the increase of alkyl chain length and the nucleophilicity of counter anion of the quaternary salt. Photochemical relation experiments revealed that the photopolymer has good photosensitivity even in the absence of a sensitizer.

Park *et al.*, [56] also synthesized other photopolymers with both pendant cyclic carbonate groups and cinnamic ester by addition reactions of poly(glycidyl methacrylate-co-styrene) with carbon dioxide and then the product with cinnamoyl chloride. Soluble quaternary ammonium salt catalysts showed good yield of cinnamoyl chloride addition to the glycidyl methacrylate groups. They observed that quaternary salt catalysts of longer alkyl chain length and of more nucleop-anion offered higher yield of cinnamoyl chloride addition. The aim of their work was to develop new materials for photoresists.

A novel utilization of polymers with pendant cinnamoyl groups was reported by Wilkes *et al.*, [57] who functionalized a copolymer of methyl methacrylate and hydroxyethyl acrylate with cinnamoyl chloride. The polymer was then used for in situ crosslinking of polymeric fibers during electrospinning. The electrospinning equipment was modified to facilitate UV light irradiation of the electrospun fibers while in flight to the collector target. Three polymers with different mol percent of the cinnamate functionality group (4, 9, and 13 mol %) were synthesized and utilized in this study. In addition to crosslinking, the irradiated copolymers were observed to form insoluble gels. As expected, it was found that the gel fraction increased with increasing mol % of the cinnamate species. The polymer can be illustrated as follows:

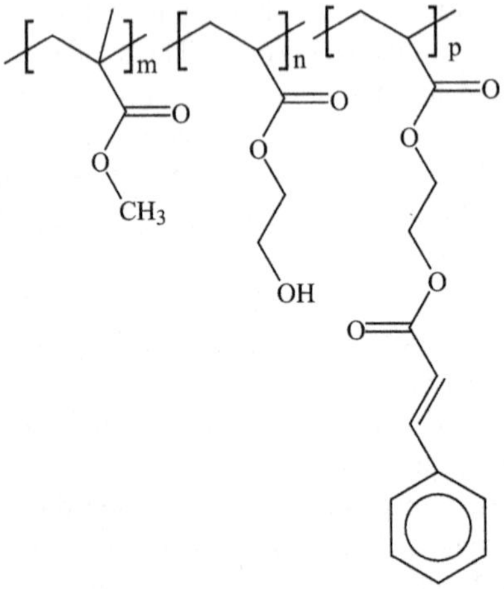

4.1.2. Polymers with Functional Chalcone Groups

Pendant chalcone groups on polymers behave similarly to pendant cinnamate groups. Thus, photocrosslincable polymer can be formed, for instance, from poly(vinyl alcohol) by a reaction with 4'-substituted-4-caboxychalcone in homogeneous dimethyl formamide solution, using 2,4,6-trinitrochlorobenzene as the condensing agent.[58]

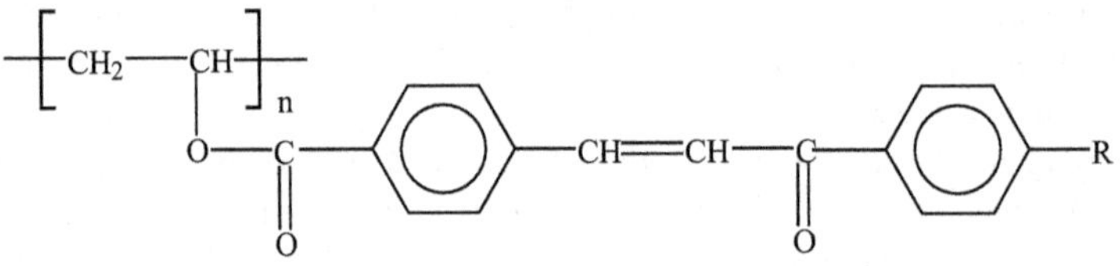

The photosensitivity of this polymer is in the range of 1 to 5 mJ/cm^2, according to Watanabe and coworkers.[58] When the R group shown above is p-Br, m-NO$_2$, or (CH$_3$)$_2$N, the crosslinking is via formation of biradicals derived from the double bonds of the cinnamoyl groups and an abstraction of protons from the neighboring methyne or methylene groups. This reaction of dimerization can be illustrated as follows [59]:

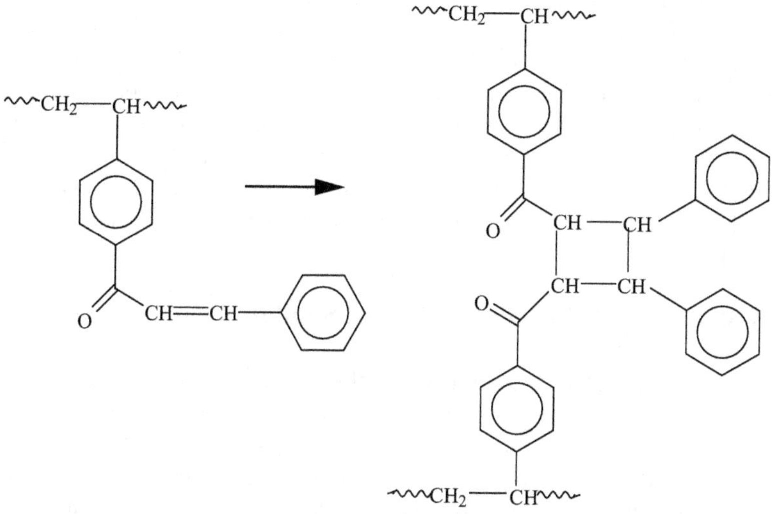

Formation of polymers by reaction of free-radical polymerizable methacryloyl groups and a photocrosslinkable pendant chalcone units was also reported. [60] The photocrosslinking reaction takes place in thin films and in solutions upon irradiation with high pressure mercury lamps. The reaction can be monitored through changes in the UV absorption spectrum. The rates of photocrosslinking in solutions were reported to be faster than in films. Photo-crosslinking in the presence of a triplet photosensitizer shows no significant changes in the rate of disappearance of the double bonds. This particular polymer with a pendant chalcone unit was found to photocrosslink at a high rate even in the absence of a photosensitizer and would be of interest as negative photoresist. [60]

A bifunctional epoxy chalcone monomer was synthesized by reacting 4,4'-dihydroxychalcone with epichlorohydrin. [61]

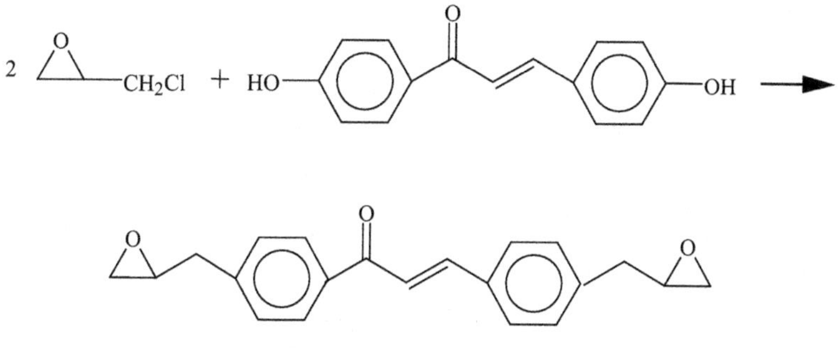

The compound has functional groups that support dimerization type crosslinking and cationic polymerization upon UV exposure (λ = 300—360 nm).. Photo-dimerization of the chalcone-epoxy compound was confirmed by UV-visible and IR absorbance changes of the C=C double bond of the chalcone unit. Additions of small amounts of onium salts will also photoinitiate cationic polymerization of the epoxy groups present in the above chalcone-epoxy compound by exposure to UV. This ultra-violet light cured chalcone-epoxy compound was reported to possess excellent thermal stability and compares well with conventional UV-cured Bisphenol A type epoxy resins.[61] (see Chapter 3)

4.1.3. Polymers with Functional Groups Similar to Cinnamates

As stated earlier, a draw back of the photocrosslinking reactions of poly(vinyl cinnamate) is the reversibility of crosslinking when irradiated with light of proper wavelength. In addition, some intramolecular cyclizations also take place. Nevertheless, because the reactions found extensive commercial applications, a number of other polymers that resemble poly(vinyl cinnamate) were developed. Some of these functional groups are as follows [62]:

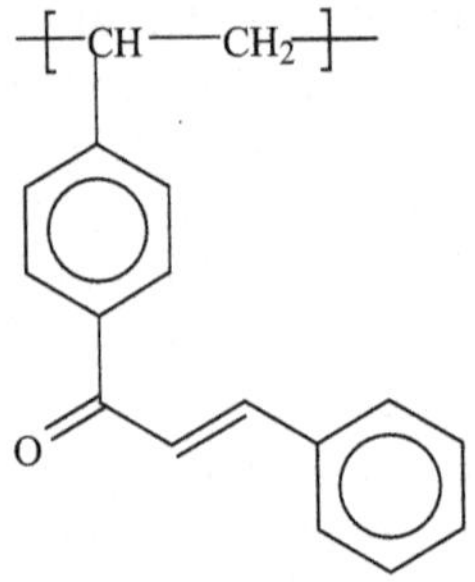

chalcone

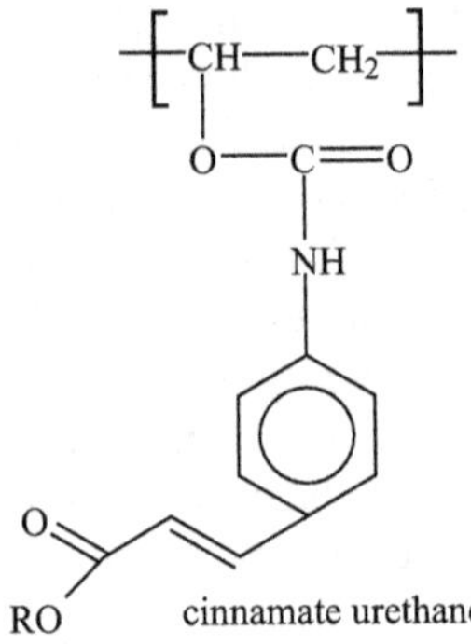

cinnamate urethane

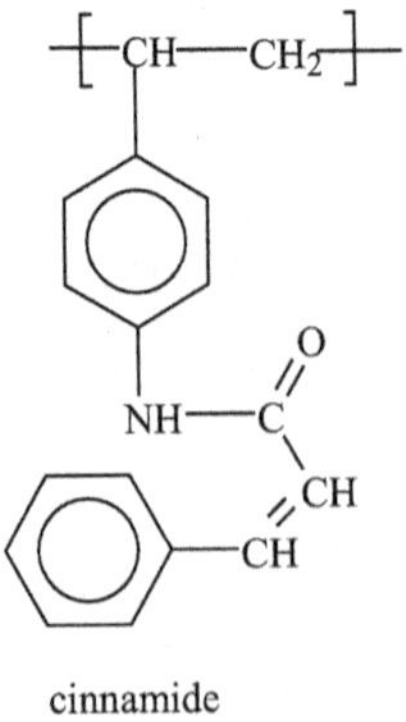

cinnamide

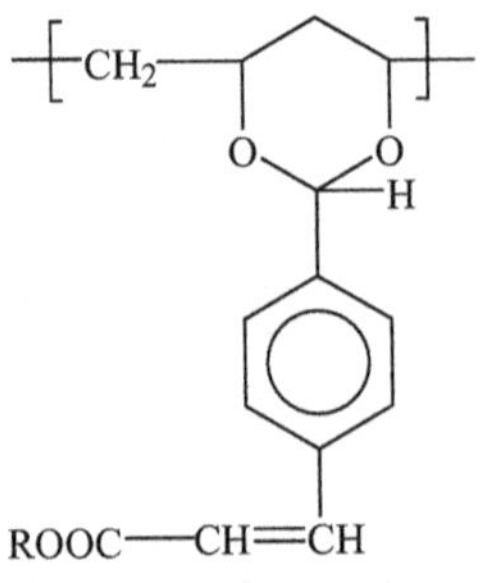

acetal

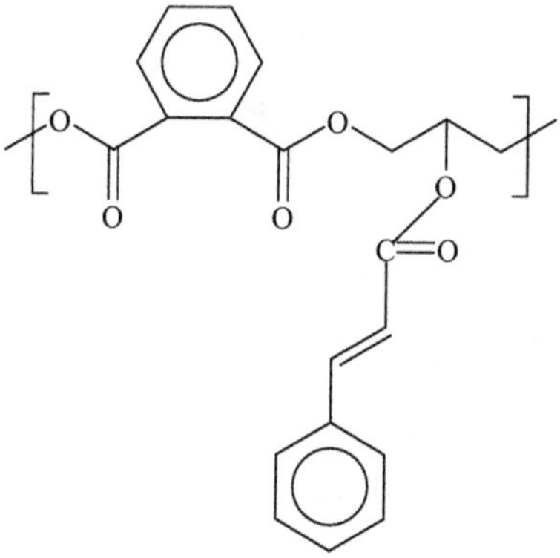

polyester

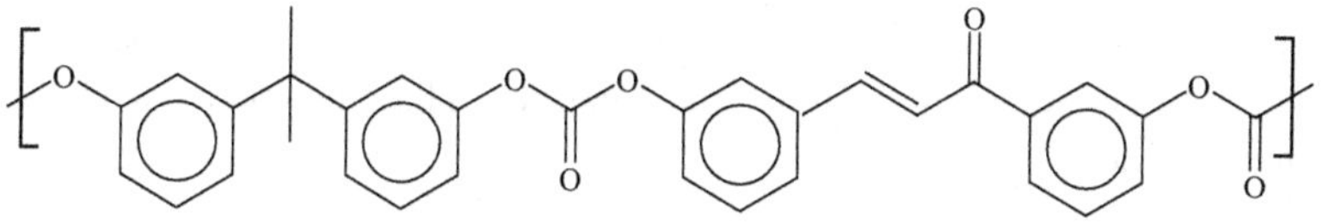

polycarbonate

4.1.4. Polymers with Pendant Furan Groups

Tsuda [63] attached 2-furylacrolein to a polymeric backbone by the following reaction:

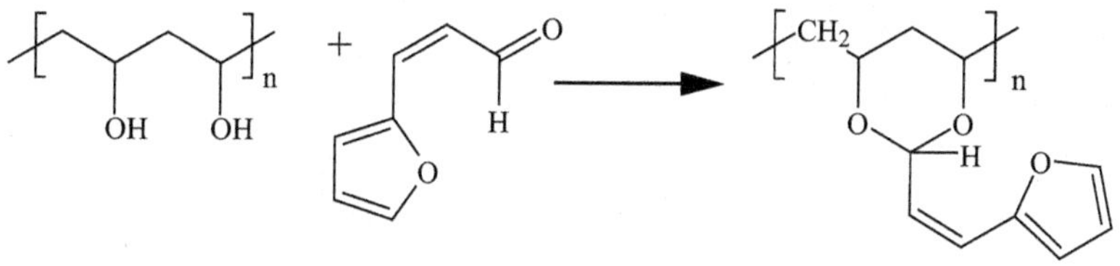

In the presence of photosensitizers this polymer was found by Tsuda to crosslink at a considerably faster rate than does poly(vinyl cinnamate).[63] He showed that the crosslinking reaction also results in formation of cyclobutane derivatives. The crosslinking is illustrated on 3-furfurylacrylic ester:

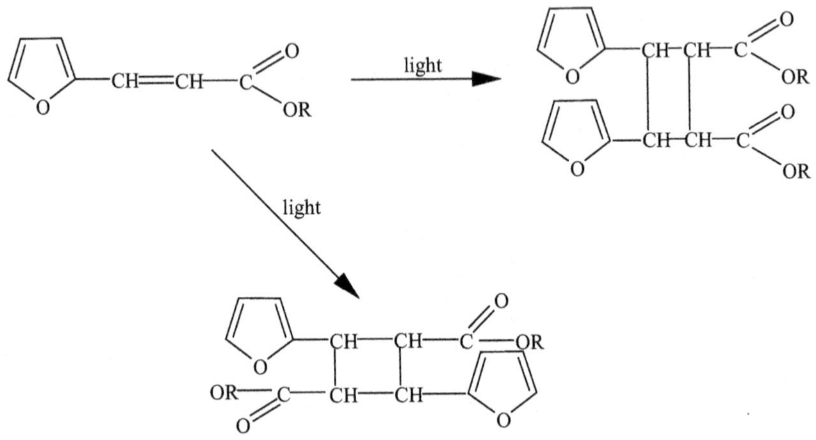

Shim and Kikuchi [64] reported preparation of a number of photocrosslinkable polymers with cinnamate and furan functionalities. The cinnamate esters were illustrated as follows:

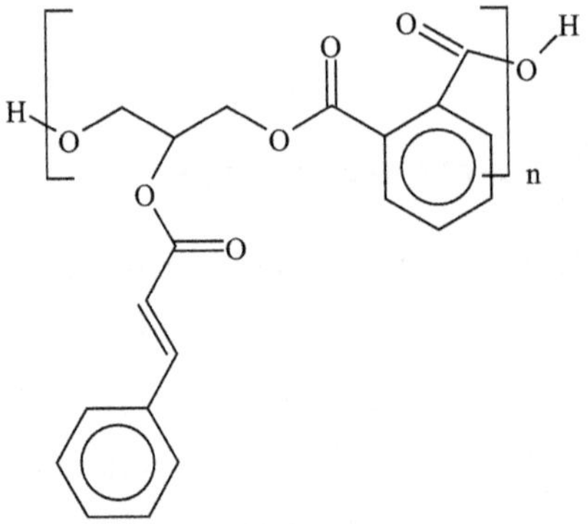

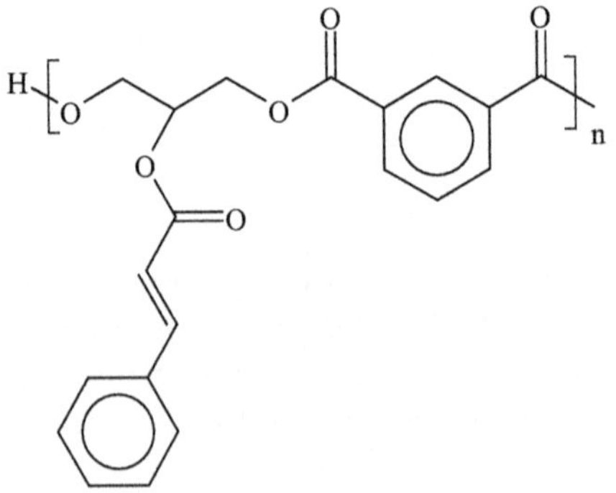

The polymers with furan functionality were shown as follows:

All materials require sensitizers. Picramide was found to be effective in this function. [64]

Tajima, Arai, and Takeuchi [65] studied the affects of singlet oxygen on photocrosslinking of poly(furfuryl methacrylate). Singlet oxygen was generated by fullerene C_{60} . The reactions were carried out in 1,1,2,2-tetra-chloroethane solutions that contained the polymer and fullerene C_{60}. Gelation occurs when exposed to visible light in the presence of photosensitizers. These solutions gel after several hours and subsequently solidify completely. [65]

It is interesting to note, that fullerines are effective in causing sensitization of poly(furfuryl] methacrylate), and also are effective in causing crosslinking upon UV irradiation. [66] This is attributed to their long-lasting ability to sensitize oxygen even when they themselves undergo oxidation The sensitivity of poly(furfuryl methacrylate) increases linearly when the concentration of C_{60} is increased from 5×10^{-5} to $I \times 10^{-3}$ mol/L. Saturation of sensitivity in the high C_{60} concentration regions was observed to be due to deficiency of oxygen molecules in the resist films. It was concluded that the dissolution rate of oxygen from the atmosphere into the resist film is lower than its consumption rate. [66]

Preparation of photocrosslinkable furan-containing polyimides was also reported. [67] It was found that the polymer crosslinks with the aid of singlet oxygen. Formations of fine pattern images can be formed. This was taken as clear evidence of successful photolithography in this photocurable system that uses C_{60} as the photosensitizer. [67]

4.1.5. Polymers with Pendant Maleimide Groups

Preparation of photocrosslinkable co- and terpolymers of N-isopropylacrylamide, 2-(dimethylmaleimido) N-ethylacrylamide as the photosensitive component, and 3-acryloylaminopropionic acid or *N-(2-(dimethylaminoethyl)-* acrylamide as ionizable comonomers was reported. [68] Here too, crosslinking takes place through formation of cyclobutane moieties:

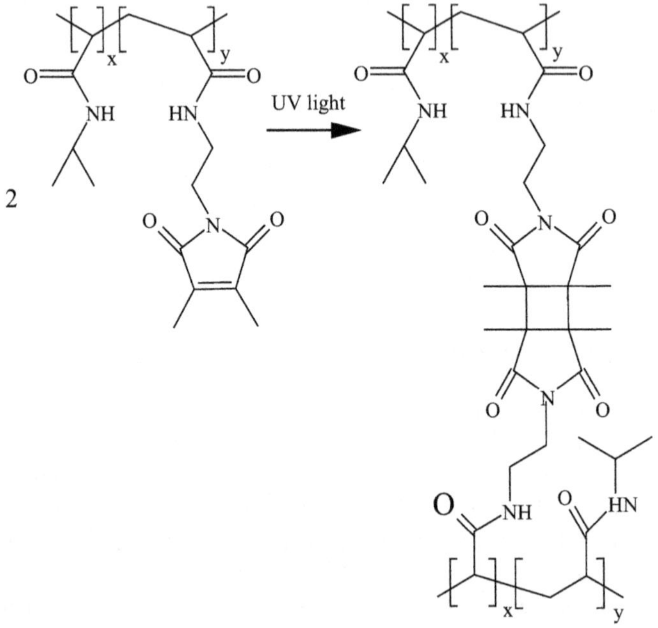

4.1.6. Polymers with Pendant Uracil, Thymene, Pyridinium, and Quinolinium Moieties

Poly(vinyl alcohol) substituted with the uracil moieties was investigated as a photocrosslinkable material. [69] It was observed that the spectra of the unirradiated material shows an absorbance peak at 265 nm in the ultraviolet region. This peak is derived from the uracil unit. This peak becomes reduced according to the progress of photodimerization when exposed to light at 265 nm. wave length. This spectral change suggests photodimerization of the uracil units in the polymer. The rate of the photodimerization for the polymer that contains long spacer units is higher than that for the polymer without long spacers. Also, the solution viscosity increases and the solubility of the polymer in water decreases with photodimerization of the uracil moiety. [69]

Also, water-soluble poly(vinyl alcohol) was modified by thymine derivatives. [69] This yielded a material that reversibly photodimerizes upon UV irradiation. Photocrosslinking experiments of this polymer were carried out in the film state on quartz plates by irradiation with light at 280 nm. The poly(vinyl alcohol) containing thymine units were found to be more light sensitive than uracil units in a similar polymer. In addition, photodimerization reaction of an ester derivative of the thymene substituted poly(vinyl alcohol) was found to be faster than that of an acetal derivatives, but the sensitivity was lower, because of higher solubility in water. Reverse reactions from photodimerization approach completion. The reversibility in solubility, however, is not complete. It was suggested that this indicates that hydrogen bonding within poly(vinyl alcohol) forms during the photodimerization reactions. [69]

The photocyclodimerization of thymine was illustrated by Guillet as follows [59]:

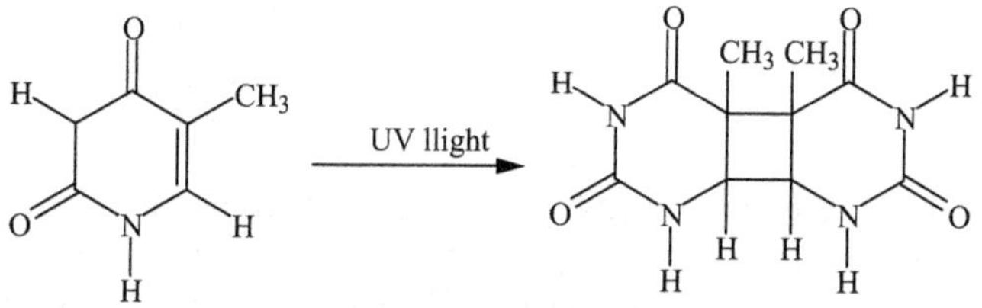

Presumably, when the thymines units are pendant groups on the polymer backbone, they dimerize by a similar mechanism.

Ichimura reported the photocrosslinking behavior of polymers with pendant styrylpyridinium, styrylquiolinium, and oxyquinolinium groups. [70] The materials can be illustrated as follows:

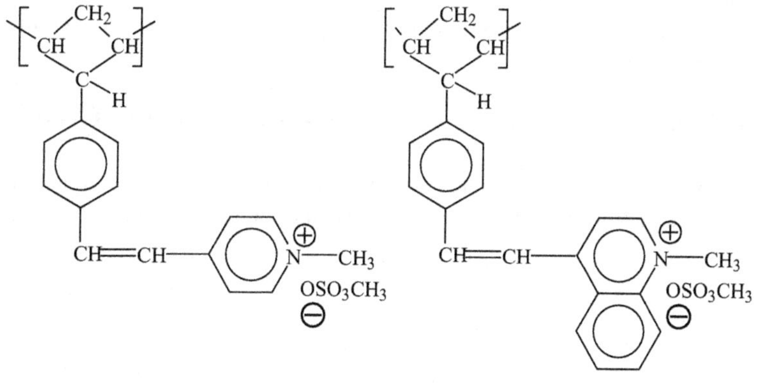

when they are formed from poly(vinyl alcohol) [70] they can be illustrated as follows:

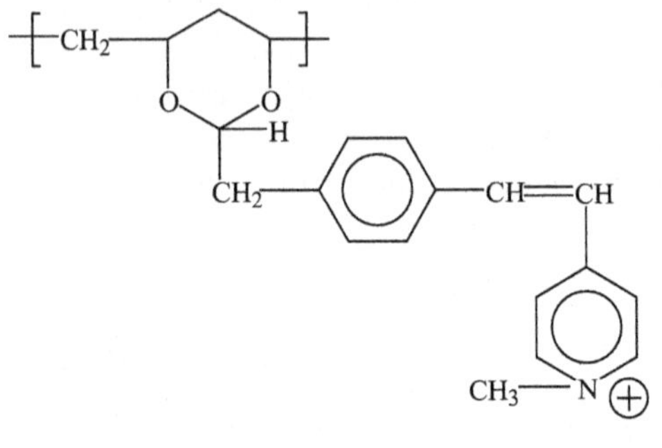

These polymers were found to exhibit anomalously high photosensitivity, even if the content of photosensitive groups is low. Irradiation with monochromatic light indicates that the reaction obeys Charlseby's equation and that the absorbed light energy leads specifically to intermolecular photodimerization. [70]

Photosensitive polymers were prepared by free radical polymerizations of oxypyridinium and oxyisoquinolinium functionalized methacrylate and styrene derivatives. [71] Also, polymerizable hydroxvpyridinium and hydroxyisoquinolinium salts were formed from vinylbenzyl chloride or glycidyl methacrylate that was reacted with 3-hydroxypyridine, 4- or 5-hydroxyiso-quinoline, and 8-hydroxyquinoline. Radical homo- and copolymerization with styrene or Me methacrylate of these salts produced copolymers. [71] The photosensitive dipolar oxypyridinium or oxyisoquinolinium betaine structures were generated in solutions with triethylamine from the low molecular weight and polymeric salt precursors. [71]

4.1.7. Polymers with Pendant Abietate and Dibenzazepine Groups

Poly(vinylbenzyl abietate) was prepared by the polymer reaction of poly(vinylbenzyl chloride) (I) with sodium abietate in chlorobenzene. [72]

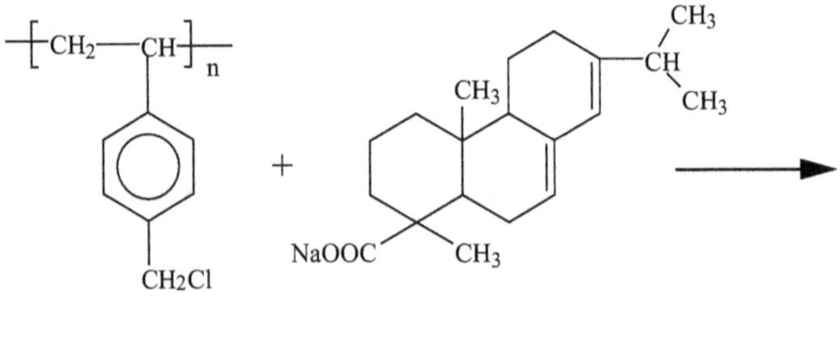

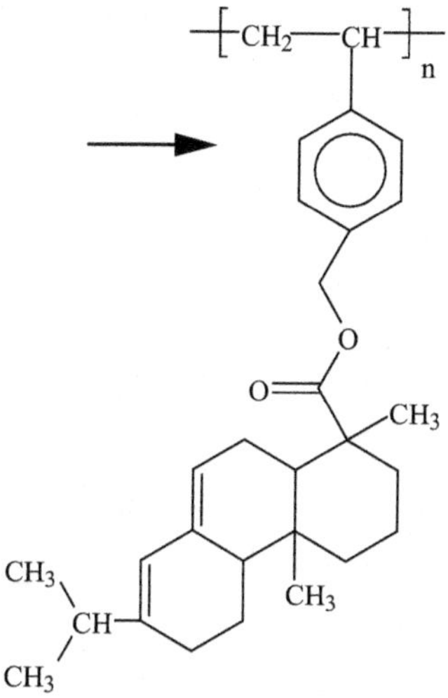

Poly(vinylbenzyl abietate) in the film state is crosslinkable via photo-dimerization of the conjugated carbon-carbon double bonds of the abietic acid moieties.[72]   What the photodimerization product looks like is not clear. Formation of a cyclobutane rings in photodimerization of steroids was represented by Trecker [14] as follows:

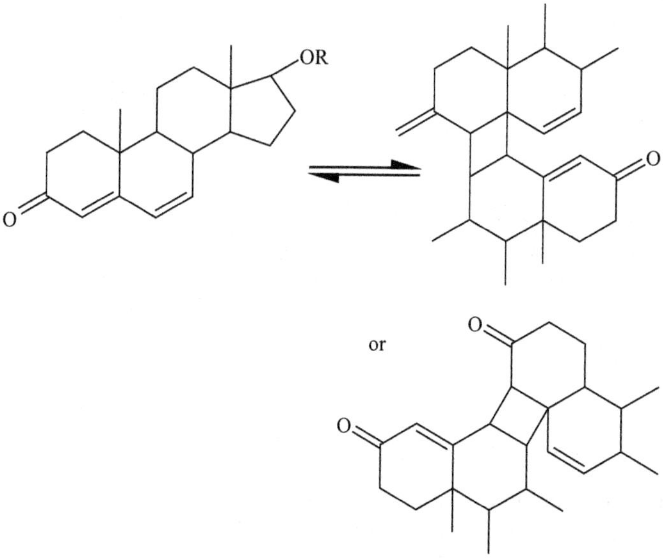

Also, judging from the photoreaction of cholestra-3,5-diene [73] one might expect some internal cyclization:

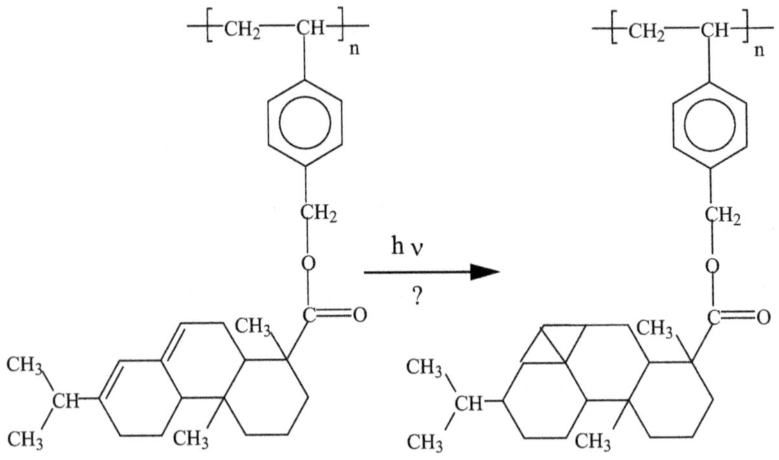

The cyclodimerization reaction of dibenzazepine was reported by Hyde *et al.,*[74] When attached to a polymer the cross-linking reaction can be visualized as follows:

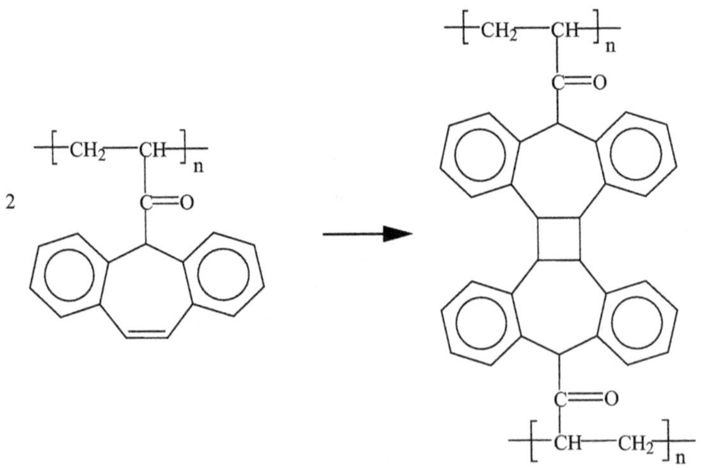

The cross-linking can result from either direct absorption of the light or in the presence of triple sensitizers like Michler's ketone. [74] The $2\pi + 2\pi$ addition leads to a formation of a cyclobutane crosslink.

Coumarin photodimerization is a known reaction[75] :

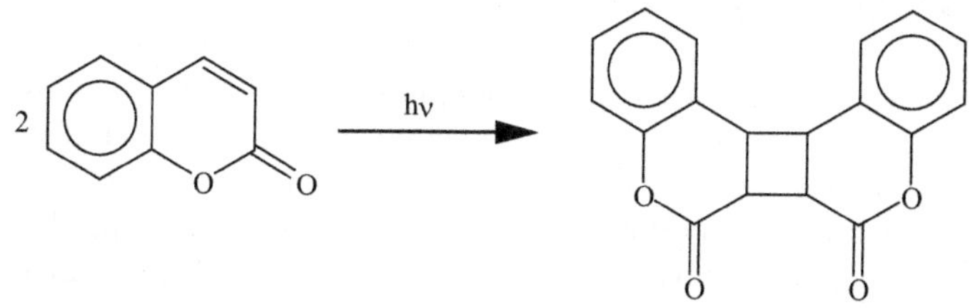

This was utilized by Tian *et al.*,[76] to prepare a new class of liquid crystal homopolymers of poly{ll-[4- (3-ethoxycarbonyl- coumarin- 7-oxy)-carbonyl-phenyloxy]-undecyl methacrylate} containing a coumarin moiety as a photo-crosslinkable unit. The preparations included polymers of various chain lengths. Also, liquid crystalline-coil diblock and liquid crystalline-coil-liquid crystalline triblock copolymers with polystyrene as the coil segment were formed. The polymers were reported to have been synthesized with the aid of atom transfer radical polymerization. The dimerization of the coumarin moieties takes place upon irradiation with light of $\lambda > 300$ nm to yield crosslinked network structures.

Lee *et al.*,[77] reported the preparation of new soluble and intrinsically photosensitive poly(amide-co-imide)s containing p-phenylenediacryloyl moiety. The copolymers were formed from p-phenylenediacryloyl chloride, aromatic dianhydrides and two equivalents of aromatic diamines. The products were subsequently imidized by reactions with the poly(amide-co-amic acid), acetic anhydride, and pyridine. The polymers were stable up to 350 °C, showed good solubility in polar aprotic solvents, and became insoluble after the irradiation due to the photodimerization of phenylenediacryloyl moiety. The photoreactivity increases with the irradiation temperature.[77]

4.2. Polymers that Crosslink by Dimerization of Nitrenes and by Other Combinations of Free-Radicals to Form Covalent Bonds

The aromatic azide groups photodecompose into nitrenes when irradiated with UV light:

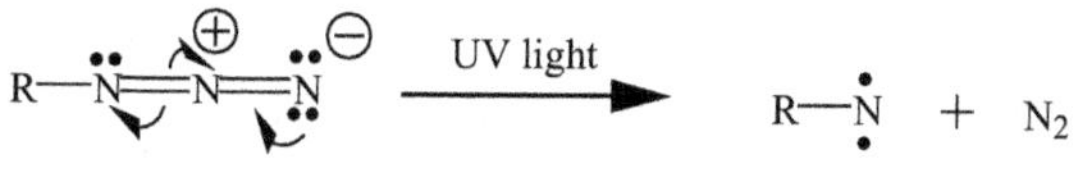

The nitrenes, that form, possess two unpaired electrons, similarly to carbenes, and dimerize readily into azo groups:

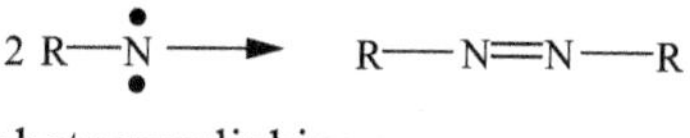

This reaction is utilized in photocrosslinking.

4.2.1. Polymers with Pendant Azide Groups

Both azide and sulfonyl azide groups are photoreactive and decompose into active nitrene groups and nitrogen upon irradiation. The dissociation of the azide moiety follows almost every transition from of an excited $n \longrightarrow \pi^*$ state to a high vibrational level of the ground state. [78-80] Introductions of pendant azide and sufonyl azide groups into polymeric structures are possible in a variety of ways and many publications describe different approaches. Thus, Tsuda and Yamaoka[81] introduced azide groups into a phenolic novolac resin by a route shown in the following scheme:

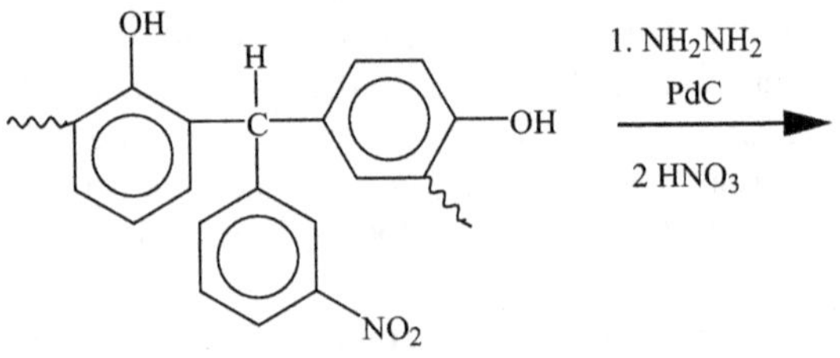

portion of a novolac

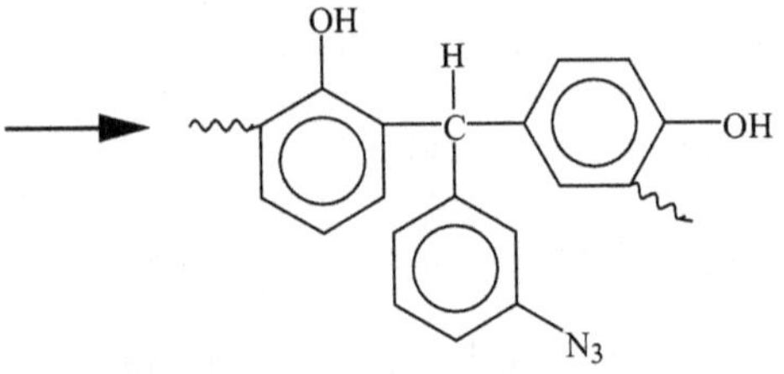

The same authors also condensed [81] formyl-1-naphthyl azide with poly(vinyl alcohol). All steps in the synthesis were not disclosed. The preparation, however, was illustrated as follows:

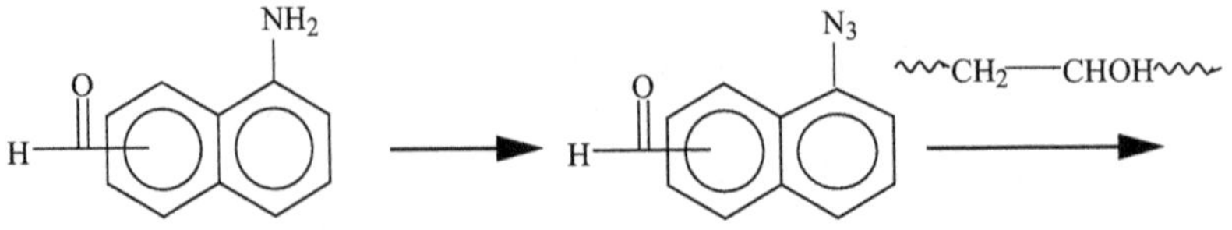

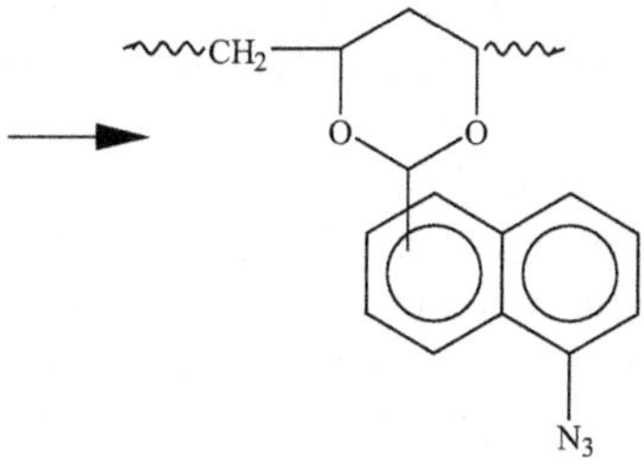

Preparations of azide derivatives from styrene-maleic anhydride copolymers, cellulose, and gelatin by attaching aromatic azide compounds are described in the literature. [50] Most of the resultant polymers crosslink rapidly when exposed to light of 260 mμ wavelength. Also, as much as 90% of the hydroxy groups of poly(vinyl alcohol) can be esterified with p-azido-benzoyl chloride. These reactions must be carried out in mixtures of chloroform and aqueous sodium hydroxide. [82] Earlier, Merril and Unruh[83] described formation of derivatives from poly(vinyl alcohol), and attachment of aromatic azide groups as follows:

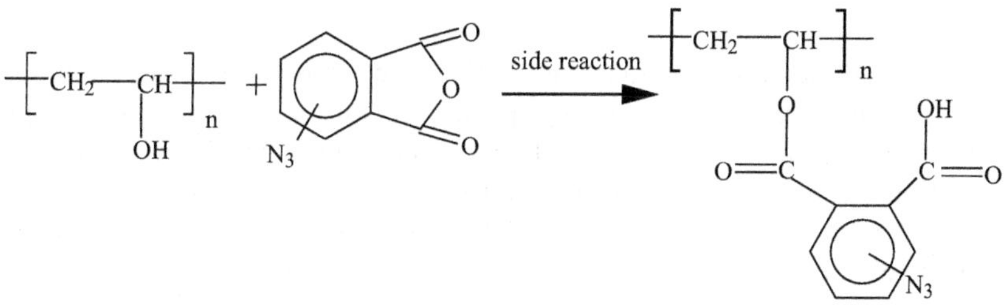

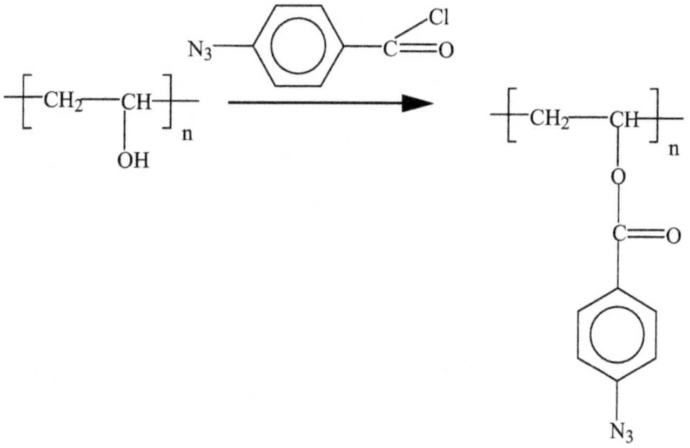

Most of these azide polymers photocrosslink at a faster rate than does poly(vinyl cinnamate), when exposed to light of 260 mμ. In addition, they responded well to photosensitization. Also, it was observed [50] that the 4-isomer of azidophthalate shows greater speed increase than does the 3-isomer. In general, the poly(vinyl alcohol) derivatives were reported to exhibit higher crosslinking speeds than do other azide functionalized polymers. [50]

The reaction product of p-azidobenzoyl chloride with polyvinyl alcohol was investigated by Tsuda and coworkers. [87] In the polymer studied by them over 90 per cent of the hydroxy groups were esterified:

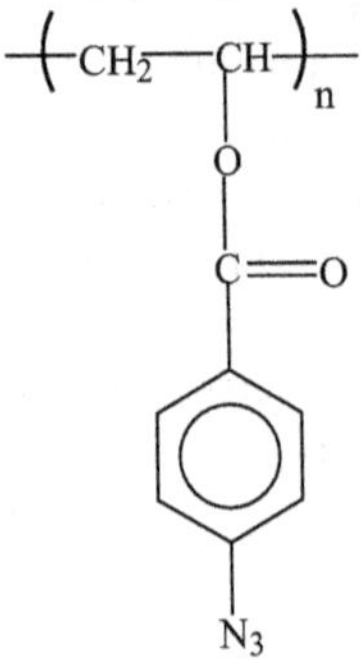

The photocrosslinking reaction was followed by observing changes in the ultraviolet and infrared absorption spectra. It was shown that the simple photochemical reaction occurs stochiometrically upon irradiation. Also, an absorption band was observed at 1500 cm^{-1} in the infrared region of the irradiated and crosslinked polymer. This band is due to N=N stretching vibration of the azo group. Based on that, Tsuda concluded the crosslinking reaction takes place by dimerization, as expected [87]:

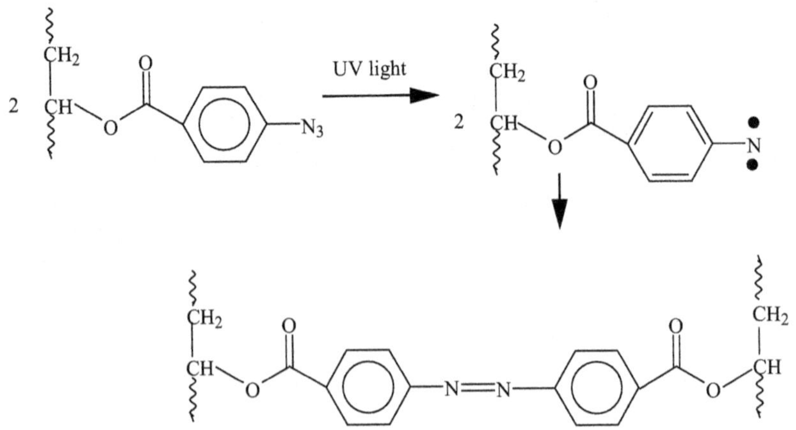

There are reports in the literature that photosensitive azide polymers can also be formed from polymers substituted with isocianate groups.[84,85,55,5,6]

The preparation and properties of many other polymers containing pendant aryl azide groups were described by Delzenne and Laridon[86]. The polymers were prepared by interfacial polycondensation of azido-substituted acid chlorides with diols and diamines. Also, in one experiment, a cinnamate moiety was combined with an azide group, together in one pendant functional structure [86]:

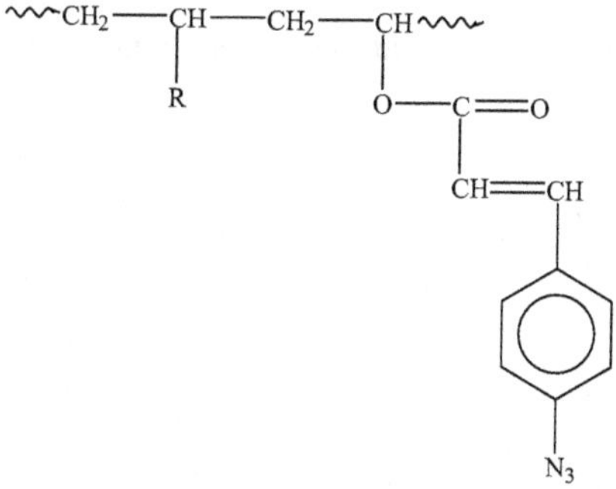

It was found, however, that addition of the azide groups to the cinnamate side chain does not increase reactivity. A marked wavelength dependence was observed on $\pi \longrightarrow \pi^*$ and $n \longrightarrow \pi^*$ transitions that occurs in both functional groupus.[86,87]

4.2.2. Polymers with Pendant Thiadiazole and Episulfide Groups

1,2,3-thiadiazoles. are photosensitive.and decompose into free-radicals upon irradiation [52] Preparation of polymers with pendant thiadiazoles was, therefore, carried out. [88] The introduction of pendant groups was carried out as follows:

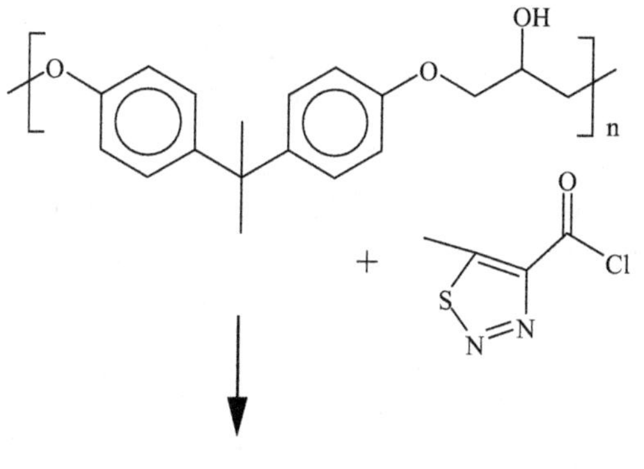

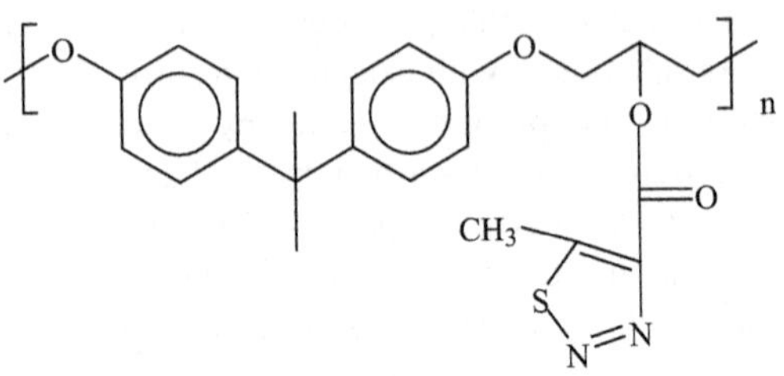

Other pendant thiadiazole groups that were also attached to polymers in a similar manner, were [88]:

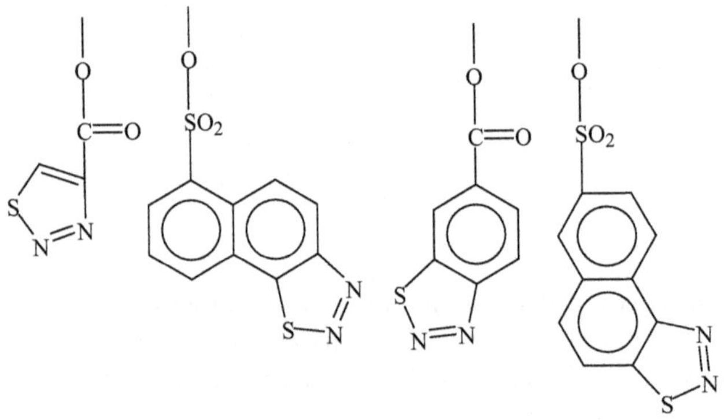

All these pendant groups require the use of proper photosensitizers. Compounds with carbonyl groups appear to be most active. The sensitization process was attributed [52] to a likely triplet energy transfer to the thiadiazole. The photolytic decomposition of 1,2,3-thiadiazoles was pictured to occur as follows [84]:

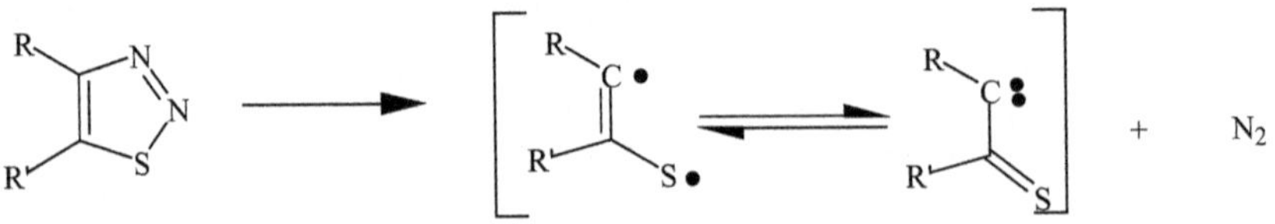

The diradicals can dimerize by several different paths. This has not been fully elucidated. Following are some of the paths that were suggested [52]:

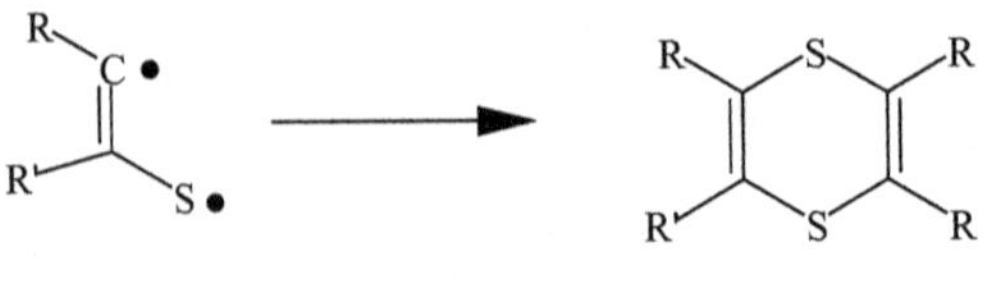

Also, a thioketocarbene can undergo a Wolff rearrangement and then dimerize:

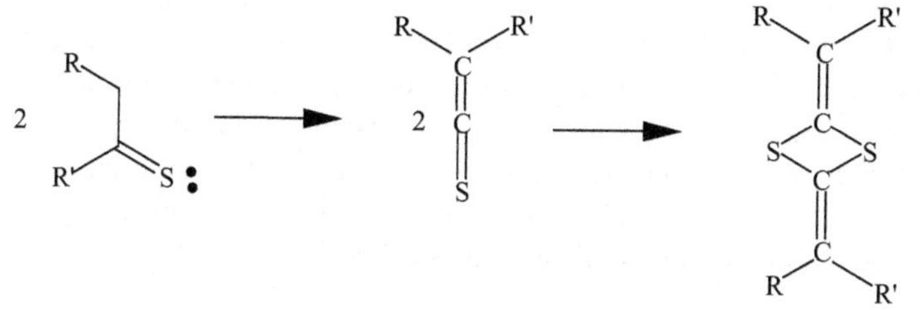

A third type of combination is also conceivable:

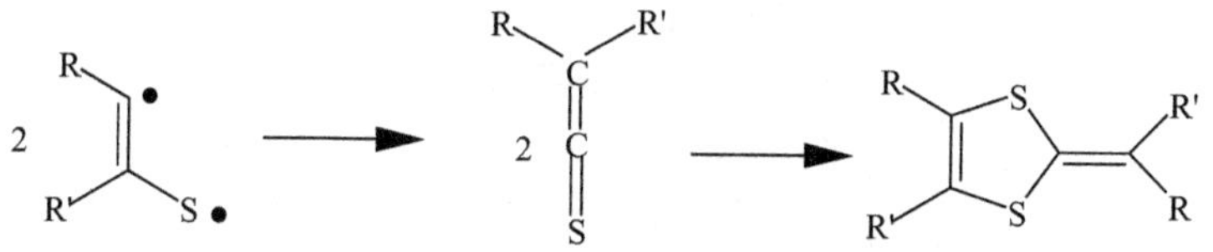

All these combinations of fragments attached to the polymeric backbone suggest that there may be various paths for photocrosslinking.

Polymers with pendant episulfide groups can be photocrosslinked by a process of photooxidation. This was shown by Tsunooka and coworkers [88] who photocrosslinked poly(2,3- epithiopropyl methacrylate).

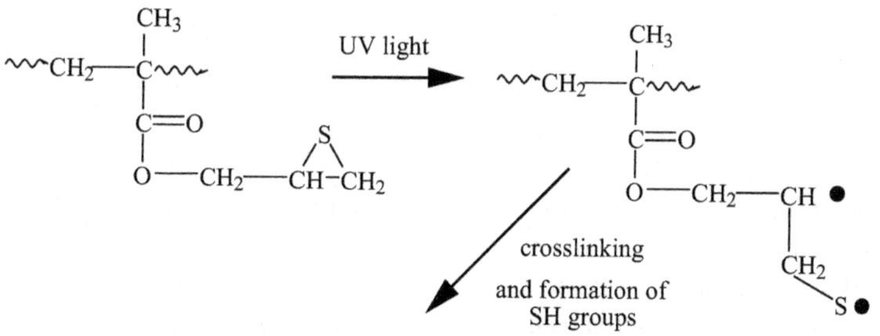

Infrared data suggests that the crosslinking takes place via oxidation of the sulfides by singlet oxygen. Also, it was observed that traces of protic components in the film (residual solvent) greatly enhance the crosslinking rate. The conclusion was that such materials promote singlet oxygen-oxidation. [88]

4.2.3. Polymers with Acrylic Functionality

Aromatic dimethacrylates are used in many applications in dentistry, optical eye wear, fiber optics, and microelectronics.[89] In addition, carbazole based polymers have demonstrated good photorefractive, optical, and charge-transporting properties, combined with ease of processing.[89] By incorporating carbazole into novel aromatic methacrylates and dimethacrylates, materials with enhanced optical properties, such as refractive index are obtained. Such photocrosslinkable oligomers can be illustrated as follows:

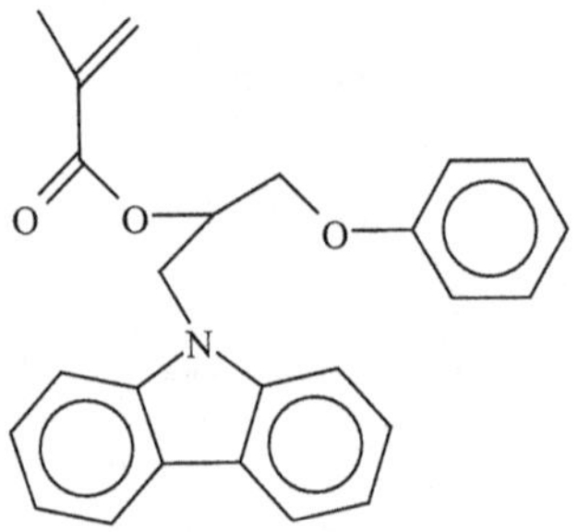

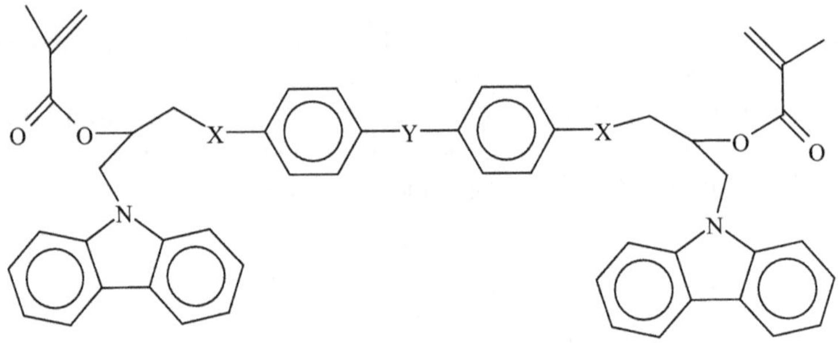

It is claimed that the materials are easy to form and photocrosslink readily when free-radial photoinitiators are used.[89]

Two new photosensitive end-capping agents, i.e., 6-(4-aminophenoxy)hexyl methacrylate and di[2-(methacryloyloxyethyl)] 5-aminoisophthalate, for polyimides were prepared.[90] These agents were used along with 2,2-bis(3,4-dicarboxyphenyl)hexafluoropropane dianhydride and 2,2'-bis(trifluoromethyl)-4,4'-diaminobiphenyl to prepare a series of methacrylate end-capped imide oligomers. The polymers required photoinitiators for crosslinking. A second approach involved the synthesis of a diamine monomer, i.e., 2,2'-dimethacryloyloxy-4,4'-diaminobiphenyl, in which methacrylate moieties were attached to the 2- and 2'-positions of a biphenyl structure. The monomer was polymerized with commercially available dianhydrides like 4,4'-oxydiphthalic anhydride. The polymers yielded line patterns 10-30 μm wide and 5-20 *μm* thick and did not develop color or shrink during UV exposure and subsequent thermal treatments. [90]

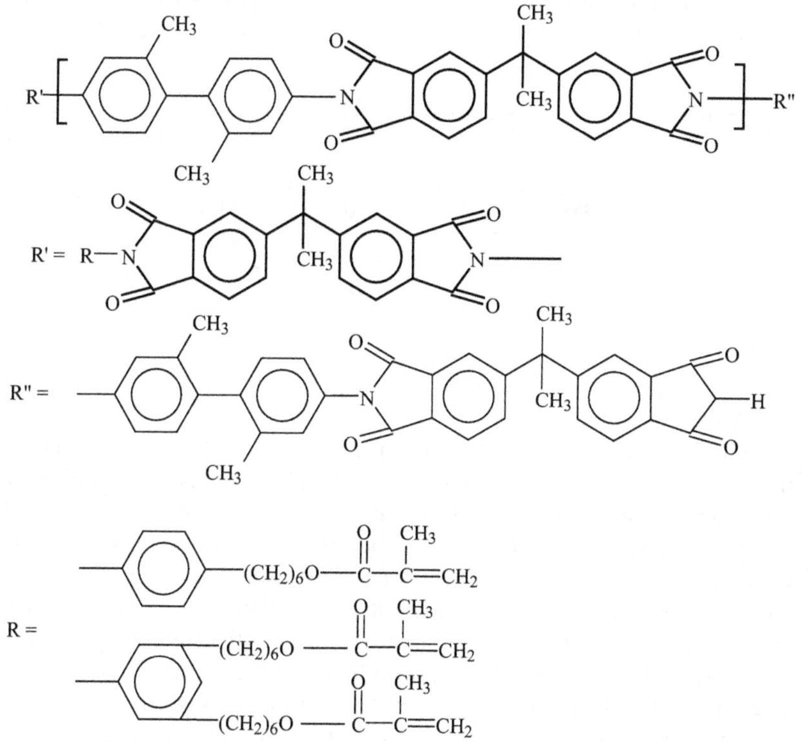

Ray and Grinstaff [91] reported preparation of methacrylated photocrosslinkable triblock copolymers of ethylene glycol and glycerol. The materials can be illustrated as follows:

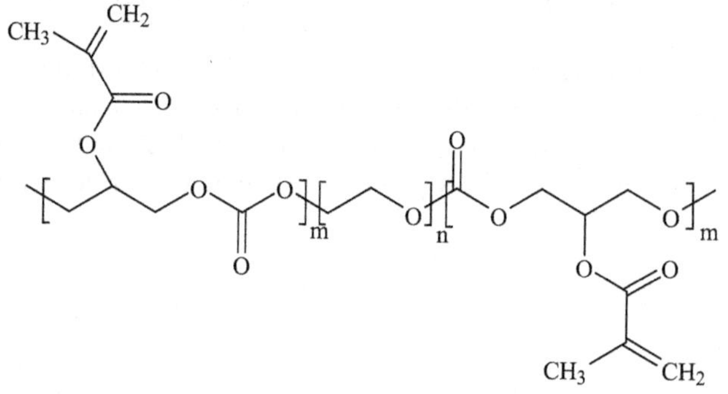

4.2.4. Ketone Containing Polymers

Polymers with pendant ketone groups can be expected to cleave by the Norrish reaction as a result of irradiation by light of a proper wave-length. This was demonstrated on films of poly(p-vinylacetophenone) [92] that were irradiated by long wave-length ultraviolet light (>300 nm). The polymer cleaved by the Norrish I type reaction:

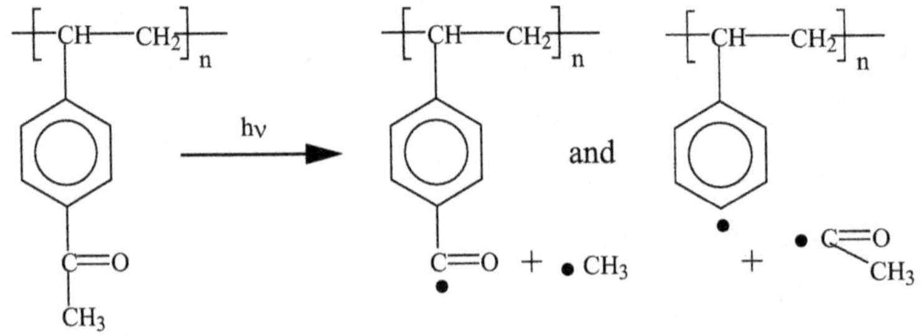

and crosslinked slowly.

Guillet studied the kinetics and mechanisms of the Type II photo-elimination reaction of some styrene-vinyl aromatic ketone copolymers in solution.[59] This included observation of the effects of solvent, ketone content, and ketone structure on the lifetimes of the chromophores, and the quantum yield of polymer chain scission. Solvent polarity was found to affect strongly the quantum yield of chain scission of a copolymer of styrene with phenyl vinyl ketone. When the irradiation was carried out in the presence of *tert*-butylhydroperoxide , the singlet and the triplet $n \rightarrow \pi^*$ states of the ketones are quenched by the peroxide at diffusion-controlled rates. The quenching constants and the Arrhenius parameters for the triplet quenching are indicative of bimolecular collision process. The inverse dependence on temperature of the singlet quenching constant indicates a possibility of reversible exciplex formation in the quenching process.

Two soluble polyimides were synthesized from aromatic diamine monomers containing benzophenone moieties, 3,3',5,5'-tetramethyl-4,4'-diaminodiphenylmethane and 3,3',5,5'-tetramethyl-4,4'-diamino-diphenyl ketone) and 3,3',4,4'-benzophenone tetracarboxylic dianhydride. [100] The polymers were reported to be photocrosslinkable and exhibited self-hypersensitizing properties. They preparation can be illustrated as follows:

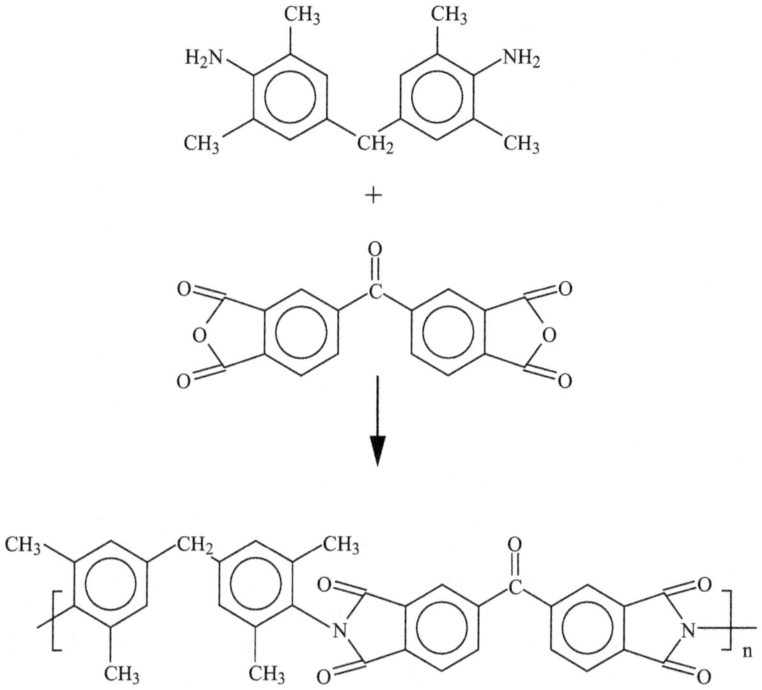

4.3. Miscellaneous Photocrosslinkable Polymers

Two interesting papers were published on synthesis of poly(benzoin acrylate) and poly(furoin acrylate)[90] for use in photocrosslinking:

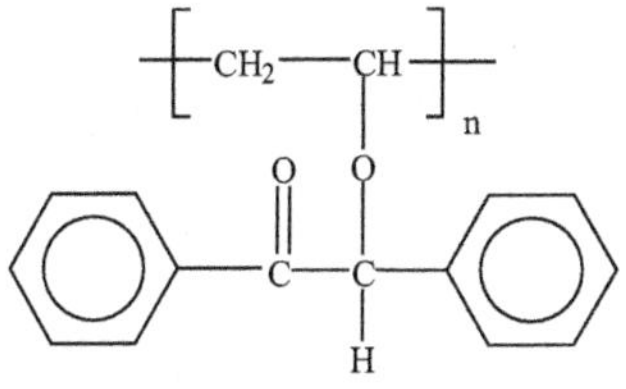

poly(benzoin acrylate) 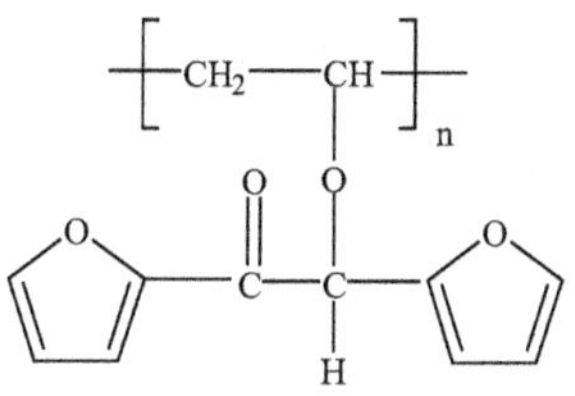   poly(furoin acrylate)

Both polymers were formed from their vinyl monomers. Pendant benzoin groups can be introduced into the polymers by other techniques. One can start, for instance, with a copolymer of styrene and maleic anhydride[31] and form ester groups:

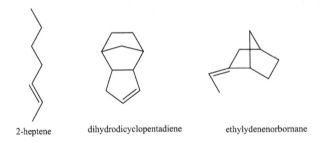

Photocrosslinking studies carried out on all these above polymers suggest that furoin structures exhibit greater photosensitivity than do the benzoin ones.

Among some other miscellaneous photocrosslinkable materials are graft copolymers of acrolein on polycarbanionic backbones.[91] In addition, various industrial method were developed for crosslinking of polyethylenes, ethylene-propylene elastomers and linear unsaturated polyesters rapidly, using UV light for initiation. The crosslinking is achieved by incorporating into these materials small amounts of UV-absorbing initiators and grafting functional groups to acts as crosslinkers. The functional groups used can be illustrated as follows :

2-heptene   dihydrodicyclopentadiene   ethylydenenorbornane

The synthesis and characterization of a photocrosslinkable liquid-crystal polymethacrylate with both biphenyl unit as a mesogen and styryl-2-pyridine unit as a photosensitive center in the same side-chain was described.[92]

Unsaturated polyamide-imides are light sensitive[93] and can be crosslinked by UV radiation. The polymers consist of polyamide backbones and dimethylmaleimidyl end groups and can be prepared from amine terminated polyamides:

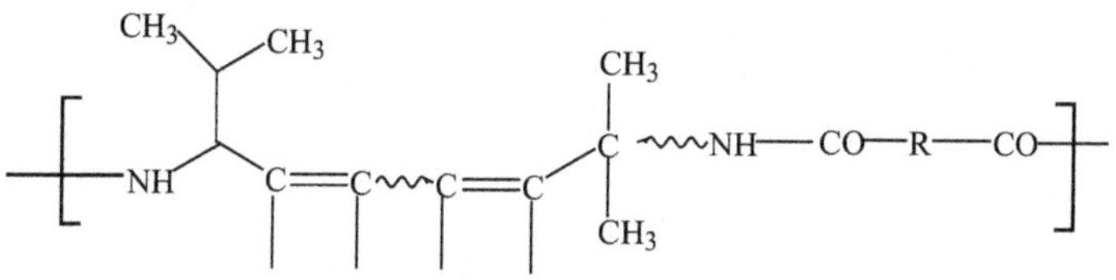

The terminal amine groups are reacted with dimethylmaleic anhydride.

Also, a photocrosslinkable polyester, designed for electro-optical application (see chapter 6) was synthesized by Roz et al.[94] The material can be illustrated as follows:

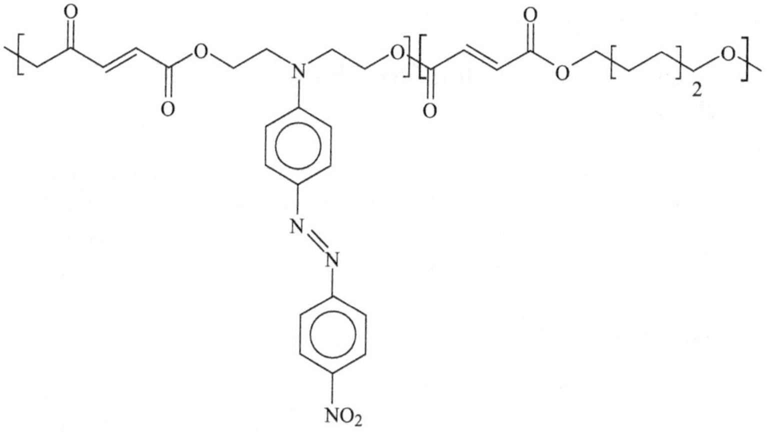

No information was provided on conditions for photocrosslinking

4.4. Polymers Designed to Crosslink Upon Irradiation with Laser Beams

To accommodate the needs of photolithography some polymers are now being developed that crosslink upon irradiation with an Ar-F excimer laser source at 193 nm.[95] To this end , for instance, Reichmanis et al [95] synthesized cycloolefin-maleic anhydride alternating copolymers. This material was formed by free-radical copolymerization of norbornene and maleic anhydride:

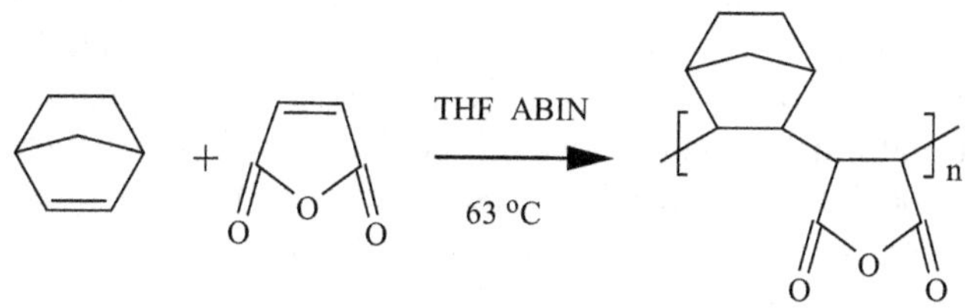

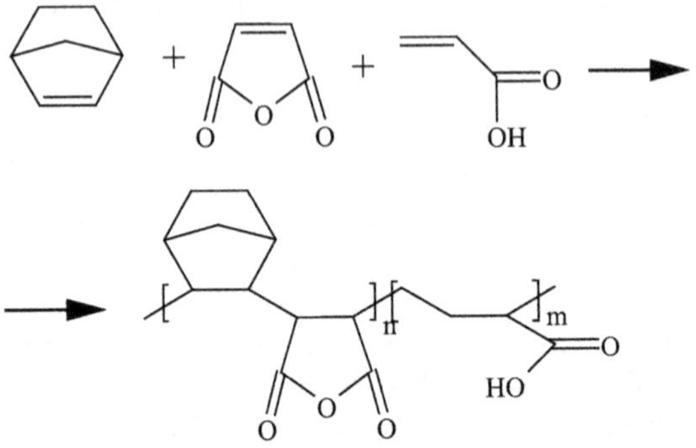

To render the polymers soluble in aqueous base acrylic acid terpolymerization was also carried out as shown in (B) above

4.5. Polymeric Materials that Crosslink by an Ionic Mechanism

While a large majority of the photocrosslinkable polymers that were investigated over the years crosslink by free-radical mechanism, utilization of ionic mechanism has received some attention. This includes polymers with pendant cyclic ethers and alkyne groups.

4.5.1. Polymers with Pendant Cyclic Ether Groups

It is not difficult to form polymers with pendant epoxy groups, because monomers, like glycidyl acrylate, glycidyl methacrylate and other epirane group carrying monomers are available commercially. These monomers can be polymerized or copolymerized readily by a free-radical mechanism to yield polymeric materials with pendant epoxy groups:

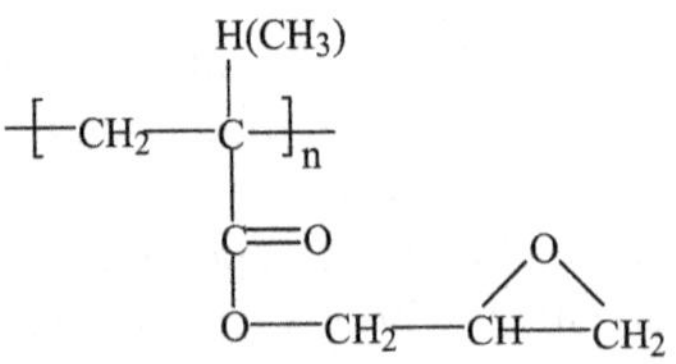

Monomers like vinylbenzyl glycidyl ether, vinylbenzyl 2-methyl glycidyl ether, vinylphenylethyloxirane, 2-methyl-2-vinylphenylethyloxirane and vinylbenzyl 3,4-epoxycyclohexyl methyl ether were prepared and polymerized to afford the corresponding polymers and copolymers with styrene.[96] In addition, poly[[2-(3,4- epoxycyclohexyl)-ethyl] dimethyl-4-vinylbenzyl silane] was also synthesized as a homopolymer and copolymer with styrene.

The polymers with cyclohexene oxide units connected to aromatic rings by -SiCH$_2$CH$_2$-linkage show highest crosslinking reactivity in the presence of photo-acid generators. The polymers with cyclohexene oxide units and ether linkages in pendant group show the lowest activity. Polymers with 2-Me glycidyl structure at pendant group crosslink faster than those containing the corresponding glycidyl structure. Ether linkages between epoxy group and aromatic rings lower the reactivity of the epoxide group by comparison to the polymers without such linkages.[96]

Shirai et al.,[97] prepared polymers that have both thermally degradable epoxy-containing moieties and sulfonic acid ester moieties in the side chain. Upon irradiation with UV light, the polymer films that contain photoacid generators become insoluble. The crosslinked polymer films become soluble again, however, in water, after a bake treatment at 120-200°C.

Inomata et al.,[98] prepared novel multifunctional photopolymers with both pendant epoxy and phenacyl ester groups by a one-pot method. Initially, poly(methacrylic acid) was reacted with epibromohydrin. This was followed by a reaction with phenacylbromide in the presence of a condensing agent. The photochemical reactions of the resulting polymers were evaluated by UV and IR spectroscopy. Pendant phenacyl ester groups photocleave to give corresponding carboxyl groups. The carboxyl groups that form in turn react with pendant epoxy groups. A baking process promotes crosslinking The photochemical reactivity of the resulting polymers is affected by the structure of the phenacyl ester group.[98]

A series of patents were issued to Mammino [99] on various resins like modified polysulfones and polysiloxanes that can be photocrosslinked when Lewis acids are added. In addition, light sensitive polysulfone copolymers can be formed through interfacial polycondensation.[104,105] Also, the reactions can be carried out on mixed bisphenols and aromatic disulfonyl chlorides in the presence of Lewis acids.

The synthesis of photocrosslinkable polysiloxanes containing. gem di-oxaalkylene styrenyl groups and gem di-urethane-α-Me styrenyl groups was reported. [100] This was accomplished by copolycondensation of α,ω—dihydroxy polydimethyl siloxanes and dichlorosilanes bearing either cyclic acetal groups or Si-H groups (onto which the cyclic acetal groups were further added) and with dichlorosilanes bearing alkyl groups. The introduction of styrenyl groups was then achieved by hydrolysis of the acetal groups into the corresponding alcohols followed by reaction with chloromethyl styrene or with 3-isopropenyl-α,α-dimethylbenzyl isocyanate. These products crosslinked under UV irradiation in the presence of a cationic photoinitiator.

4.5.2. Polymers with Pendant Alkyne Groups

Monomers with acetylene functionality are known to polymerize upon light irradiation in the presence of some metal carbonyls, like $Cr(CO)_6$, $Mo(CO)_6$, and $W(CO)_6$ and some other group VI metal complexes. A variety of Lewis acids, like $SiCl_4$ or $GeCl_4$ are also effective catalysts for the polymerization of various acetylenes in nonchlorinated solvents.

Badarau and Wang [101] reported preparation of a photocrosslinkable alkyne group containing polymers. The materials can be illustrated as follows:

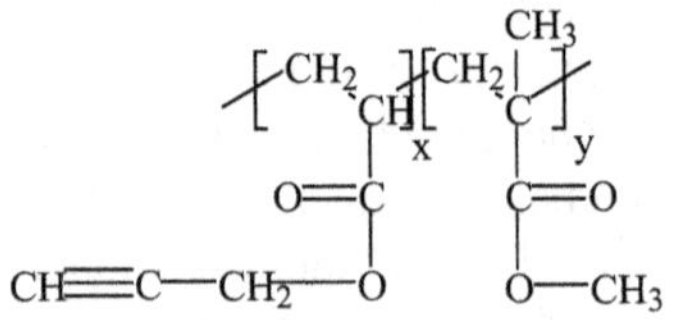

The above shown polymer photocrosslinks when irradiated in the presence of small amounts of $W(CO)_6$ using a light beam at 320-390 nm. [101] UV photolysis of $W(CO)_6$ in the presence of an alkyne results in the formation of a η^2-alkyne tungsten pentacarbonyl complex, $[(\eta^2\text{-}RC\equiv CR')W(CO)_5]$, that rearranges to a vinylidene derivative $[R(H)C=C=W(CO)_5]$ and is an intermediate in the initiation step in the polymerization of alkynes. [101]

In addition Badarau and Wang [101] reported preparation of a series of diacetylene-containing polymers that were formed by oxidative coupling polymerization of 9,9-bis(4-propargyloxyphenyl)fluorene and dipropargyl 4,4'-(hexafluoroisopropylidene)diphthalimide.

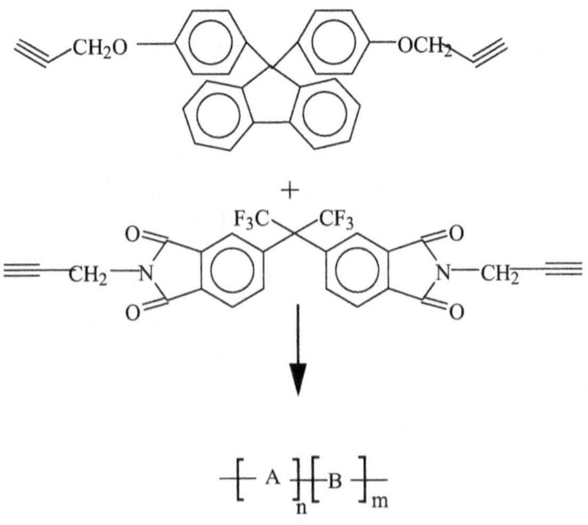

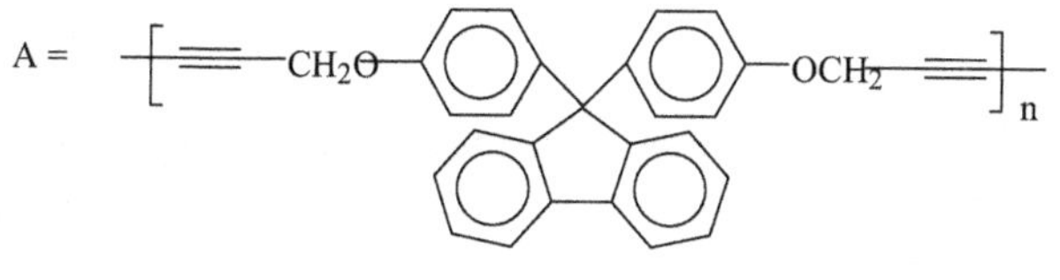

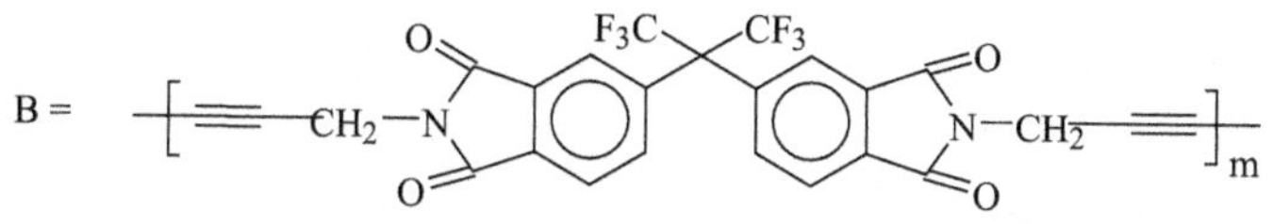

A =

B =

The molecular weights of the polymers are in 42 000-235 300 range and the polydispersity indices are 1.1-1.8. All the polymers show typically amorphous diffraction patterns, good solubility in organic solvents, and good film-forming capability. These polymers were reported to readily cross-linkable upon exposure to UV light at ambient temperature and can also be thermally cured. The alkyne groups are involved in both thermal and photocuring processes. [101]

In another paper, Wang et al., [102] also reported preparation of polymers containing pendant propargyl groups that can also be effectively crosslink upon UV irradiation in the presence of a small amounts of $W(CO)_6$. The polymer, like the ones above, crosslinking is a result of conversion of the acetylenic units to polyene and diacetylene moieties in the crosslinked system. [103]

Greenfield et al., [103] obtained a patent on polymerizable mesogenic or liquid crystalline compounds with cinnamic acid residue and alkyne groups.

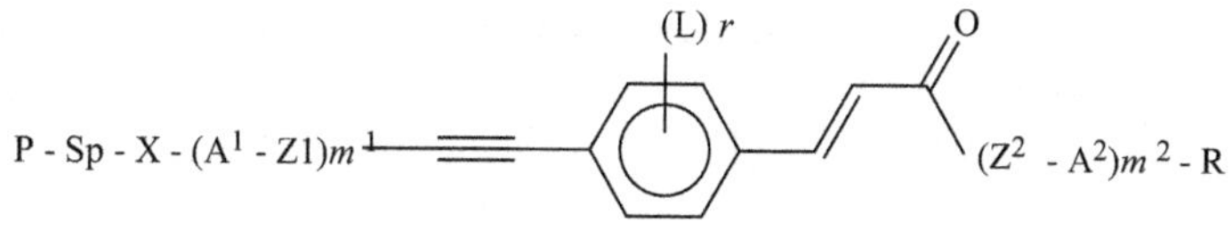

$P - Sp - X - (A^1 - Z1)m^1$

where, P is a polymerizable group, Sp is a spacer group or a single bond, X, Z^1 and Z^2 are each independently various residues or a single bonds, A are each independently aliphatic or aromatic, optionally substituted carbocyclic or heterocyclic groups, m^1 and m^2 are each independently 0, 1, 2 or 3 with $m^1 + m^2 < 5$, R is hydrogen. Preferably, the polymerizable group P is a vinyl group, an acrylate group, a methacrylate group, a propenyl ether group or an epoxy group.

4.5.3. Metallopolymers

Manners *et al.*, [29] reported synthesis of a photocrosslinkable polyferrocenyl-silane with methacrylate side groups:

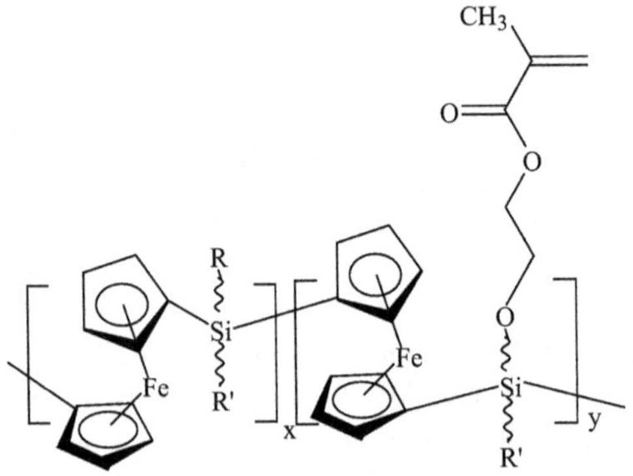

where R are methyl or ethyl groups and R" can be methyl or phenyl. The polymers photocrosslinks through the methacrylate group.

References

1.   F. Gersheim, "History of Photography", Oxford Press, London, 1966
2.   W.I. le Noble, *Highlights of Organic Chemistry*, Chapter 14, Marcel. Dekker, Inc. New York, 1974
3.   C. Decker, *Polymer International* **2002**, *51*(11), 1141-1150
4.   M. Wen, L.E. Scriven, and A.V. McCormick, *Macromolecules* **2003,** *36*(11), 4140; ***ibid.***, 4159
5.   P.J. Flory, *J. Am. Chem. Soc.*, **1941**, *63*, 3083, 3091, 3096
6.   P.J. Flory, "Principles of Polymer Chemistry," Cornell University Press, Ithaca, N.Y., 1953
7.   W.H. Stockmayer, *J. Polym. Sci.*, **1952**, *9*, 69; *ibid.*, **1953**, *11*, 424
8.   R.C. Dolby and R.H. Engebrecht, Canadian Pat. # 1,106,544 (1981)
      British Patent # 438,960 (Nov. 26, 1935)
      U.S. Patent # 1,965,710 (July 10, 1934)
      French Patent # 1,351,542 (Feb. 7, 1964)
      U.S. Patent # 3,066,117 (Nov. 27, 1962)
      Brit. Patnent # 822,861 (Nov. 4, 1959)
      U.S. Patent # 2,948,706 (Aug. 9, 1960)
      U.S. Patent # 2,738,745 (Dec. 27. 1955)

9.   W.G. Herkstroeter and S. Farid, *J. Photochem.*, **35**, 71 (1986)

10. M. Kato, *J. Polymer Sci.*, **1969**, *B-7*, 605

11. M.J. Farrall, M. Alexis and M. Trecarton, *Polymer*, **24**, 114, (1983); A.O. Patil, D.D. Deshpnde, and S.S. War, *Polymer*, **22**, 434 (1981); O. Zimmer and H. Meier, *J. Chem. Soc. Chem. Commun.*, 481 (1982); M.P. Stevens and A.D. Jenkins, *J. Polym. Sci., Chem. Ed.*, **17**, 3675 (1979); M. Kata, M. Hasegawa, and T. Ichijo, *J. Polym. Sci., B-8*, 263 (1970); G.E. Green, B.P. Stark, and S.A. Zahir, *JK. Macromol Sci., Rev. Macromol Chem.* **C21**, 187 (1991-1982); G. Oster, and N. Yand, *Chem. Rev.*, **68,** 125 (1968); D.R. Randall, ed., *Radiation Curing of Polymers,* CRC Press, Boca Raton, 1987

12. F. Wilkinson, *J. Phys. Chem.*, **1961**, *66*, 2569

13. H.L. Backstrom and K. Sandros, *Acta Chem., Scand.,* **1962**, *16*, 958

14. D.J. Trecker in "Organic Photochemistry", vol 2, O.L. Chapman, ed., M. Dekker, New York, 1969

15. L.M. Minsk, J.G. Smith, Van Devsen, and J.F. Wright, *J. Appl. Polym. Sci.*, **1959**, *2*, 302

16. Bertram and Kvrsten, J. Prakt. Chem., 51,323 (1895)

17. C.N. Ruber, Ber., 35, 2415 (1902)

18. G.M.J. Schmidt, J. Chem. Sec., 1964, 2014

19. J. Bergman, G.M.J. Schmidt, and F.I. Sonntag, J. Chem. Sec., 1964, 2021

20. M.D. Cohen, G.M.J. Schmidt, and F.I. Sonntag, J. Chem. Sec., 1964, 2000

21. H. Tanaka and E. Otomegawa, *J. Polymer. Sci., Chem. Ed.* **1974**, *12*, 1125

22. H.G. Curme, C.C. Natale, and D.J. Kelley, J. Phys. Chem., 71,767(1967)

23. F.I. Sonntag and R. Srinivasan, S.P.E. Conference on Photopolymers, Nov. 6-7, 1967, Elenville, N.Y.

24. C. Libermann and M. Zsuffa, Ber., 44, 841 (1911)

25. K. Nakamura and S. Kikuchi, Bull. Chem. Sec., Japan, 40, 1027 (1967)

26. W.M. Moreau, Am. Chem. Sec. Preprints, 10(1), 362 (1969)

27. K. Nakamura and S. Kikuchi, Bull. Chem. Sec., Japan, 41, 1977 (1968)

28. M. Tsuda, *J. Polymer Sci.*, **1969**, *7*, 259; K. Nakamura and S.K. Kuchi, *Bull. Chem. Soc. Japan*, **1968**, *41*, 1977

29. M. Tsuda and S. Oikawa, Chapt. 29 in *Ultraviolet Light Induced Reactions in Polymers*, Am. Chem. Soc. Symposium Series #25, Washington, 1976, 87

30. E.D. Kvasnikov, V.M. Kozenkov, and V.A. Barachevsky, *Zh. Nauchn. Prikl.* Fotogr. *Kinematogr.* **1979**, *24*, 222; *Chem. Abstr.* **1979**, *91*, 99887h.

31. B.A. Barachevsky, *SPIE* **1991**, *1559*, 184.

32. M. Schadt, H. Seiberle, A. Schuster, S.M. Kelly, *Jpn. J. Appl. Phys.* **1995**, *34*, L764.

33. M. Schadt, H. Seiberle, A. Schuster, and S.M. Kelly, *Jpn. J. Appl. Phys.* **1995**, *34*, 3240

34. A.G. Dyadyusha, T.Y. Marusii, Y.A, Reznikov, A.I. Khizhnyak, and V.Y. Reshetnyak, *JETP Lett.* **1992**, *56*, 17

35. T.Y. Marusii, and Y.A. Koznikov, *Mol. Mater.* **1993**, *3*, 161

36. A.G. Dyadyusha, A. Khizhnyak, T.Y. Marusii, V.Y. Reshetnyak, Y.A. Reznikov, and W.S. Park, *Jpn. J. Appl. Phys.* **1995**, *34*, L 1000.

37. Y. Akita, H. Akiyama, K. Kudo, Y. Hayashi, and K. Ichimura, *J. Photopolym. Sci. Technol.* **1995**, *8*, 75.

38. Y. Iimura, T. Satoh, and S. Kobayashi, *J. Photopolym. Sci. Technol.* **1995**, *8*, 257.

39. H. Tomita, K. Kudo, and K. Ichimura, *Liq. Cryst.* **1996**, *20*, 171.

40. M. Schadt, K. Schmitt, V. Kozinkov, and V. Chigrinov, *Jpn. J. Appl. Phys.* **1992**, *7*, 2155

41. K. Ichimura, Y. Akita, H. Akiyama, K. Kudo, and Y. Hayashi, *Macromolecules*, **1997**, *30*, 903-911

42. Y. Kawanishi, T. Tamaki, M. Sakuragi, T. Seki, and K. Ichimura, *Mol. Cryst. Liq. Cryst.* **1992**, *218*, 153.

43. K. Ichimura, Y. Hayashi, and N. Ishizuki, *Chem. Lett.* **1992**, 1063

44. K. Ichimura, Y. Hayashi, H. Akiyama, T. Ikeda, and N. Ishizuki, *Appl. Phys. Lett.* **1993**, *63*, 449

45. K. Ichimura, H. Akiyama, N. Ishizuki, and Y. Iawanishi, *Makromol. Chem., Rapid Commun.* **1993**, *14*, 813.

46. H. Akiyama, K. Kudo, and K. Ichimura, *Macromol. Chem. Rapid Commun.* **1995**, *16*, 35.

47. M. Schadt, K. Schmitt, V. Kozinkov, and V. Chigrinov, *Jpn. J. Appl. Phys.* **1992**, *7*, 2155.

48. C.-H. Zhang, Z.-H. Yang, and M.-X.Ding, *Liquid Crystals* **2003**, 30(1), 65-69

49. S.W. Lee, S.I. Kim, B. Lee, W. Choi, B. Chae, S.B. Kim, and M. Ree *Macromolecules* **2003**, *36*, 6527-6536

50. M. Nagata and S. Hizakae, *Macromolecular Bioscience*, **2003**, *3*(8), 412

51. H.R. Allcoci and C.G. Cameron, Macromolecules, **1994**, *27*, 3125, 3131

52. G.A. Delzenne and U. Laridon, S.P.E. Conference on Photopolymers, Nov. 6-7, 1967, Ellenville, N.Y.

53. Y. Shindo, T. Sugimura, K. Horie, and I. Mita, *Polymer J.*, **1986**, *22*, 859

54. S.Y. Park, H.Y. Park, H.S. Lee, S.W. Park, and E.W. Park, *Optical Materials (Amsterdam, Netherlands)* **2003**, 21(1-3), 331-335; W.-S. Kim, J.-W. Lee, Y.-W. Kwak, J.-K. Lee, Y.-T. Park, and S.-D. Yoh, *Polym. J.* **2001**, *33*(8), 643-646

55. D.-W. Lee, J.-H. Hur, B.-K. Kim, S.-W. Park, D.-W. Park, *Journal of Industrial and Engineering Chemistry,* **2003**, 9(5), 513-51

56. H.-Y. Park, D.-W. Lee, H.-S. Lee, D.-O. Lim, D.-W. Park, *Reaction Kinetics and Catalysis Letters* **2003**, 79(2), 245-255

57. P. Gupta, S.R. Trenor, T.E. Long, and G.L. Wilkes, *Macromolecules,* **2004**, *37*, 9211

58. S. Watanabe, S. Harashima, and N. Tsukada, *J. Polymer. Sci., Chem. Ed.,* **1986**, *24*, 1227

59. J. Guillet, p.317, *Polymer Photophysics and Photochemistry,* Cambridge University Press, Cambridge, 1985

60. K. Subramanian, V. Krishnasamy, S. Nanjundan, and A.V. Kami Reddy, *Ear. Polym. .J.* **2000**, *36*(11), 2343-2350

61. D.H. Choi, S.J. Oh, H.B. Clia, and J.Y. Lee, *Eur. Polym. J.* **2001**, *37*(10), 1951-1959

62. J.L.R. Williams, S.P.E. Photopolymers Meeting, Nov. 6-7, Elenville, N.Y. 1967

63. M. Tsuda, J. Polymer Sci., A-1,7, 259 (1969)

64. J.S. Shim and S. Kikuchi, *Kogyo Hagaku Zasshi,* **1965**, *68*(2), 387, 393; from *Chem. Abstr.,* **1965**, *63*, 10119c

65. Y. Tajima, H. Arai, and K. Takeuchi, *Kagaku Kogaku Ronbunshu* **1999**, *25*(6), 873-877

66. H. Arai, Y. Tajima, and K. Takeuchi, *J. Photopolym. Sci. Technol.* **1999**, *12*(1), 121-124.

67. Y. Tajima, Y. Shigemitsu, H. Arai, W. He, E. Takeuchi, and K.Takeuchi, *J. Photopolym. Sci. Technol.* **1999**, *12*(1), 125-128

68. M. E. Harmon, D. Kuckling, and C.W. Frank, *Macromolecules* **2003**, *36*,162-172

69. Y. Ohtani, Y. Inaki, and M. Miyata, *J. Photopolym. Sci. Technol.* **2001**, *14*(2), 295-296

70. K. Ichimura, Makromol. Chem., 188, 2973 (1987) ; *J. Polymer Sci., Chem. Ed.* **A-1,20**, 1411 (1982)

71. V. Goertz, and H. Ritter, *Macromolecules* **2002**, *35*(11), 4258-4265.

72. W.-S. Kim, H.-S. Jang, K.-H. Hong, and K.-H. Seo, *Macromol. Rapid Commun.* **2001**, *22*(11), 825-828

73. J. March, "Advanced Organic Chemistry", Wiley, New York, 1985, p. 216

74. P. Hyde, L.J. Kricka, and A. Ledwith, *J. Polymer. Sci., Lett. Ed.,* **1973**, *11*, 415

75. N.J. Turro, *Moleular Photochemistry,* W.A. Benjamin, New York, 1967

76. Y. Tian, X. Kong, Y. Nagase, and T. Iyoda, *Journal of Polymer Science, Part A: Polymer Chemistry* **2003**, *41*(14), 2197-2206

77. M.-H. Lee, Y.S. Chung, and J. Kim, *Polymeric Materials Science and Engineering* **2001**, *84*, 613

78. A. Reiser, H.M. Wagner, R. Marley, and G. Bowes, Trans. Faraday Sec., 63, 2403 (1967)

79. A. Reiser and R. Marley, *Trans. Faraday Sec.*, **1968**, *64*, 1806

80. A. Reiser, F.W. Willets, G.C. Terry, V. Williams, and R. Marley, *Trans. Faraday Soc.*, **1969**, 65, 3265

81. T. Tsunoda and T. Yamaoka, S.P.E. Conference on Photopolymers, Nov. 6-7, 1967, Elenville, N.Y.

82. T. Tsunoda, T. Yamaoka, G. Nagarnatsu, and M. Hirohashi, S.P.E. Conference on Photopolymers, Oct. 15-16, **1970**, Elenville, N.Y.

83. S.H. Merrill and C.C. Unruh, J. Appl. Polymer Sci., 7, 273 (1963)

84. H. Holtschmidt and G. Oertel, *Ang. Makromol. Chem.*, **1969**, *9*, 1

85. Q. Lin, J.P. Gao, and Z.Y. Wang, *Polymer* **2003**, 44(19), 5527-5531

86. G.A. Delzenne and U. Laridon, J. *Polymer Sci.*, **1969**, *C-22*, 1 149

87. A. Yabe, M. Tsuda, K. Honda, and H. Tanaka, *J. Polymer Sci.*, **1972**, *A- 1*, *10*, 2376

88. M. Tsunooka, S. Tanaka, M. tanaka, H. Egawa, and T. Nonaka, *J. Polymer Sci., Chem. Ed.*, **1985**, *23*, 2495

89. H.C. Ng and J.E. Guillet, Macromolecules, 18, 2294, 2299 (1985);

90. L. Yang, Z. Li, G. Cheng, K. Yao, *Gongneng Cailiao* **2000**, 31(2), 196-199

91. W.C. Roy, III and M.W. Grinstuff, *Am. Chem. Soc., Polymer Preprints*, **2002**, *43*(2), 768

92. Y. Kurusu, H. Nishujama, and M. Okawara, *Makromol. Chem.*, **1970**, *138*, 49

93. M. Schadt, H. Seiberle, A.Schuster, and S.M. Kelly, *Jpn.J. Appl. Phys.* **1995**, *34*, 3240.

94. J. Berger, *Polymer Communications*, **1985**, *26*, 11

95. Z. Roz, S. Maaref., and S.-S. Sun, *Am. Chem. Soc. Polymer Preprints*, **2003**, *44*(1), 86

96. F.M. Houlihan, T.I. Wallow, O. Nalamasu, and E. Rechmanis, *Macromolecules*, **1997**, *30*, 6517

97. T. Oyama, T. Yamashita, T. Suzuki, K. Ebitani, Hoshino, T. Motoki and M. Tornoi, *Reactive & Functional Polymers,* **2001**, *49*(2), 99-116

98. K. Inomata, S. Kawasaki, A. Kluneyama, and T. Nishikubo, *J. Pulym. Sci., Part A: Polym. Chem.* **2001**, *39*(4), 530-538

99. M. Shirai, A. Kawaue, H. Okamura, M. Tsunooka, *Chemistry Letters* **2002,** (9), 940-941

100. J. Manimino, U.S. Patents # 3,408,181-3,408.189 (Oct. 29, 1968)

101. L. Ab-dellah, B. Boutevin, G. Caporiccio, and F. Guida-Pietrasanta, *Europ. Polymer J.,* **2003** *39*(1), 49, 56

102. C. Badarau and Z.Y. Wang, *Macromolecules*, **2003**, *36*, 6959; *Macromolecules*, **2004**, *37*, 147

103. Z. Wang, F.R.M. McCourt, and D.A. Holden, *Macromolecules*, **1992**, *25,* 1579 (1992)

104. S. Greenfield, R. Harding, J. Vaughan-Spickers, A. Smith, I. Hassall, and C.Dunn, **Brit. Pat. 2,388,600** (Nov 19, 2003)

105. P.W. Cyr, D. A. Rider, K. Kulbaba, and I. Manners, *Macromolecules*, **2004**, *37*, 3959

Chapter 5

Photoresponsive Polymers

Photoresponsive polymers are materials that are able to respond to light irradiation by undergoing reversible changes in their chemical structures and/or their physical properties. Also, photochromism refers to the photoinduced reversible transformations in chemical compounds between two electronic states characterized by different absorption spectra. [1] There are many ways in which the photoresponsiveness of polymers can manifest itself. One might observe changes in viscosity of polymeric solutions, in contraction of polymeric chains, in sol-gel transitions, in electrical conductivity, or even in color changes as a result of irradiation with light of an appropriate wavelength. Another interesting manifestation of photoresponsiveness in some special polymers is a change in the permeability to gases in films. These changes can and are utilized in many ways. Thus, for instance, structural changes due to isomerization are employed to align liquid crystals and photoconductivity is utilized in xerography. Over the last two decades, the photoresponsive materials have grown in practical and scientific importance, because such materials are useful in many applications.

5.1. Polymers for Harvesting Sun's Energy

The goal of harvesting light energy has led to research in polymeric materials that could potentially mimic photosythesis. In such materials, the choice of chromophores is the most critical variable. The location of the chromophores on the polymeric chains and the tacticity of the polymers are also very important. Weber points out, for instance, that among a number of chromophores attached to polymeric chains naphthalene and carbazole form very stable eximers, while phenanthrene and diphenyl anthracene do not. [2] At present, many polymeric materials are known and utilized in the vast areas of nonsilver based imaging, information storage, remote sensing, electroresponsive materials for displays and others. Fox and coworkers concluded, however, that noncompare in efficiency to naturally occurring photon harvesting polymers for photosynthesis. [3]

5.1.1. Polymers with Norbornadiene Moieties

A different approach to harvesting light energy, however, is to utilize pendant groups that reversibly absorb light energy, rearrange, and then release this absorbed energy in a rearrangement back to the original structure. To that end, research is going on in various laboratories to develop systematically derivatized polymer arrays than can collect and convert light energy. Among these, photorearrangements from norbornadiene to quadricyclane are back are of considerable interest, because photoenergy can be stored as strain energy (about 96 kJ/mol) in a quadricyclane molecule and later recovered [4]

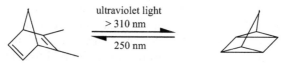

This photoisomerization reaction is referred to as a valence isomerization. It is a reaction in which electron reshuffling occurs and the nuclei move to make or break new π and σ bonds. A number of polymers were, therefore, prepared with the norbornandiene moieties either in the backbone or as pendant groups. Among them are polyesters that were synthesized with donor-acceptor norbornadiene residues in the main chain [53] by polyaddition of 5-(4-methoxyphenyl)-1,4,6,7,7-pentamethyl-2,5-norbornadiene-2,3-dicarboxylic acid or 5,6-bis(4-methoxyphenyl)-7,7-dimethyl-2,5-norbornadiene-2,3-dicarboxylic acid, with bis(epoxide)s. This preparation of such polymers and the accompanying photorearrangements can be illustrated as follows:

The photorearrangements of the norbornadiene residues in the resulting polyesters were observed to proceed smoothly to the quadricyclane groups. Also, it was found that the norbornadiene residues in these polyesters show resistance to fatigue in repeated cycles of the interconversions. [4]

The goal to utilize photochemical valence isomerization between norbornadiene and quadricyclane for solar energy collection and storage was reported by others. [5] Nagai and coworkers synthesized five different polymers with trifluoromethyl-substituted norbornadiene moieties in the side chains and in the main chain [5]:

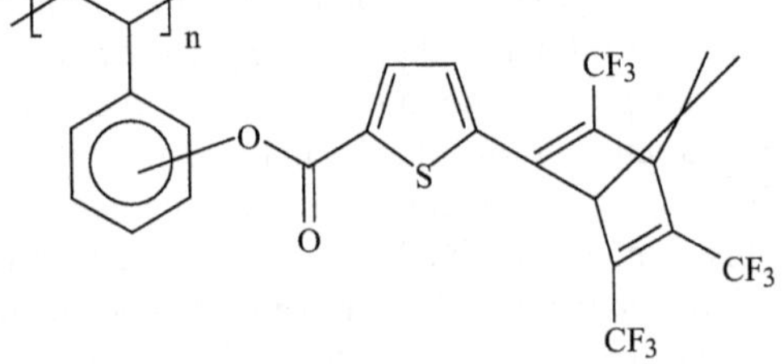

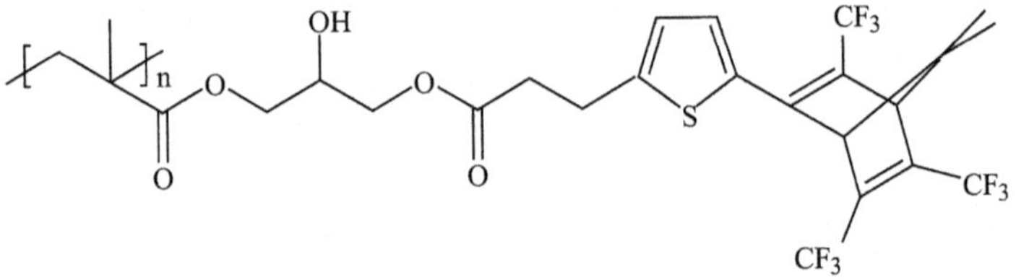

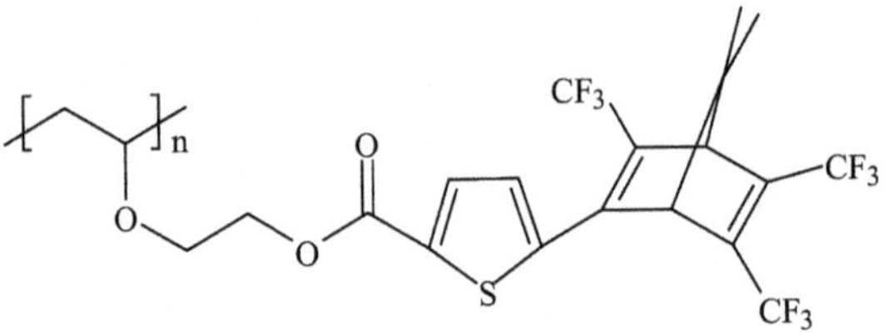

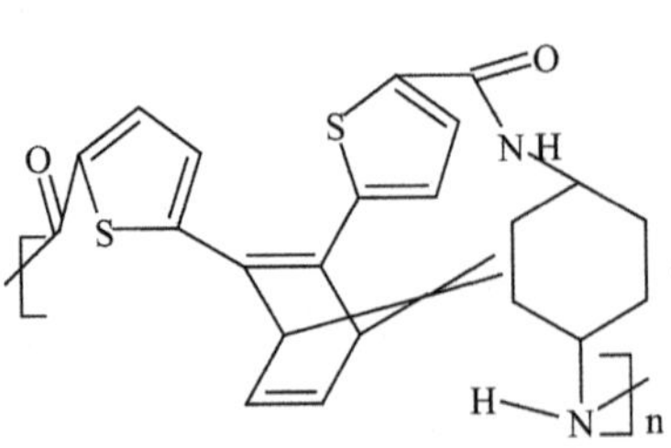

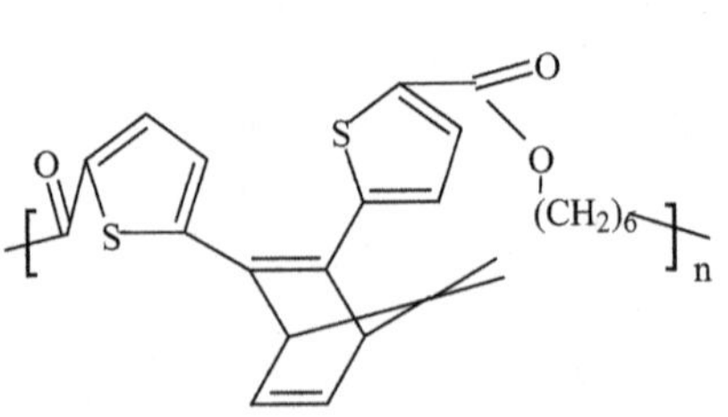

All or the above polymers exhibit large absorption bands in the visible region of the spectra and the norbornadiene moieties in these polymers isomerize very rapidly. In addition, the norbornadiene moieties also exhibit efficient fatigue resistance.[5]

Kawatsuki *et al.*[6] synthesized styrene polymers with pendant norbornadiene groups attached via amide linkages:

where R is a methoxy or a ring substituted aniline group attached through the nitrogen or through another position. These pendant groups also undergo reversible conversions into quadricycline units in polymer films when irradiated by ultraviolet light of two different wavelengths. The materials exhibit high photosensitivity as well as a large red shift in the absorption spectrum upon irradiation.

Sampei *et al.*,[7] reported that polyaddition of 2,5-norbornadiene-2,3-dicarboxylic acid diglycidyl ester to adipoyl chloride gave a polyester containing norbornadiene residues in the polymer backbone and in the pendant groups. When a photochemical rearrangement of norbornadiene residues took place in polymer films, the rate of the photochemical reactions in the polymer backbones was higher than that in the side chains.[7]

Kawashima, *et al.*,[8] reported preparation of donor-acceptor type norbornadiene carboxylic acids compounds with carbamoyl groups, such as dipropylcarbamoyl, methylphenylcarbamoyl, propylcarbamoyl, and phenyl-carbamoyl. Benzyl esters were also prepared. Addition of these groups to polystyrenes formed polymers with pendant donor-acceptor type norbornadiene. Some were formed with 100% substitution. It was found that the polymers containing phenylcarbamoyl groups exhibits especially high photoreactivity. In addition, the rate of the photochemical reaction in films of these norbornadiene polymers increases efficiently by an addition of 4,4'—bis(diethyl-amino) benzophenone photosensitizer. As a result, all the norbornadiene groups of the polymers isomerize to the quadricycline groups in as little as 20 seconds. The stored thermal energy of the irradiated polymers was found to be 32-52 kJ/mol.

5.1.2. Polymers with Pendant Spiropyran Groups

An entirely different approach was taken by Rodriguez-Cabello and coworkers [5] They prepared a molecular-scale machine that converts light into mechanical work. The system uses a polypeptide modified by spiropyran and undergoes transitions from reversible coil to α-helix when exposed to light. Irradiating this polymer with UV light causes ring-opening reactions of the spiropyran group and formation of charged species. The reactions are reversible in sunlight to reform the spiropyran ring structure. Attaching as few as 2.3 spiropyran groups per 100 amino acid residues allows clear photomodulation of the material. The reactions can be illustrated as follows:

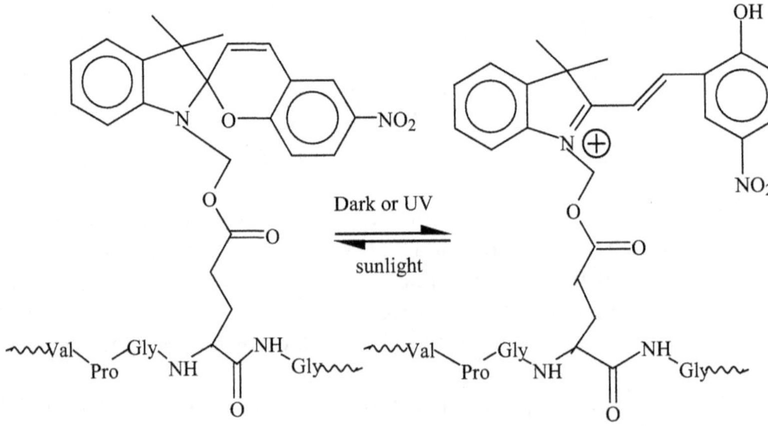

5.1.3. Novel Approaches to Light Energy Harvesting Polymers

Groups, like 6-oxo-1,6-dihydropyrimidin-3-ium-4-oleates, are light sensitive and undergo intramolecular photocyclization to bis-β-lactams when irradiated with UV light between 320 and 490 nm. [9].

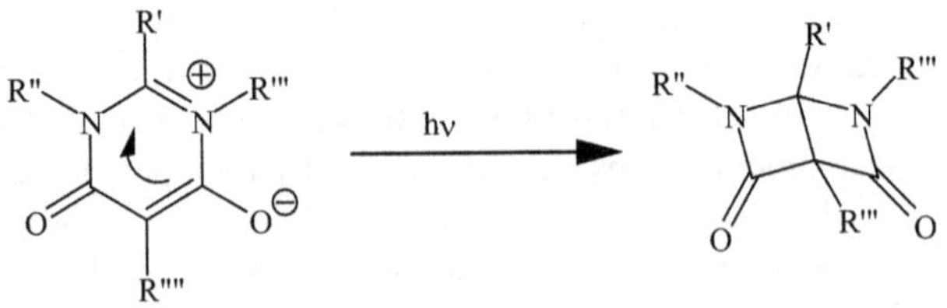

This was utilized to form copolymers of methacrylates with colesteric monomers. [10] An example of such copolymers and the rearrangement is shown below [10]:

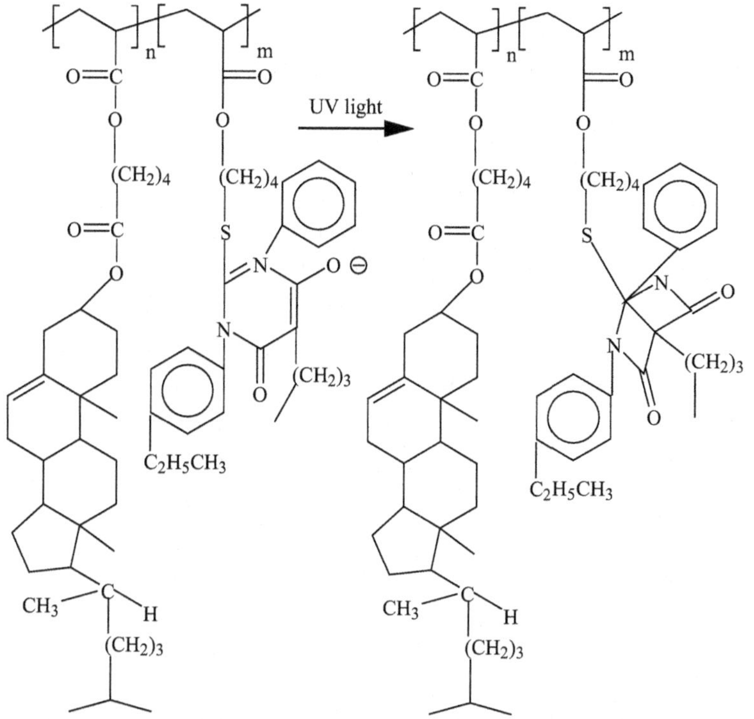

It was found that the rate of photorearrangement is affected by the length of the alkyl side group in the 5 position of the pyrimidinium oleate and by the embedding of a mesoionic chromophore, as shown above. [10]

In photosynthesis, antenna pigments harvest energy and transfer it into a reaction center for redox reactions. [11] Different antenna chromophores that surround the reaction center are arranged morphologically in an order of energy gradient. [12-14] Such an arrangement allows the energy to be sequentially transferred and efficiently funneled into reaction centers over small distances in the direction of decreasing band gaps. Considerable research has been carried out on means to develop the sequential multistep energy transfer systems. This was done not just to mimic the natural light-harvesting process, [12-14] but also for possible applications and use in optoelectronic and biological systems. [15] It is speculated that a conjugated polymeric backbone, with well-designed interruptions of conjugation by insulating spacers, might allow tuning the emission properties and by providing an alternate model. [16]

Based on this concept, Krebs and coworkers [17] reported a synthesis of a light harvesting material that consists of three structural domains. Two of them are conjugated homopolymers that are linked with a zinc porphyrin:

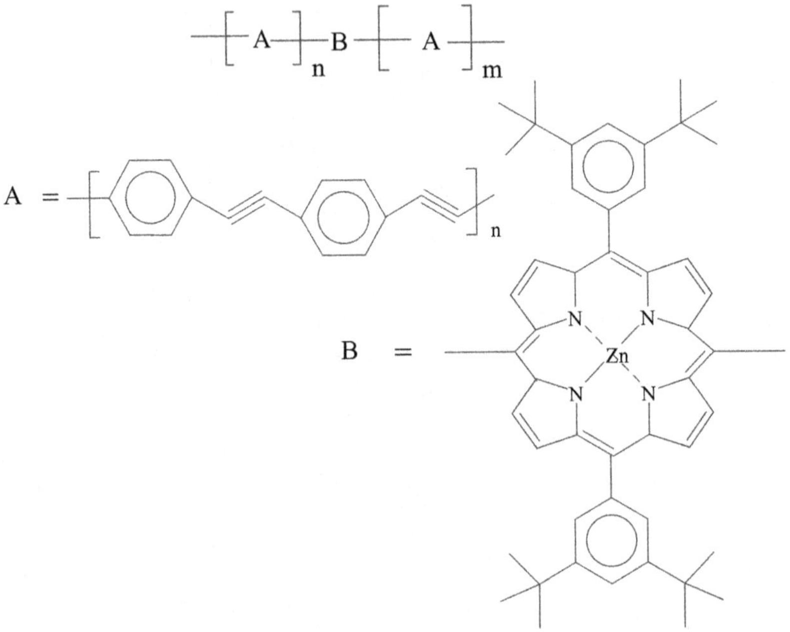

Sections A and B of the above block copolymer are illustrated above. The polymer has a constant ratio of the zinc porphyrin to the incorporated monomer units, regardless of the molecular weight. The ratio of zinc porphyrin to the polymer blocks can be varied in the material by varying the size of blocks A. Studies of energy transfer from the polymer to the zinc porphyrin showed that there is actually very little energy transfer when the material is in solution. On the other hand, there is quantitative energy transfer in the solid state. Also, it was observed that the light-harvesting properties of the three-domain structures depend on the chain lengths of the conjugated polymers.

Cheng and Luh reported that they are also trying to develop polymers that would mimic natural photosynthesis with synthetic polymers.[18] They point out that silylene moieties have been used extensively as insulating spacers.[18] In general, when the silylene spacer contains only one silicon atom, no conjugative interactions between the π systems and the silicon moiety is observed.[19] They believe, therefore, that introduction of an energy gradient with three well-designed chromophores into a silylene-spaced polymeric chain may lead to sequential energy transfer. To achieve this goal, they carried out preparations of regioregular silylene-spaced copolymers composed of energy gradients with three different chromophores. One of the polymers prepared in this way can be illustrated as follows:

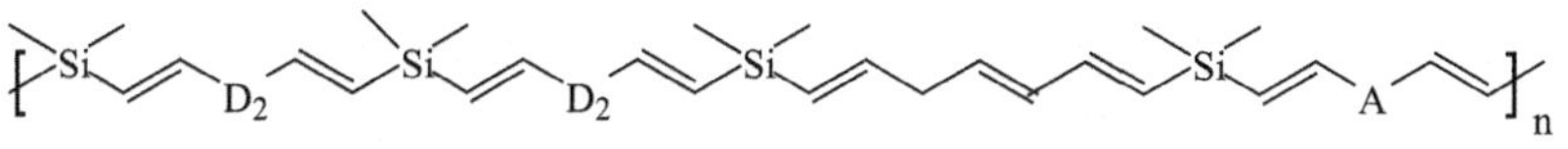

Their synthesis utilized rhodium-catalyzed hydrosilylation of bis-vinylsilanes and bis-alkynes. The ratio of the three chromophores in the above polymer is 1:2:1, corresponding to D1, D2, and A chromophores, respectively. Cheng and Luh found that upon excitation of the donor chromophore D1 only emission from the acceptor A was observed. This is quite encouraging. [19] The chromophores and the acceptor were illustrated as follows:

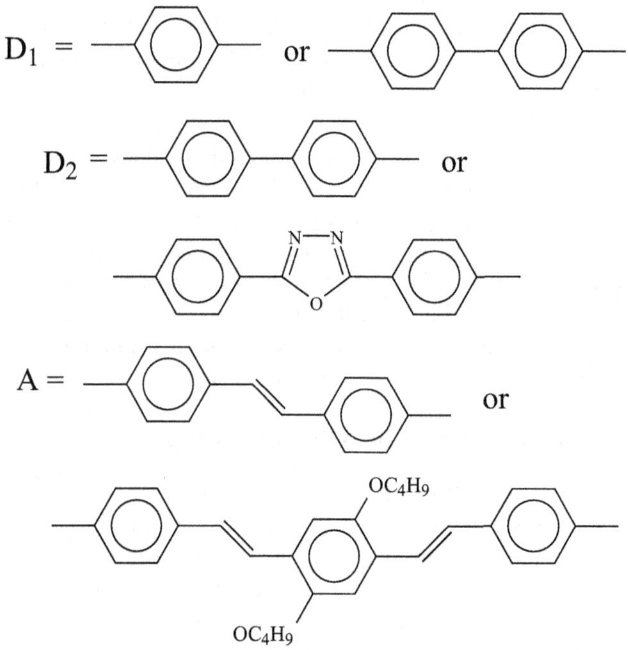

It has not been disclosed yet, at this time, how successful this whole approach is to actually mimicking photosynthesis, but it appears to be a very interesting concept.

5.2. Photoisomerization Reactions in Polymeric Materials

A common photoisomerization reaction is intramolecular rearrangement that takes place in the nitro groups in poly(β-nitrostyrene) and its dimeric model compound upon irradiation with UV light [20]:

This reaction, however, is accompanied by molecular weight loss and the molecular weight of the photoproduct is considerable lower than that of the starting polymer prior to irradiation. This may, perhaps, be due to hydrolysis of the product with a trace amount of water in methyl alcohol/ammonia medium that is used during the irradiation. [20] Whether this photoisomerization reaction has practical utility, however, is not certain.

Among all the photorearrangements, a *cis-trans* isomerization reaction is the most useful one. A well known example one is that of *cis*-stilbene to the *trans* isomer. This reaction has been described in many books. [21] The isomerization reaction takes place because many olefins in the excited singlet and triplet states have a perpendicular instead of a planar geometry. This means that in the excited state the cis-trans isomerism disappears. Upon return to the ground state, S_0, it is possible for either isomer to form. The return, however, usually takes place to the more stable form. Generally, photoisomerization of chromophores in dilute solutions is a first order reaction.

5.2.1. Photoisomerization of the Olefinic Group

An example of olefinic groups rearrangement is work by Onciu *et al.*,[22] who formed three bis(trimellitimide)s by condensing three aromatic diamines with trimellitic anhydride. This was followed by preparation of two series of photoreactive copoly(amide-imide)s by direct polycondensation of the bis(trimellitimide)s and 1,4-phenylenediacrilic acid with either 4,4'-diphenyl-methanediisocyanate in one case or with 1,6-diisocyanatohexane in another case, respectfully. All of the copoly(amide-imide)s were found to be soluble in polar aprotic solvents and to yield transparent, flexible and tough films. [22] When the polymers are irradiated in solution the p-phenylenediacryloyl units undergo *trans-cis* photoisomerization and $(2 + 2)$ photocycloadditions. [22] The fully aromatic polyamides, also undergo a photofries rearrangement. The photo-fries reaction, however, is completely suppressed in polymers that contain an aliphatic amide moiety. [22] The same processes are also observed in the polymer films. [22]

Polymers prepared by condensation of 4,4'-diacetylstiblene as the bis(ketomethylene)monomer with 4,4'-diamino-3,3'-dibenzoylstilbene, a bi(amino ketone), exhibit photoviscosity effects in dilute solutions due to *cis-*

trans isomerization. [23] The preparation of the polymers and the photo-rearrangements can be illustrated as follows:

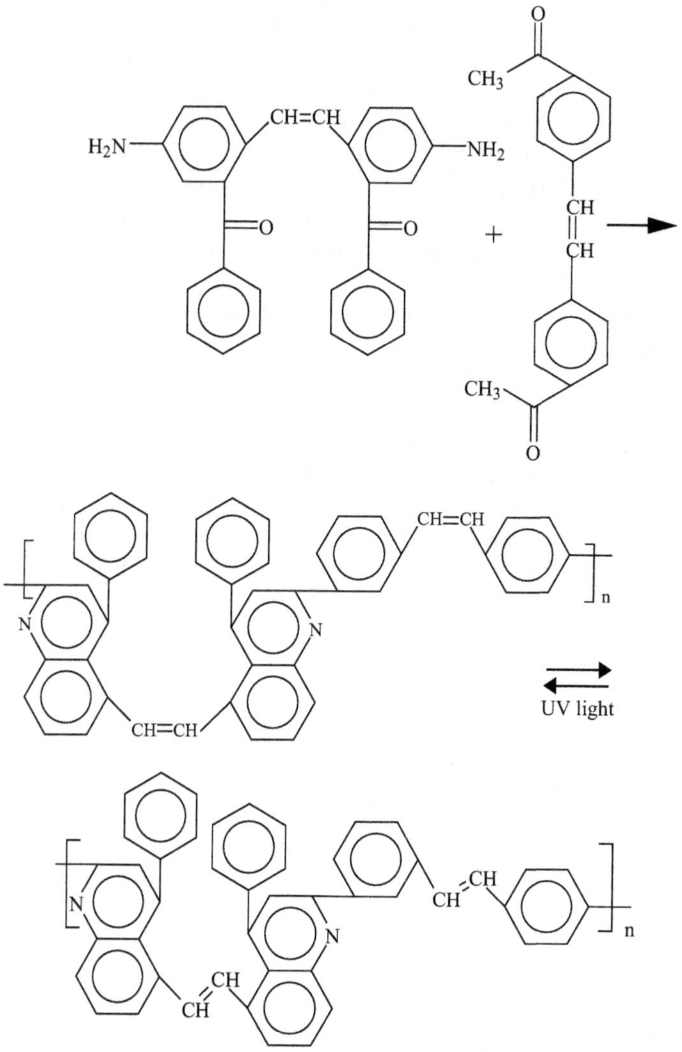

The changes in viscosity can vary from 2 to 23% as a result of irradiation.

5.2.2. Photoisomerization of the Azo Group

Azobenzene is a well-known photoresponsive chromophore, and its photoinduced and thermal geometric isomerizations have been extensively explored. [21] Azobenzene and its derivatives assume both *trans* and *cis* conformations with respect to the azo linkage. Azo $\pi \rightarrow \pi*$ excitation and azo n $\rightarrow \pi*$ excitation trigger *trans*-to-*cis* and *cis*-to-*trans* isomerizations. [24-26] The azo linkage normally exists in the more stable *trans* form. Also, the *trans* isomer of azobenzene exhibits an intense absorption around 320 nm due to the $\pi \rightarrow \pi*$ transition, while the *cis* isomer shows a weak absorption of the n$\rightarrow \pi*$ transition, around 430 nm. [27] Reversible isomerizations between *cis* and *trans* structures are due to these transitions. Photoisomerization can proceed almost quantitatively. [28] By comparison, the thermal isomerizations from *cis* to *trans* configurations take place due to low activation energy of the *cis*-to-*trans* process. Isomerizations of the azo chromophore in compounds are often accompanied by drastic changes in a number of properties such, as for instance, changes in the dipole moments. [29] The isomerization back to the *trans* configuration can be readily carried out either thermally, or by visible light irradiation. Because, the isomerizations have a strong effect on physical properties, the azobenzene units are potentially useful components for photo-control of polymer structures and properties. As a result, the azobenzene functionality has been incorporated in wide variety of polymers. Changes in the molecular structure, such as *cis*-*trans* isomerization in polymers can induce contraction and expansion of the polymeric chains on both microscopic and macroscopic scale. This was demonstrated on a polymer with azo linkages. Exposure from dark to light can result in a contraction of as much as 0.5 % [30]:

Some interesting reports on photoisomerization of azo-polymers are presented below, while the subsequent sections describe specific physical changes and possible utilizations.

Thus, photochemical conformation changes were demonstrated in a copolymer of styrene, where azobenzene structures are attached in the co-monomer to the benzene portion through amide linkages [31] :

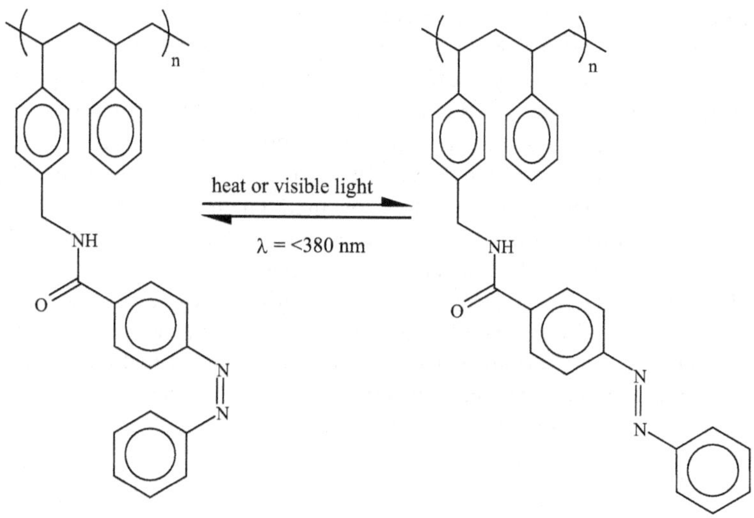

After 10 minutes of irradiation with ultraviolet light the photostationary state is reached, consisting of 79% of the *cis* isomer. Back isomerization to *trans* of the sample is slow in the dark(less than 10% in 60 min), but is much faster when exposed to visible light. [31]

A Japanese patent [32] describes preparation of isophthalic type polyesters that include monomers with pendant azo groups:

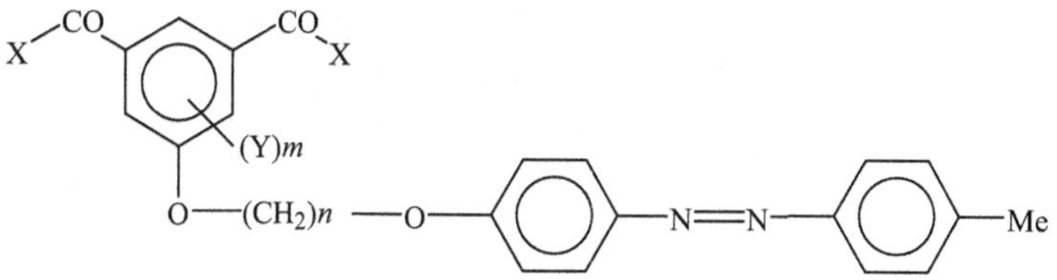

where Y is a hydrogen or a lower alkyl group; m = 1-3; n = 2-18. Polyesters based on this monomers are claimed to be useful for optical recording media such as hologram recording media with low light absorption or loss and wide range of working wave lengths. The preparation of the polyester consisted of first heating diethyl-5-hydroxyisophthalate with 4-(6-bromohexyloxy)-4'-methyl-azobenzene in acetone at reflux to give diethyl-5-{6-[4'-(4"-methylphenyl-azo)phenoxyhexyloxy}isophthalate. This compound is subsequently reacted with 4,4'-- bis(6-chlorohexyloxy)diphenyl ether in the presence of Zn acetate in a vacuum at 160° to form a polyester. The patent claims that coating the polyester on a glass plate and recording with either 532 nm or 515 nm incident light gave a hologram. [32]

    Izumi et al., [33] prepared poly(p-phenyleneethynylene) and hetero-aromatic-containing poly(phenylene)-based conjugated polymers with photo-isomerizable azobenzene units in the main chain . These were formed by using palladium-catalyzed cross-coupling reaction of 4,4'-diiodoazobenzenes with 1,4-diethynylbenzenes and bis(trimethylstannyl)heteroaromatics, respectively. The product, a polymer with a low degree of main-chain conjugation shows reversible photoisomerization of azobenzene units. This is reported to be accompanied by changes in electrochemical properties.

    In addition, Izumi and coworkers, in a earlier study, also carried out similar preparations of conjugated polymers with azobenzenes in the main chain. [34-36] Application of various palladium-catalyzed coupling methods such as the Suzuki coupling and the Heck reactions allowed formation of poly(p-phenylene)- and poly(phenyl vinylene)-based polymers:

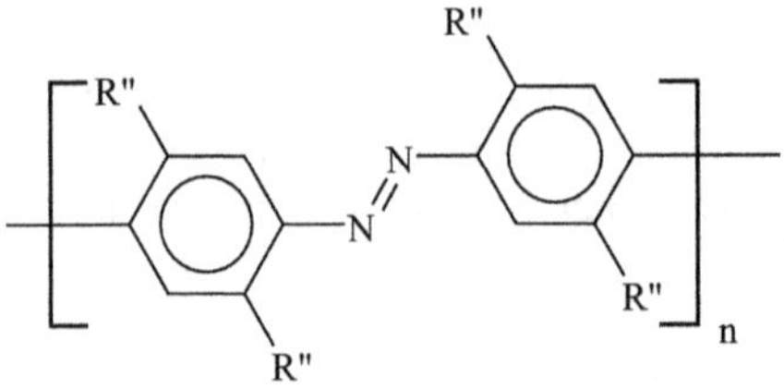

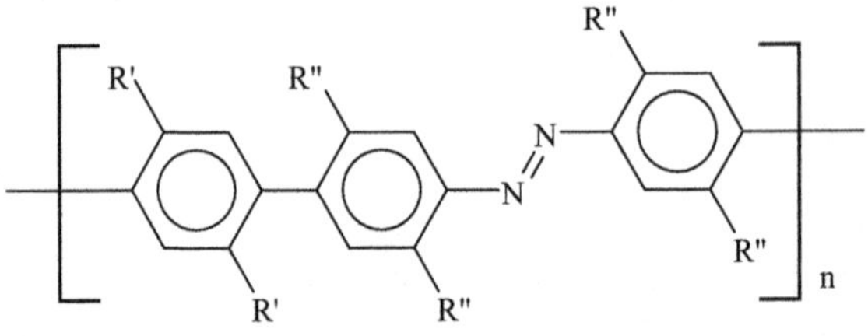

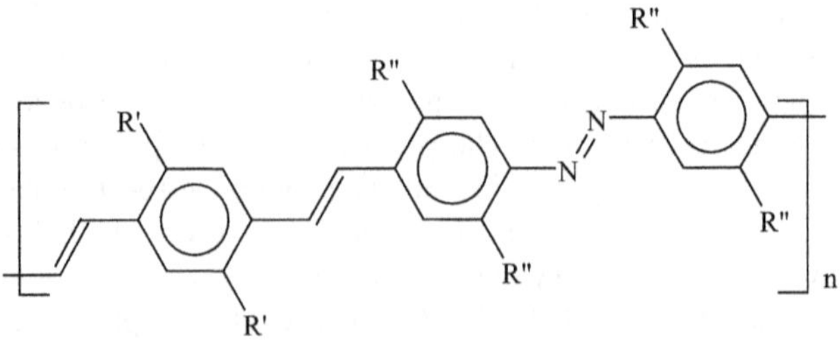

where R',R" = H or n-C$_6$H$_{13}$

The preparation of the poly(p-phenylene)-based conjugated polymers with photoisomerizable azobenzene units in the main chain can be illustrated as follows[37]:

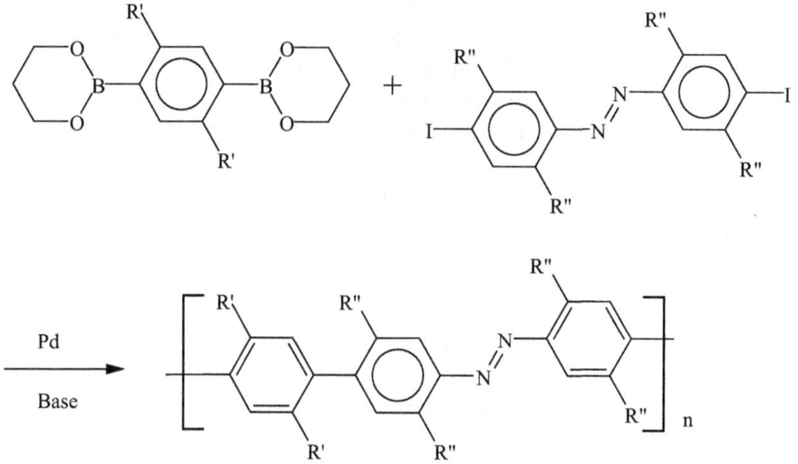

where R' and/or R" can be hydrogens or n-C$_6$H$_{13}$. Despite their stiff structures, these polymers are soluble in common organic solvents, such as chloroform, toluene, and tetrahydrofuran, when either or both of the monomers possess n-hexyl side chains on the aromatic rings. The resulting polymers were found to possess high molecular weights $(M_n > 7 \times 10^3)$ and to be thermally stabile (to > 350 °C), Also, they exhibit fluorescence. Yet, photochemical *trans*-to-*cis* isomerizations in the solid state were found to take place at 20 °C. These isomerization processes are also accompanied by changes in the three-dimensional hydrodynamic volumes of the polymers.[37]

Synthesis of a series of azobenzene-cored polyphenylene dendrimers was reported.[38] The photoresponsive properties of these dendrimers were investigated by optical spectroscopy and gel permeation chromatography The polyphenylene dendrimers exhibit a photoresponsive behavior upon UV and visible light irradiation, resulting in a reversible and appreciable changes of the dendrimer structures.[38]

Actually, formation of polyphenylene dendrimers with azobenzene structures in the core was reported earlier.[39] The materials were also demonstrated to exhibit a photoresponsive behavior upon UV and visible light irradiation, resulting in reversible and appreciable changes in the dendrimer structures.

Junge and McGrath[40] reported studying photoresponse of dendrimers containing three interior azobenzene moieties that are radially configured about the core. With each azobenzene capable of *cis-trans* isomerization, the dendrimers can exist in four discrete states, *cis-cis-cis, cis-cis-trans, cis-trans-trans,* and *trans-trans-trans*. All states are detectable in solution and exhibit different physical properties. ^{1}H NMR spectral data indicates independent, rather than simultaneous isomerization of azobenzene units. Dark incubation in CH_2Cl_2 solutions of the dendrimers maximized the $n \to \pi^*$ transition at 352-nm, and irradiation with 350-nm light resulted in photoisomerization, as evidenced by a decrease in the absorbance at 352-nm and an increase in the $n \to \pi^*$ transition at 442-nm. The presence of sharp isobesic points in the absorption spectra during isomerization can be expected on the basis of the independence of the chromophores. [40]

Nozaki *et al.*,[42] synthesized photoresponsive γ-polyketones with azobenzene side chains:

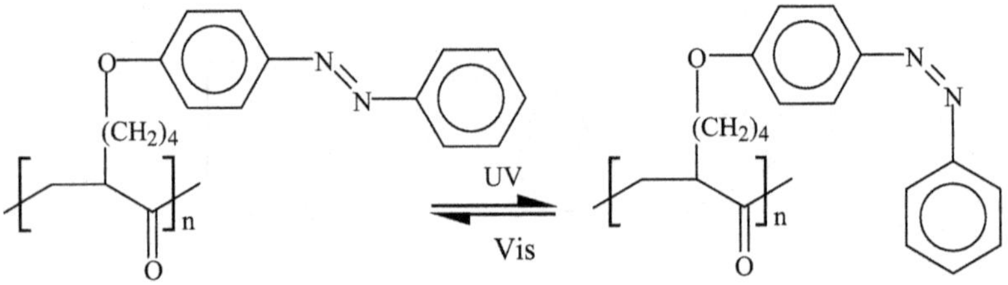

and a ketone/spiroketal one:

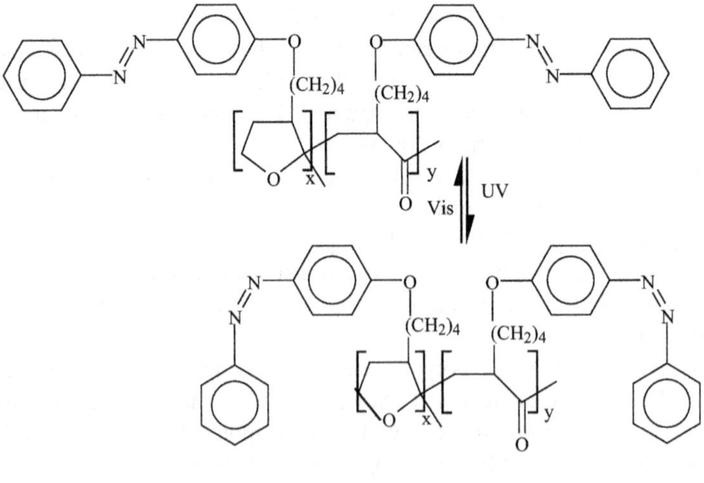

They found that the conformations of the polymers that contain the ketone units only are not affected by the *cis-trans* isomerization. On the other hand, the keto/spiroketal polymers drastically change in the main-chain conformations. [42]

Condensation of acryloyl chloride with 4-(phenylazo)aniline was used to form a monomer p-(phenylazo)acrylamilide. The monomer was polymerized to yield poly[p-(phenylazo)acrylanilide] that undergoes photochemical interconversions between *trans* and *cis* isomers under UV or visible light irradiation. This reaction is reversible and reproducible for the homopolymer and for a copolymer with methacrylic acid [43]:

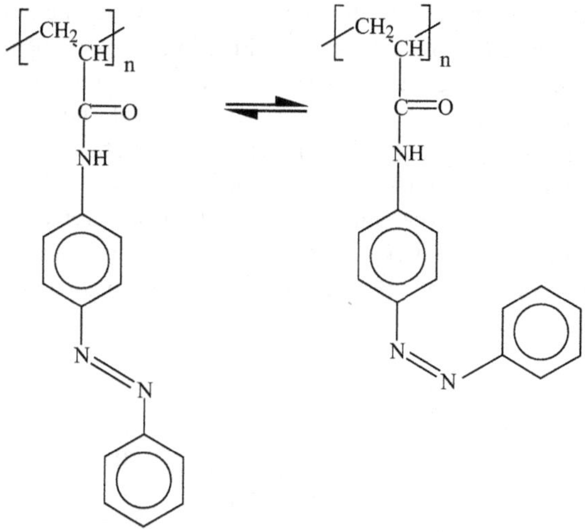

5.2.2.1 Changes in Viscosity and Solubility of Polymeric Solutions

Changes of viscosity in polymeric solutions that are associated with photoinduced conformational changes of the macromolecules were observed by Lovrien in 1967. [43] He reported that solutions of a copolymer of methacrylic acid and N-(2,2'-dimethoxyazobenzene)acrylamide exhibit an increase in specific viscosity when irradiated with UV light. He also observed a decrease in the viscosity of a poly(methacrylic acid) and chrysophenine solution in water under the influence of UV light. [43] This was followed by various reports of photoviscosity effects in solutions of azobenzene based polymers. Matejka and Dusek [44] studied a copolymer of styrene and maleic anhydride with azobenzene in the side chains. U.V. light irradiation of a solution of this polymer in 1,4-dioxane causes a decrease in specific viscosity between 24 to 30 % and in tetrahydrofuran between 1 and 8%. They also observed that this decrease in viscosity is reversible. The magnitude of the effect was found to be related to the quantity of azobenzene linkages present in the polymer.

Hallensleben and Menzel [45] found that irradiation of poly(5-(4-phenylazobenzyl)-L-glutamate in 1,4-dioxane solution with UV light ($\lambda > 470$ nm) decreases the viscosity by 9% . Here too this change in viscosity is accompanied by a *trans* to *cis* isomerization that was estimated to be 23%. With additional irradiation by 360 nm UV light, the viscosity decreases an additional 9% and the isomerization to *cis* reaches 89 %.

Irie *et al.*, [46] synthesized a number of polyamides with azobenzene groups in the backbone. All the polymers exhibit photoviscosity effects. In solutions in N.N'-dimethylacetamide, a 60% reduction in specific viscosity can be achieved by UV light irradiation (410 > λ > 350 nm). The initial viscosity is regained by storage in the dark at room temperature for 30 hours. Changes in viscosity of solutions in dimethylsulfoxide of a range of polyureas with azobenzene groups in the polymer backbone were reported. [47] The irradiations were carried out at 35 °C with UV light between 410 and 350 nm. It was observed that the intrinsic viscosity is about 40% lower during UV irradiation than in the dark. Also, toluene solutions of polydimethylsiloxane with azobenzene residues were shown to exhibit 20 % lower viscosity under UV light irradiation than in the dark. [48] Recently, Fernando and coworkers [49] prepared a series of copolymers of trans-4-methacryloyloxyazobenzene with methyl methacrylate.

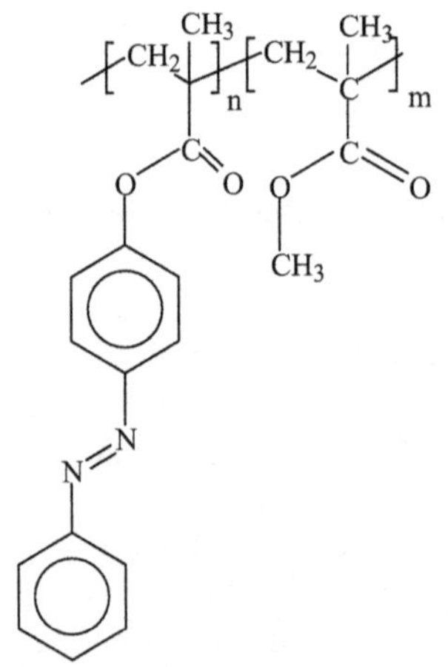

All the polymers exhibit a photoviscosity effect when exposed to UV irradiation at $\lambda = 365$ nm. This effect was attributed by them to conformational contraction of the polymer chains due to dipole-dipole interaction between neighboring chromophores. [40] This conformational change might possibly be illustrated as follows:

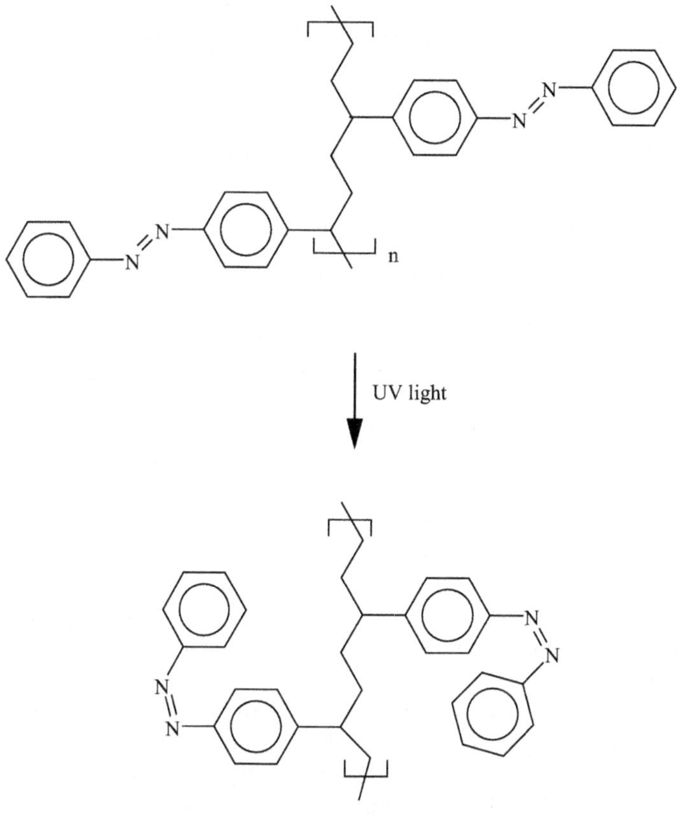

Also, when copolymers of polystyrene and 4-(methacryloyl-amino)azobenzene containing 2.2 to 6.5% of the latter are irradiated in a cyclohexane solution with 15 flashes of 347 nm. of light. the polymeric chains contract. [50] This occurs at a high rate per second as a consequence of isomerization. At a later stage, several hundred seconds after the flash there is evidence of polymer aggregation and precipitation. [50]

In addition, when azobenzene residues are introduced into the main chain of poly(dimethylsiloxane), reversible solution viscosity changes can be obtained by irradiation with ultraviolet light. [51] The conformational changes of the polymer can be illustrated as follows:

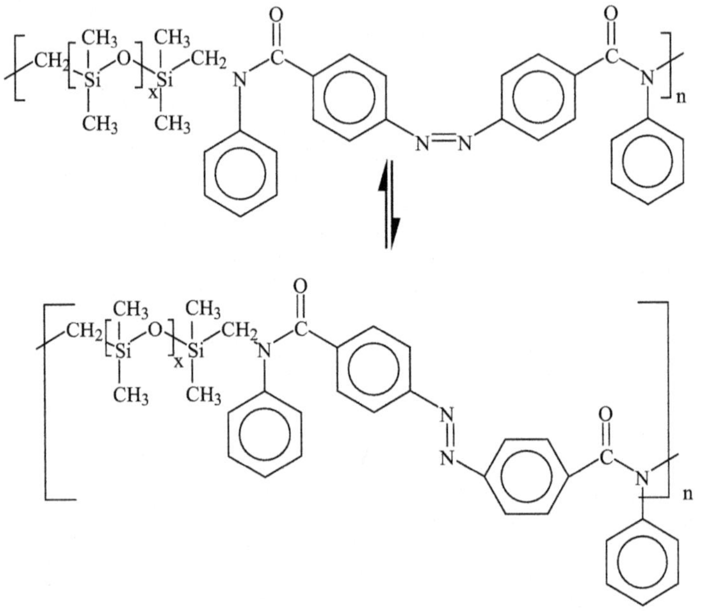

A shown above, the conformational changes are confined to the isomerization of the azo group.

Isomerization from *cis* to *trans* and back of azo groups, however, is not the only mechanism that can affect photoviscosity change in polymeric solutions. Thus, reversible solution viscosity changes were also observed [48] in solutions of poly(dimethylacrylamide) with pendant triphenylmethane leucohydroxide in methanol. This can be illustrated as follows:

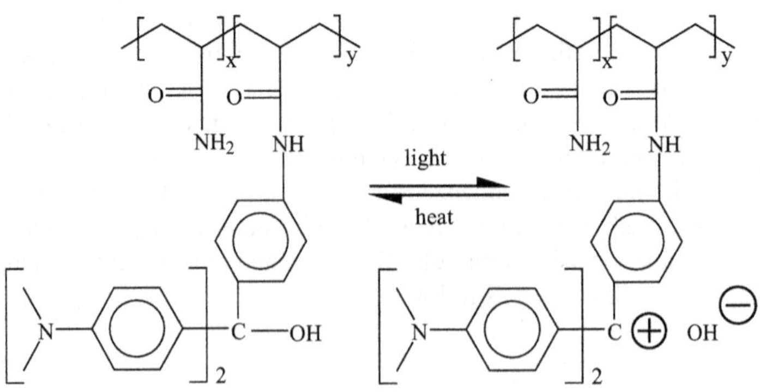

So, as shown, the viscosity changes are due to positive charges that form on the pendant groups.

Also, when a terpolymer that consists of vinyl pyrrolidone, styrene that is *p*-substituted with 9,10-diphenyl anthracene, and a flluorescein derivative of acrylamide, is irradiated with light, there is a viscosity increase [52]:

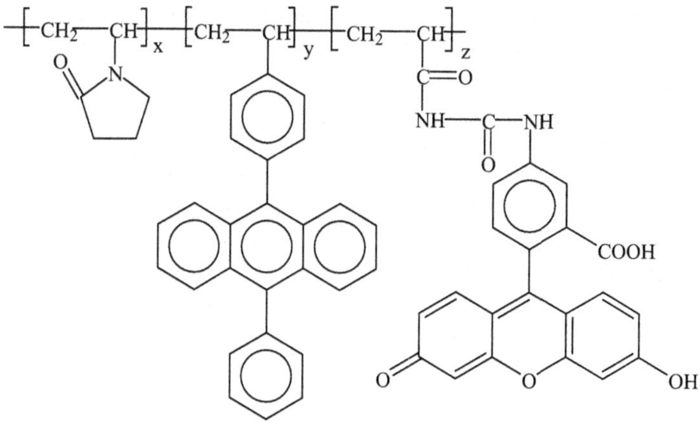

The absorption of light by the diphenylanthracene was found to result in efficient intracoil sensitization by fluorescene. The quantum efficiency of this process was determined to be 0.4 in methanol and 0.8 in water. [52] This increase corresponds to a decrease in polymer coil size in water. Analysis of the fluorescence decay also demonstrates that the intracoil energy transfer is essentially a static process and that aggregation can result in nonexponential fluorescence decay that is interpreted as a dynamic equilibrium that takes place between diphenylanthracene and a nonfluorescent dimer state [52]:

The solubility of a copolymer of styrene in cyclohexane was found to change reversibly upon ultraviolet light irradiation when the copolymer contains small amounts (~2 mol %) of spirobenzopyran among the pendant groups. [53] This is believed to be due to photoisomerization of the pendant spirobenzopyran groups to the polar merocyanine form with the resultant decrease in polymer-solvent interaction and subsequent precipitation of the higher molecular weight fractions of the polymer. A copolymer with a high content of spirobenzopyran groups (12.3 mol %) performs as a negative photo-resist with high contrast. [53]

5.2.2.2. Sol-Gel Reversible Conversions

Gel systems that respond to external light stimuli are of considerable interest in many fields, from potential use in paints and coatings, to personal care items and drug delivery systems. Applications might also include photonics, optical storage, and photoswitchable materials.

Ire and Iga demonstrated that photostimulated sol-gel transitions are possible with gels of a polystyrene copolymer in carbon disulfide solution that carry some azobenzene groups. [54] The transitions are reversible and take place upon ultraviolet light irradiation. The polymer undergoes *cis* to *trans* isomerization of the pendant azobenzene groups with a 62% conversion for polymers containing as much as 10.5% of azo groups:

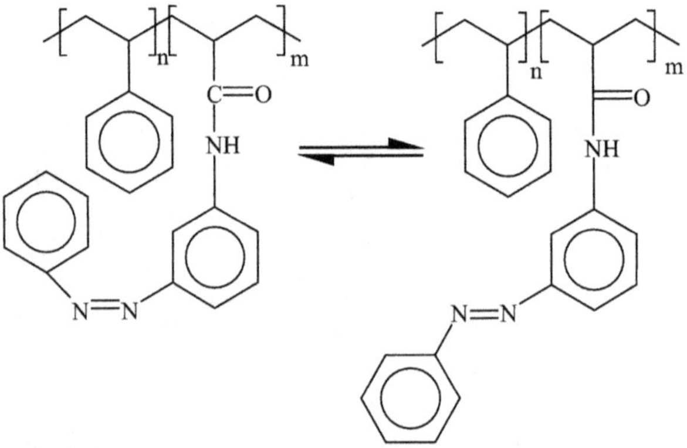

Irradiation with light of wave length between 310 and 400 nm converts the solution to a gel and visible light (>450 nm) reverts the gel back to the solution The process is a result the *trans* isomerizing to the *cis* form upon irradiation with UV light. [55] It suggests that *cis* azobenzene moieties in the pendant groups make interpolymer crosslinking junctions that are more stable than the interactions of the *trans* forms.

Lee, Smith, and Hatton observed [56] that the viscosity and gelation of mixtures of hydrophobically modified poly(acrylic acid) and a cationic photosensitive surfactant can be controlled reversibly by switching irradiation of the sample between UV and visible.

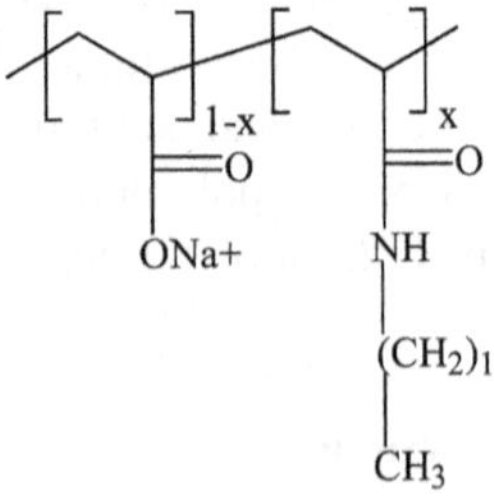

hydrophobically modified poly(acrylic acid)

$$CH_3CH_2-\underset{}{\bigcirc}-N{=}N-\underset{}{\bigcirc}-O(CH_2)_4\overset{\oplus}{N}(CH_3)_3 \ Br^{\ominus}$$

surfactant

The planar *trans* form of the surfactant, that is attained by visible light irradiation, is more hydrophobic than the nonplanar *cis* conformation of the material that is attained by ultraviolet light irradiation. At the critical aggregation concentration of the surfactant, micellar aggregates form on the polymer and solubilize the alkyl side chains grafted on the hydrophobically modified poly(acrylic acid), leading to physical crosslinking and gelation. The hydrophobic *trans* (visible light) form, with a planar azobenzene group in the surfactant tail, has a lower critical aggregation concentration than the more polar *cis* (UV light) form, resulting in gelation at lower surfactant concentrations under visible light. Reversible viscosity changes of up to 2 orders of magnitude are observed upon exposure to UV or visible light. Observed viscosity maxima of 5.6×10^4 cPa for the *trans* form and 2.2×10^3 cPa for the *cis* form of the surfactant suggest that the *trans* form surfactant micelles are more effective at solubilizing the alkyl side chains than are the more hydrophilic *cis*-form micelles.[56]

Harada and coworkers [57] reported forming a photoresponsive hydrogel system by combination of α-cyclodextrin, dodecyl-modified poly(acrylic acid), and a photoresponsive competitive guest, 4,4'-azodibenzoic acid. An aqueous solution of dodecyl-modified poly(acrylic acid) exhibits a gel-like behavior, because polymer chains form a network structure via hydrophobic associations of C_{12} side chains. When α-cyclodextrin is added to the gel-like aqueous solution, the gel is converted to a sol mixture because hydrophobic interactions of C_{12} side chains are dissociated by the formation of inclusion complexes of α-cyclodextrin with the C_{12} side chains. Upon addition of 4,4'-azodibenzoic acid to the binary sol mixture of the modified poly(acrylic acid) and α-cyclodextrin, the cyclodextrin interacts predominantly with 4,4'-azodibenzoic acid:

$$\underset{O}{\overset{HO}{\diagdown}}C-\underset{}{\bigcirc}-N{=}N-\underset{}{\bigcirc}-C\underset{OH}{\overset{O}{\diagup}}$$

+

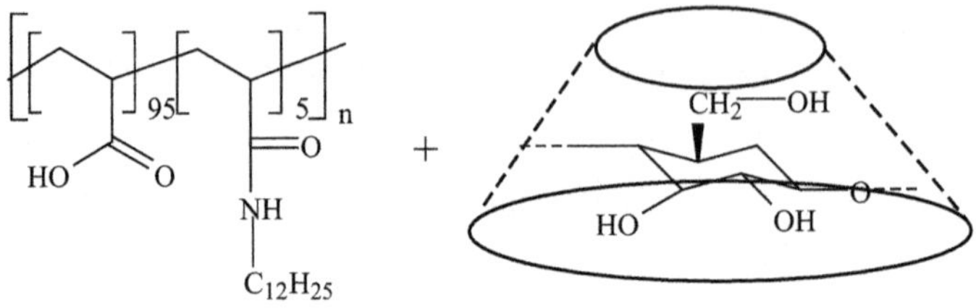

Hydrophobic associations of the C_{12} side chains are restored, resulting in a sol-to-gel transition. When this ternary gel mixture is irradiated with UV light, the azobenzene compound isomerizes from *trans* to *cis*. As a result, the mixture undergoes a gel-to-sol transition, because α-cyclodextrin formed inclusion complexes more favorably with C_{12} side chains than with *cis*-azobenzoic acid. On the other hand, when the ternary sol mixture is irradiated with visible light, the azo group isomerizes back from *cis* to *trans* and the mixture undergoes a sol-to gel transition. Harada and coworkers [57] reported that these gel-to-sol and sol-to-gel transitions occur repeatedly by repetitive irradiations of UV and visible light.

5.2.2.3. Changes in Birefringence and Dichroism

Over the last twenty years there were reports that exposure to linearly polarized light of some photoreactive films can generate optical anisotropy in these films through an anisotropic polarization axis. This phenomenon received more attention recently [58-60] If the axis-selective reaction is accompanied by molecular reorientation, a large anisotropy can form. Thus, Kawatsuki *et al.*, [131] reported that in axis-selective triplet phtotosensitized reactions of photo-crosslinkable liquid crystalline polymeric films achieve molecular reorientation. Irradiation is done with linearly polarized light at 405 nm. Also, the molecular reorientation is thermally enhanced. Four photosensitizers were used in this work. These are as follows:

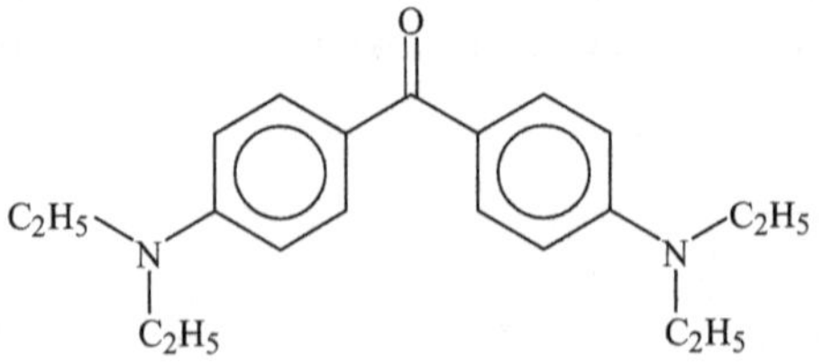

4,4'-(N.N'-bisdiethylamino)benzophenone

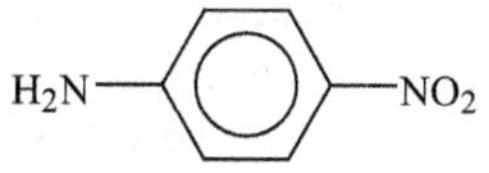

p-nitroaniline

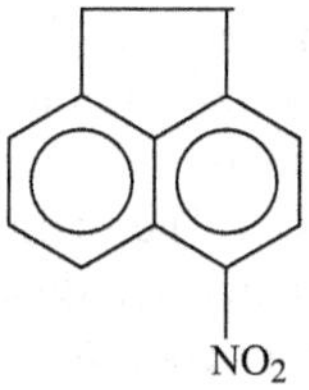

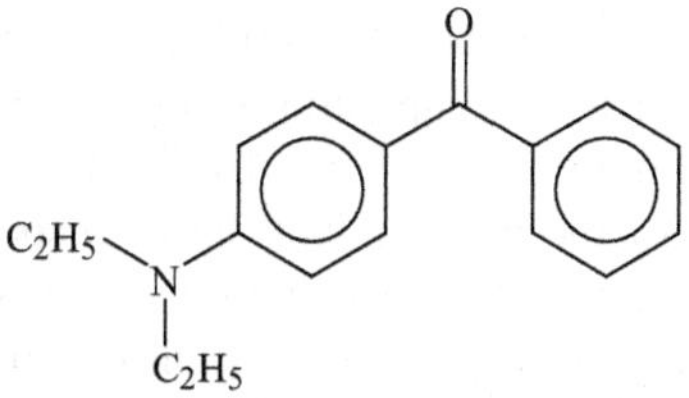

4-nitroacenaphthene   4-(N-dimethylamino)benzophenone

Also, several types of polymeric films were used in this work that are liquid crystalline polymethacrylates with photocrosslinkable mesogenic side groups:

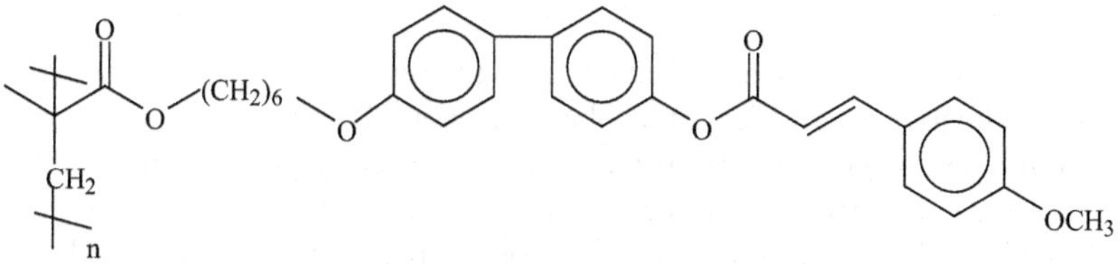

In all cases the photoreactivity of the films was generated with less that 9% doping. The axis-selective photoexcitation of the triplet sensitizers and the polarization-preserved energy to the photoreactive mesogenic groups determined the photoinduced optical anisotropy as well as the thermal enhancement of the molecular orientation.

Amorphous polymers with pendant azobenzene groups were demonstrated to be good candidates for optical data storage and other electrooptic applications. [61] One such material can be illustrated as follows:

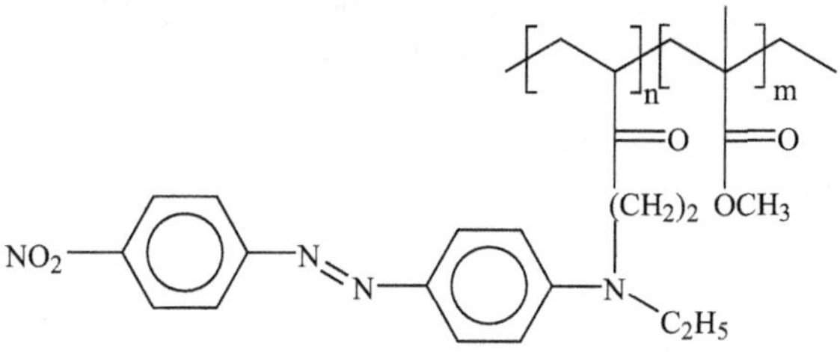

This is due in part to a mechanism that allows anisotropy to be optically induced at room temperature in these materials. [62] The induced anisotropy manifests itself in birefringence and in dichroism of long-term stability. The anisotropy can, however, be eliminated by either optical or thermal means. The mechanism by which anistoropy is induced involves selective photoexcitation followed by a spatial reorientation of the azo-aromatic side chain dipoles via *trans-cis-trans* isomerization. By selectively exciting only the dipoles possessing a sufficient dipolar component, or by absorption probability in a single direction, a perfectly stable dipolar orientation is created that is perpendicular to this direction. [62] When the external electric dc field poling is done (see Chapter 6) at temperatures above the glass transition temperature, the orientation of dipoles takes place (by the typical theoretical description of the process) in one of the independent dipole responses. [63] The literature indicates, however, that the resultant orientation order is not linearly dependent on the chromophore content, especially at high chromophore concentration. [64, 65] The conclusion is that there is some form of dipolar interaction between the azo groups on the side chains that are next to each other on the polymer backbone. Absorption maxima shifts to higher energy with increasing azo concentration. indicate that the dipoles are associated in an anti-parallel fashion. This interaction reduces the mobility of the side chain but enhances the orientation stability.

It was demonstrated by Natansohn and coworkers [64, 65] how the *trans-cis-trans* isomerization cycles of azobenzene bound in high glass transition temperature polymers could be used to induce birefringence in the polymer films. Linearly polarized light orients an excess of the azobenzene group perpendicularly to the polarization direction, while circularly polarized light restores the original disorder of orientations, thus "erasing" the induced birefringence. This means that irradiation with an argon laser at 514 or 488 nm will photochemically activate both the *trans-cis* and *cis-trans* isomerization process. The *cis* isomers are relatively short lived even in the absence of irradiation. Thus, after the laser is switched off, only the trans isomers remain present in the film. [65]

In addition, it was also reported that circular dichroism and circular birefringence can be induced in thin films of achiral polymer liquid crystals containing 7 to 15 mol % azobenzene chromophores by irradiation with circularly polarized light. [66] The polymer can be illustrated as follows:

Circular dichroism was observed not only for the azobenzene moieties but also for the nonphotoactive mesogens, cyanobiphenyl groups, due to a cooperative motion. Circularly polarized light of the opposite handedness produces enantiomeric structures, and a chiroptical switch can be achieved by alternating irradiation with left and right circularly polarized light. The level of photoinduced chirality depends upon incorporation methods (doped or chemically bound) of the azobenzene chromophores, the amount of azobenzene units, and the sample temperature. The photoinduced circular dichroism can be erased by heating the films above the clearing temperatures or by annealing the films in the liquid-crystalline phase.

Two newly synthesized polyelectrolytes functionalized polymers with branched side chains bearing electron donor-acceptor type azobenzene chromophores were used as polyanions to build up multi-layer films through an electrostatic sequential adsorption process [67]:

where ,Y is NO_2 in one polymer and COOH in the other. Poly(diallyldimethyl-ammonium chloride) was used as the polycation. When dissolved in an anhydrous dimethylformamide and also in a series of dimethylformamide-water mixed solvents of different ratios, both azo polyelectrolytes formed uniform multilayer films. This was done through the layer-by-layer adsorption process. Altering the water content of the dipping solutions of both azo polyelectrolytes was found to dramatically change the thickness of the sequentially adsorbed bi-layers, chromophore orientation, and surface roughness of the multi-layer films. On the other hand, the solvent effect on the hydrogen-aggregation in the multi-layer films was determined by the structural details of the azo polyelectrolytes. After the irradiation with a linearly polarized Ar+ laser beam at 488 nm, significant dichroism was induced in the multi-layer films that were prepared from the dimethylformamide dipping solution. Upon exposure to an interference pattern of the Ar+ laser beam at modest intensities, optically induced surface modulation on the multilayer surfaces were observed. [67]

Enzymatic synthesis, employing horseradish peroxidase, was used to prepare a photoactive azo polymer, poly-[4-(phenylazo)phenol], from 4-(phenylazo)phenol. [68] Spectroscopic data show that the coupling reaction occurs primarily at the ortho positions, with some coupling at the meta positions of the phenol ring as well. This results in the formation of a branched polyphenylene backbone with pendant azo functionalities on every repeat unit of the macromolecules.

This enzymatically synthesized azophenol polymer has an extremely high dye content (nearly 100%) and is soluble in most polar organic solvents. It forms good optical quality thin films. Polymer solutions show reversible *trans* to *cis* photoisomerizations of the azobenzene groups with long relaxation time. The poly(azophenol) films also exhibit photoinduced absorption dichroism and large photoinduced birefringence with unusual relaxation behavior. [68]

A photoresponsive azo-polymer was grafted onto polyethylene film. [69] The monomer used in the grafting reaction is 4-[N-(acryloxyethy)-N-ethylamino]-4'-nitroazobenzene. It was found that the azo groups through rapid *cis-trans* isomerization orient themselves perpendicularly to the polarization direction of the light. When the films are irradiated with polarized light of 488 nm this resulted in formation of anisotropic films. [69]

There is considerable interest in using polymeric photorefractive materials for holographic recording. The photorefractive effect must be completely reversible so that the recorded holograms can be erased at will. Such reversibility would make these polymeric materials suitable for real time optical applications. This requirement, however, creates a real problem, because recording and reconstruction of the stored information is done by illuminating the material with a spatially homogenous light beam of the same wavelength. The light gets diffracted by the reversible index modulation that, at the same time, simultaneously erases the hologram. Earlier, it was demonstrated that polymeric composites with high diffraction efficiency, two-beam coupling gain coefficients, and ms response times can be prepated.[70,] Also, Blanche and coworkers [71] demonstrated that recording of volume holograms on such polymers with femtosecond pulses using two photon absorption coupled with a nondestructive readout using cw lasers of the same wavelength is possible. The polymer is a composite of poly(N-vinylcarbazole), N-ethylcarbazole, and benzyl butyl phthalate. A photosensitizer, an electrooptic chromophore was sandwiched between two coated transparent electrodes. It can be illustrated as follows [15] :

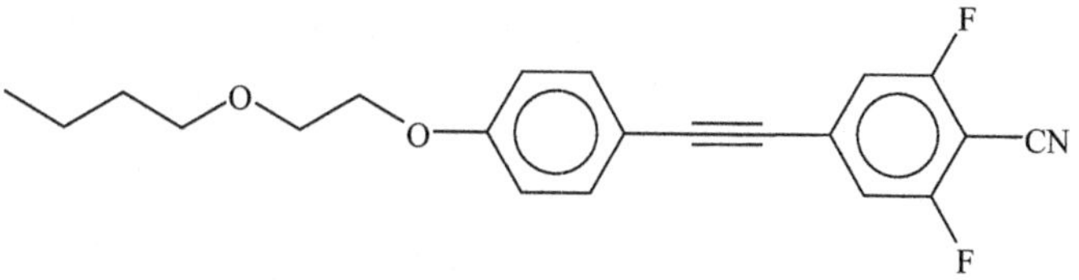

The composite was fabricated into photorefractive samples by injection molding.

5.2.2.4. Application to Optical Data Storage

Due to possible utilization of photoinduced orientation in polymeric films in optical data storage, this phenomenon and the quadratic nonlinear optical effects were extensively investigated in the last few years. [71] It was reported, for instance, that to study photoisomerization in a polymeric environment, a series of polymers containing azo dyes with large differences in the second order transition temperature were compared. [71] Particular emphasis was placed on the relationship between photoisomerization, T_g of the polymers, and their molecular structure. As a result, it was shown that light-induced non-polar orientation in very high Tg polyimides (Tg up to 350 °C) can take place even at room temperature. The polymers used in one of these studies can be illustrated as follows [71]:

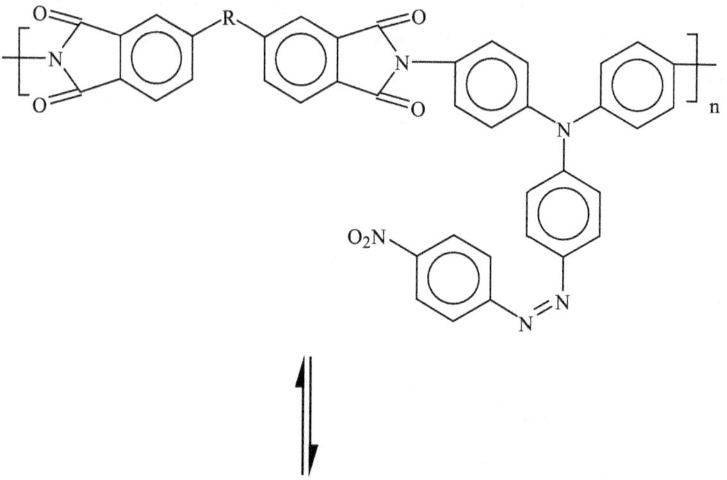

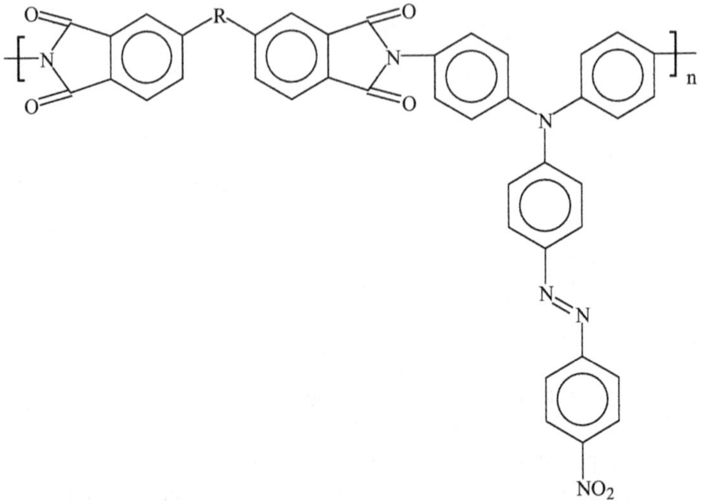

where R $=$ $(CF_3)_2$ -C < in one polymer and -COO-$(CH_2)_2$-COO- in another. From the behavior of the mean absorbance it was concluded that all the azo chromophores revert to the *trans* form on completion of a thermal back reaction. The observed increase in the dichroic ratio over the first 25 hours is believed to be due to the thermal back isomerization and not due to the relaxation of the induced orientation. [71] Heating polymers at 170 ^{0}C for one hour fails to erase the green light-induced dichroism in the samples. This dichroism is, however, completely erased on heating the samples above their T_g for 10 minutes. Irradiation of the films with incident light gives holograms. [72]

Also, polyesters were formed by melt transesterification [73] of diphenyltetradecanedioate with a series of mesogenic 2-[-4-[(-4-cyanophenyl)azo]phenoxy]alkyl]-1,3-propanediols, where the alkyl spacer is hexa, octa, or decamethylene. The molecular weights of these polyesters ranges between 5000 and 8900:

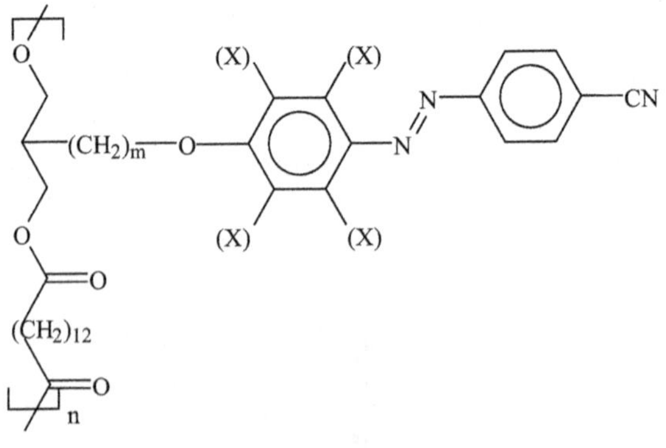

Optical storage properties of these polyesters were studied through measurements of polarization. anisotropy and holography. A resolution of 5000 lines/mm and diffraction efficiencies of about 40 % were achieved. Lifetimes greater than 30 months for information storage was obtained, even though the glass transition temperatures are about 20 °C. Complete erasure of the information can be obtained by heating the films to about 80 °C. It was also claimed that the films can be reused many times without failure. [73]

Belfield and coworkers [74] describe simultaneous two-photon absorption as a process in which the probability of a ground to excited state transition scales quadratically with incident intensity of the irradiation source. This nonlinear or quadratic dependence makes two-photon excitation attractive for use in several emerging technologies, such as two-photon fluorescence imaging, three-dimensional microfabrication, and optical power limiting. Of greatest interest is the possible development of organic three-dimensional optical data storage media and processes.

Previously, Belford et al., [75] reported the synthesis and characterization of organic fluorescent dyes with high two-photon absorptivity. Based on their findings, they reported that they have prepared a photosensitive polymer with a high two-photon absorptivity group:

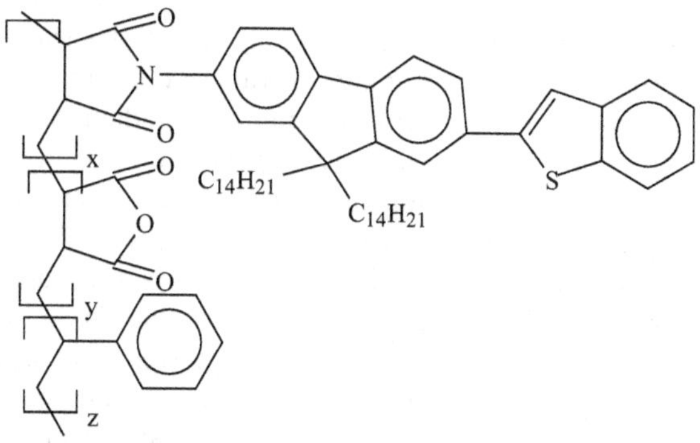

Photophysical properties indicate, the this polymer undergoes two-photon unconverted fluorescence. [74] Fluorescence properties of the polymer were modulated in the presence and absence of a photoacid generator, providing image contrast. A multi-layered assembly and bulk two-photon fluorescence lithographic imaging provided both negative" and "positive" contrast images, demonstrating the possibility of' three-dimensional optical data read-out. Furthermore, the photosensitive polymer was also responsive to two-photon induced "writing" followed by "reading", indicating the possibility of high optical data density storage in the organic material.[74]

5.2.2.5. Changes in Contact Angle of Water

Wu *et al.*, [76] reported syntheses of a series of novel azo polyelectrolytes, starting with a reactive precursor, poly(acryloyl chloride). The precursor polymer was functionalized by the Schotten-Baumann reaction with several aromatic azo compounds containing hydroxyl end groups. The degrees of functionalization were controlled by selecting suitable feed ratios between the azo reactants and poly(acryloyl chloride) The unreacted acyl chloride groups were hydrolyzed to obtain ionizable carboxylic acid structures.

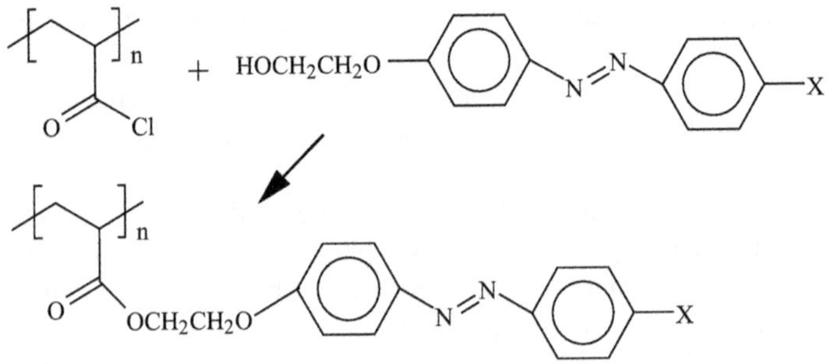

When irradiated by 365 nm UV light, azo polyelectrolytes showed a significant photochromic effect. The contact angles of water on the surfaces of spin-coated films of two of the polymers decreased upon UV irradiation. The extent of the photoinduced contact angle changes depends on the type of the azo chromophores and the degree of functionalization. Self-assembled multilayers of the azo polyelectrolytes were formed by a layer-by-layer adsorption method. A significant photochromic effect from *cis-trans* isomerization of the azo chromophores was observed for the multilayers. Photoinduced contact angle changes of water on the self-assembled multilayers were also observed. [76]

A reversible photocontrol of wettability of polymeric materials is possible by a technique developed by Irie and Iga [54] When a copolymer of butyl methacrylate and (2-hydroxyphenyl)-α-(4-vinylphenyl)benzyl alcohol is irradiated with ultraviolet light, there is a large increase in the contact angle and wettability of the material. This reverses back to the original structure in the dark [54]:

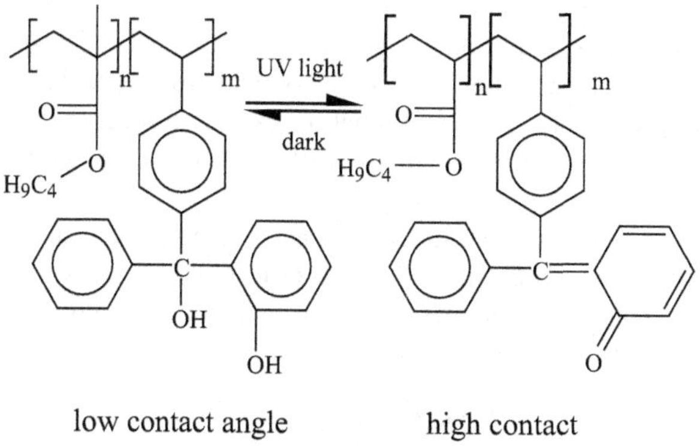

low contact angle   high contact

Polyacrylamide gels that contain triphenylmethane leucocyanide groups (1-4 mol %) exhibit reversible photostimulated dilation [72] upon irradiation with ultraviolet light of 270 nm. The weight of the gels increases as much as 13 times and the size expands approximately 2.2 times in each dimension as it swells with water. The dilated gel de-swells in the dark. The cycles of dilation and contraction of the gel by photoirradiation can be repeated several times. The gel expansion, however, is suppressed by addition of salts, such as NaCI or KBr. The salt effect and semiquantitative theoretical considerations of the behavior of ions suggests an osmotic pressure differential between the gel inside and the outer solution. This is produced by photodissociation of triphenylmethane leucocyanide groups contained in the gel network and is the main driving force of the photostimulated gel dilation angle.

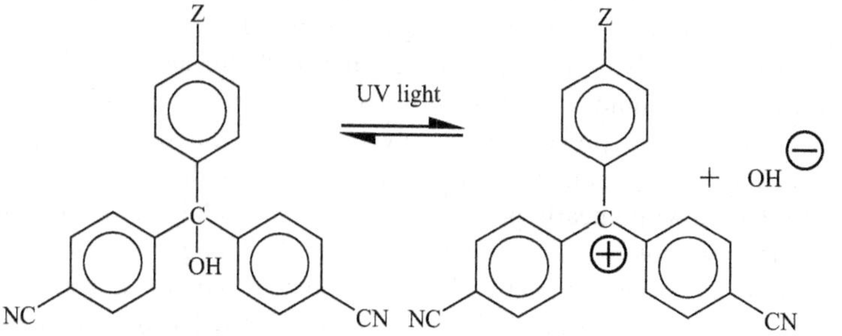

5.2.2.6. Changes in Optical Activity

Conformationally restricted copolyaramides containing a combination of 4,4'- azobenzene, 1,4-phenylene, and chiral 2,2'-binaphthylene main-chain segments were demonstrated to exhibit photoresponsive chiroptical behavior. [77] This is due to multiple *trans-cis*-isomerization reactions triggered within the polymer backbones. In contrast to the more randomly constructed counterparts, copolymers that possess periodic backbone structures undergo reversible, wavelength-dependent inversions in their optical rotations in response to multiple UV- light/visible- light illumination cycles. Similar behavior is also observed for smaller oligomers fitted with periodic arrangements of the monomer units.[77]

Also a new chiral menthone-based acrylic monomer, capable of *trans-cis* isomerization, was synthesized and used to obtain a homopolymer and a series of comb-shaped copolymers with a nematogenic monomer. [78] This monomer was illustrated as follows:

The copolymers are capable of forming chiral nematic phases.. UV irradiation significantly changes the step of the helix in copolymer samples with planar orientation.

5.2.3. Liquid Crystalline Alignment

In the past ten or more years glass transitions and mobility in confined polymer systems aroused much interest. It was influenced by need for alignment in liquid crystalline flat panel displays, because in these displays films of polyimides are widely used . The surfaces are usually treated to produce uniform alignment of the liquid crystals into suitable "pretilt" angles. The treatments consisted of rubbing process with velvet fabrics. Search for new methods, however, led to development of molecular structures that undergo alignment upon irradiation with linearly polarized UV light. [79-81] Polymer-stabilized liquid crystals are low-molar-mass liquid crystal. Their bulk alignment or their texture is stabilized by a polymer network. Such polymer network is usually in low concentration. [82, 83] Several types of polarized-light-induced liquid crystalline aligning of molecules were reported in the literature. [89]

One photoalignment material is poly(vinyl cinnamate). The polymer and its copolymers were reported to have the ability to align in thin films in the direction perpendicular to the axis of the linearly polarized ultraviolet light. This photoalignment mechanism has not been fully elucidated at present. A draw back to using poly(vinyl cinnamate) and its copolymers is a low glass transition temperature. As a result, they remain mobile after treatment and chain orientation. Other materials with higher T_g are, therefore, needed. Among the most promising ones are polyimides. They form liquid crystal alignment layers in flat-panel displays and possess good optical transparencies, adhesion, heat resistance, dimensional stability, and are good insulators.

There are various reports in the literature about preparations of soluble photoreactive polyimides with cinnamate chromophore side groups. Thus, it was reported by Lee et al., [84] that they prepared a photoreactive polyimide with cinnamate chromophores side groups :

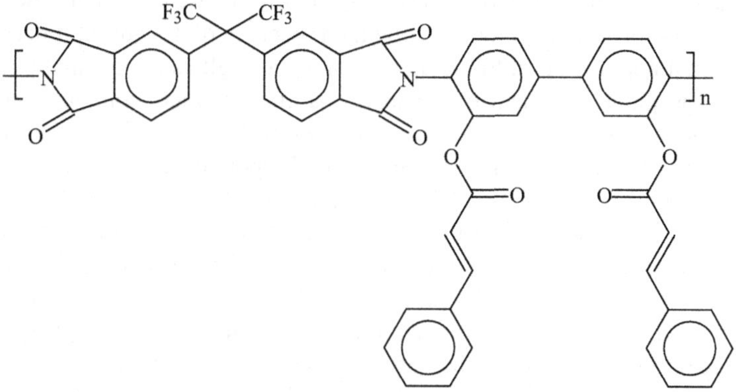

This polyimide is claimed to be thermally stable up to 340 °C and has a glass transition temperature of 181 °C. Also, it was demonstrated that the cinnamate chromophores, upon irradiation with linearly polarized ultraviolet light, undergo both photoisomerization and dimerization. In addition, the light exposure induces anisotropic orientation of the polymer main chains and of the cinnamate side groups in the film. The irradiated films align homogeneously the nematic liquid crystal molecules along one direction at an angle of 107 ° with respect to the polarization. The liquid crystalline alignment was found to be thermally stable up to 200 ° C.

It was also reported [84] that photoreactivity of side-chain liquid-crystalline polymers can align liquid crystals both in a parallel mode or perpendicularly, depending on the degree of the photoreaction of the polymers. Presumably this particular polymer can multi-photoalign the liquid crystal pattern without a change of the direction of the linearly polarized UV light. The chemical structure of such an aligning polymer is depicted as follows:

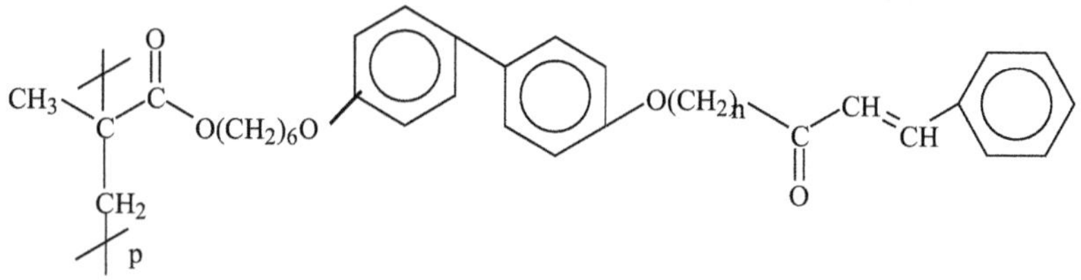

where n = 2 or 6

It was concluded, [84] therefore, that the liquid crystals align both parallel and perpendicular to the incident E direction on the photocrosslinked polymer film by changing the degree of the reacted cinnamoyl group. That can be controlled by irradiation time. A bias-tilt angle between the liquid crystals director and the substrate is also realized by controlling the irradiation angle of the light. [84]

Another approach to liquid crystalline alignment is based on photo-isomerization of azo compounds in polymeric materials or as part of the polymer structure. [85-87] In recent years, investigation of the use of azobenzene-containing polymers for liquid crystalline alignment became quite thorough because of the potential application in holographic storage as well as optical and photonic use. [88-90] The photoalignment of liquid crystalline polymers containing azobenzene groups has an advantage of local variation of the orientation order due to pixel wise irradiation. This is a process that is reported to takes place via angular dependent excitation, a series of *cis-trans* photoisomerization cycles, and rotational diffusion within the steady state of the photoreaction. This results in the photochromic side group becoming oriented perpendicularly to the electric field vector of the incident light and establishing an oblate order in the films.

In addition, an approach to photo aligning liquid crystals by a two step procedure was reported by Stumpe, Gimenez, and coworkers. [92] In this procedure, films of liquid crystalline copoly(methacrylate)s with photochromic azobenzene and benzanilide side groups and related terpolymers (containing additional dye side groups) are oriented by irradiation. Linearly polarized visible or, alternatively, UV light is used. The light-induced ordering of the azobenzene group is connected with cooperative alignment of the nonphotochromic co-mesogenic and dye side groups even below the glass transition temperature. They noted that the light-induced orientational order generated in the glassy state is significantly amplified by the subsequent annealing of the irradiated films at temperatures within the mesophases of co- and terpolymers. Amplification factors up to 30 were found in this series of materials. By this technique, the required dose or the irradiation time is significantly reduced by the optimization of light-induced and thermal processing, respectively. This procedure results in anisotropic monodomain films of nematic and smectic polymers.

Another investigation also reported photo and thermal isomerizations of azobenzene based groups that are covalently bound to copolymers of methacrylates [101]:

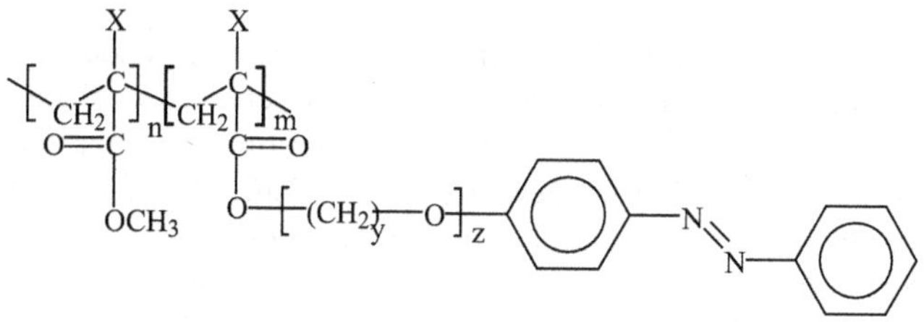

Here, kinetic data indicates that when these liquid crystalline polymers are exposed to visible light, the azo groups orient themselves from a more or less random arrangement to a perpendicular one. This is a photoselection process. [101]

It was also reported that photoinduced alignment in cast thin films of liquid crystalline polymers, of 6-[4-(4-ethylphenyl)diazenylphenyloxy]hexyl methacrylate,

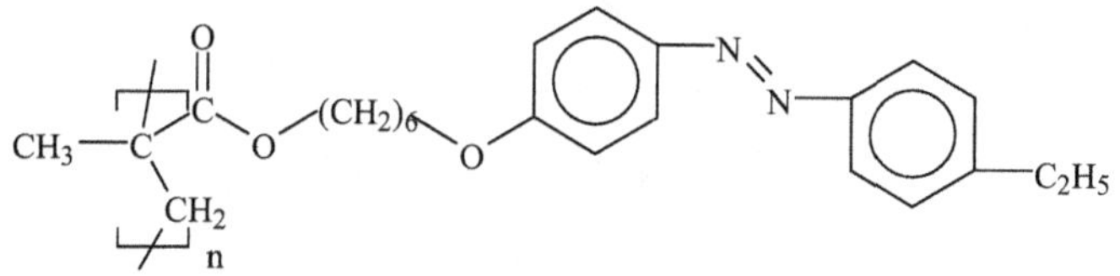

and poly[4'-[6-(methacryloyloxy)hexyloxy]-4-cyanobiphenyl] were studied. [93] When the polymer films with *cis*-azobenzenes were subjected to irradiation with linearly polarized light at 436 nm, biaxiality of the azobenzene moieties was induced during the in-plane alignment process. On the other hand, when they were exposed to unpolarized light at 436 nm, three-dimensional manipulation of the polymer liquid crystals was achieved. Examination of the three-dimensional alignment behavior by UV and FTIR spectroscopies, and by polarizing optical microscopy techniques found that the azobenzene moieties become aligned along the propagation direction of the irradiation light through repetition of *trans-cis-trans* isomerizations. The induced anisotropy is large (an approximate order parameter as 0.36), reversible, and stable (more than 5 months at room temperature). A nonphotoactive mesogen, cyanobiphenyl, also underwent three-dimensional reorientation efficiently together with the azobenzene moieties due to the cooperative motion. [93] :

Han *et al.*, [91] also studied exposure of films of liquid crystalline polymers with azobenzene side chains to linearly polarized light of 436 nm that results in successive occurrences of uniaxial in-plane orientations, followed by out-of-plane orientations of azobenzenes. [91] Two kinds of orientation modes were observed. These are possibly extreme cases, when linearly polarized light with the electric vector parallel to the xz-plane comes along the z-direction. One is the uniaxial in-plane orientation of the azobenzene with a dipole moment parallel to the x-axis from the *x*- to y-direction and the other is out-of-plane (homeotropic) one toward the z-direction. Marked dependence of photo-orientation processes on film temperatures was observed. In-plane orientation was generated in the glassy state. Photoorientation at higher temperatures, slightly below the transition temperature between smectic and nematic phases, gives rise to distinct transformations from in-plane orientation at the early state to successive out-of-plane reorientations. [91] These orientations can be illustrated as follows:

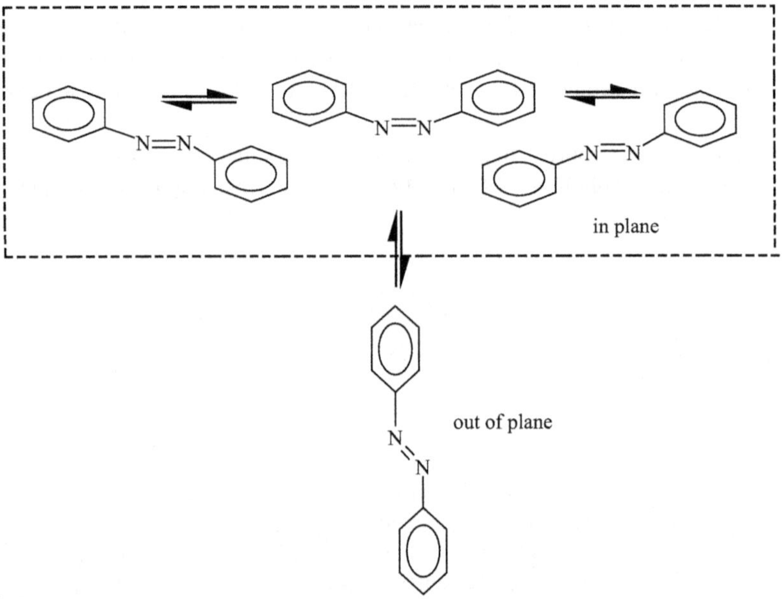

Zebger *et al.*, [94] reported that they prepared a liquid crystalline polyester with azobenzene side groups:

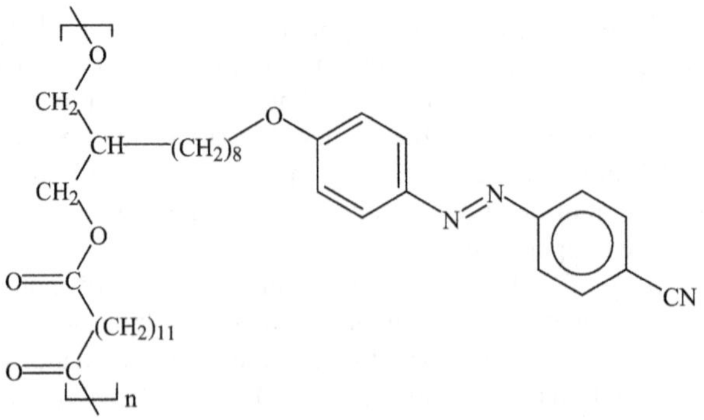

They studied the effects of irradiation of the films of this polymer with linearly polarized red light after photochemical pretreatment. Conventional photo-orientation processes are usually done with linearly polarized green, blue, or UV light of an argon laser (514, 488, or 351 nm). The found that orientation of 4-cyano-4'-alkoxyazobenzene side groups can take place parallel to the electric field vector upon irradiating of the films with linearly polarized red light. The polyester is characterized by smectic and nematic phases and a strong tendency to form J-aggregates. It requires a pretreatment by irradiation with ultraviolet light or an exposure to visible light of high power density to produce a certain concentration of the *trans*-isomer, which destroys any initial orientation order and J-aggregates. The orientation process is cooperative, whereas the light-induced orientation of the photochromic moiety causes an ordering of the alkylene spacers and even of the main-chain segments into the same direction. Zebger et al., [94] concluded that the most probable mechanism of this two-step process is the angular-selective transformation of the bulky *trans*-isomers to the rod like *cis*-isomeric formed by the red light. The aligned *cis*-azobenzene side groups become strongly J-aggregated. Very high values of dichroism of about 0.8 and birefringence of about 0.3 were generated as a result of this combination of the photoinduced orientation process and the thermotropic self-organization, which takes place simultaneously under the irradiation conditions. The process results in a uniaxial prolate order of the film, whereas conventional photo-orientation leads to a biaxial oblate order. [93] Thus, the direction of photo-induced orientation and the type of the three-dimensional orientation order can be controlled by the wavelength of the irradiating linearly polarized light. This can also be done in the same film of a smetic polyester with 4-cyano-4'-alkoxyazobenzene side group. [93]

Also, it is interesting to note that in a diblock copolymer that contains azobenzene groups [94]:

photoinduced birefringence shows microdomain structures, characteristic of diblock copolymers. These structures hinder photoalignment of the azobenzene mesogenic groups. [94] It is important, therefore, to take into consideration any possible steric hindrance that might occur in the preparation of the polymeric material.

Zhao and coworkers [95, 96] reported that an azobenzene polymer network can also optically align ferroelectric liquid crystals. This was done by dissolving two chiral dimethacrylate and one chiral diacrylate monomers containing azobenzene groups in a commercial ferroelectric liquid crystal host. The monomers were illustrated as follows:

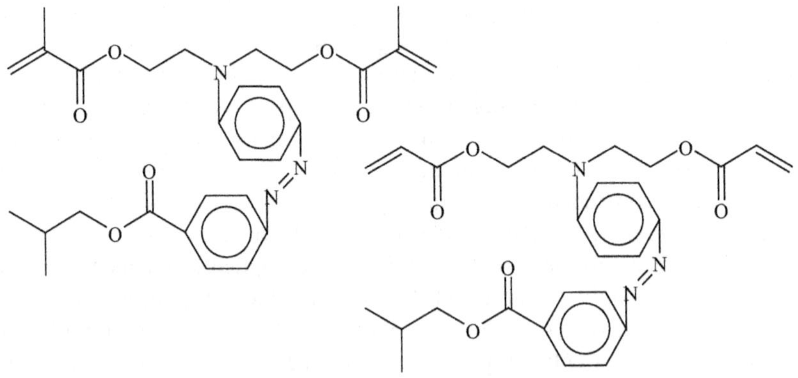

methacrylate   acrylate

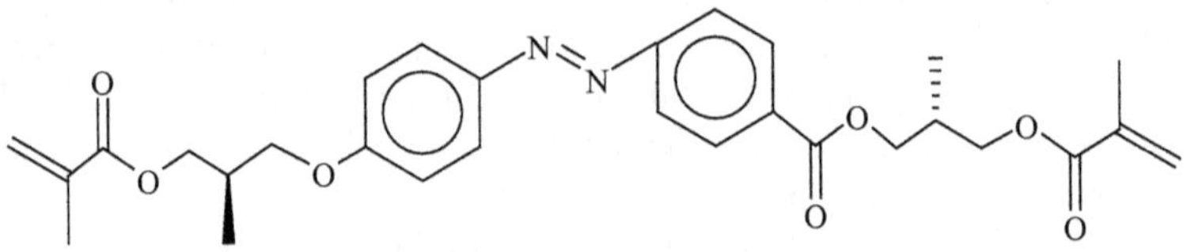

The monomers were then thermally polymerized and simultaneously irradiated with linearly polarized light. Two of the monomers were able to induce bulk alignment of the liquid crystals in direction perpendicular to the polarized light. Monomer #1 was effective in concentrations as low as 1%. It was also concluded from the experimental evidence that the photoinduced bulk alignment of the ferroelectric liquid crystals may take place by a mechanism that is different from one that takes place in achiral azopolymers.

In addition, Zhao and coworkers [97] reported photoinduced alignment of ferroelectric liquid crystals using azobenzene polymer networks of polyethers and polyepoxides. Bulk alignment was achieved by polymerizing several divinyl ethers and diepoxide monomers bearing an azobenzene moiety. Here too, thermal polymerizations were conducted in solution within the ferroelectric liquid crystals, while exposing the reaction mixture to linearly polarized irradiation. The monomers can be shown as follows:

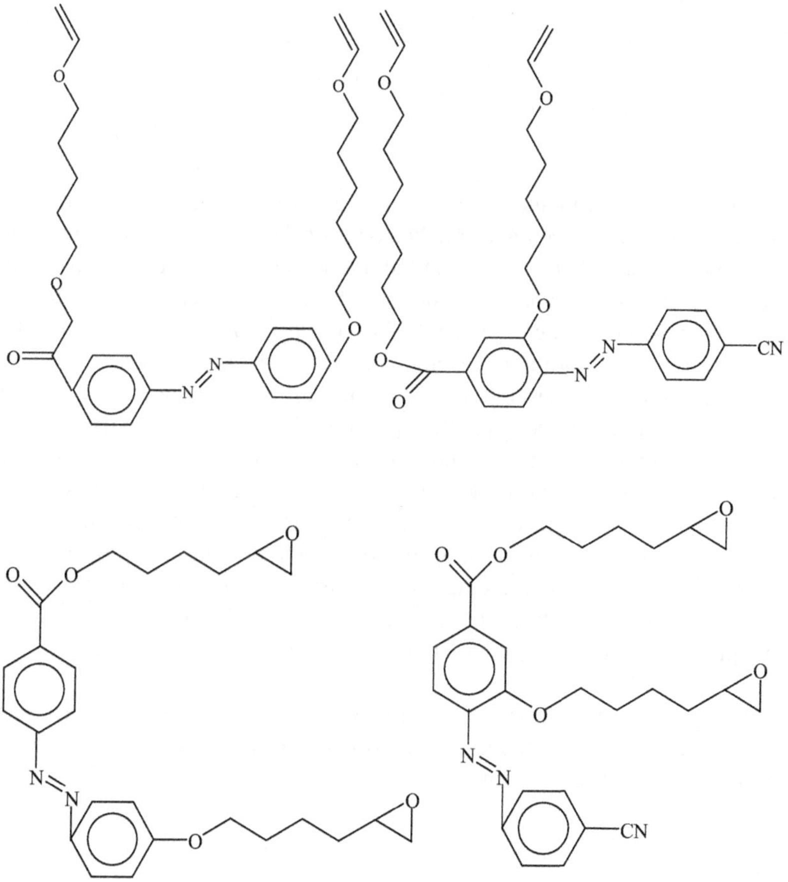

Polymerization of these monomers was achieved by cationic mechanism. The monomers were also found capable of inducing and stabilizing bulk alignment of the liquid crystals. Zhao and coworkers [97]concluded, however, that the mechanism of action might be different from the one obtained with chiral azobenzene polymethacrylates. Instead, the results suggest to them that the mechanism might be based on formation of an anisotropic azobenzene polyether or polyepoxide network.

In still another, subsequent work, Zhao and coworkers [98] used block copolymers composed of polystyrene and liquid crystalline azobenzene-containing polymethacrylate copolymers as a model system:

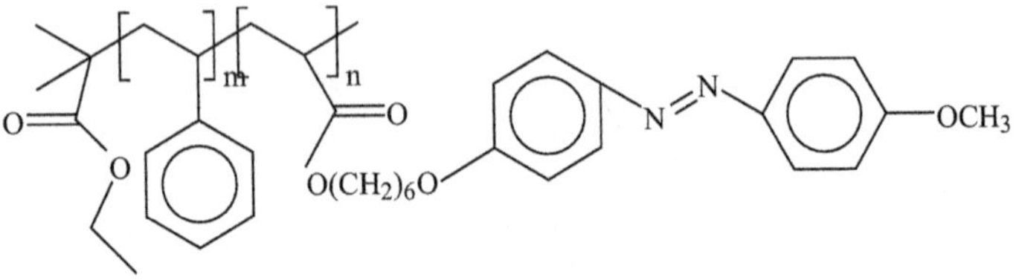

to investigate the confinement effects on the photoalignment, photochemical phase transition, and thermo-chromic behavior of the azobenzene polymer. The study showed [98] that when confined in the microphase-separated domains in the diblock copolymers, the azobenzene polymer behaves differently than it does as a homopolymer free from confinement. The confinement effects are manifested by (1) decreased photoinduced and thermally enhanced orientation of azobenzene mesogenic groups in different aggregation states, (2) slower transformation from a liquid crystalline phase to the isotropic state triggered by the *trans-cis* photoisomerization and slower recovery of the liquid crystalline phase after the thermally induced *cis-trans* back-isomerization, and (3) severely reduced and even suppressed changes in the aggregation states of azobenzene groups on heating, which is at the origin of the thermo-chromic property. The common cause of these confinement effects is the restriction imposed by the confining geometry on either an order-disorder or a disorder-order reorganization process involving the motion and rearrangement of azobenzene groups. [98]

Synthesis of liquid crystalline photoactive triblock poly(butyl acrylate) thermoplastic elastomer copolymers with methacrylate-based azobenzene-containing side-chain was reported. [99]

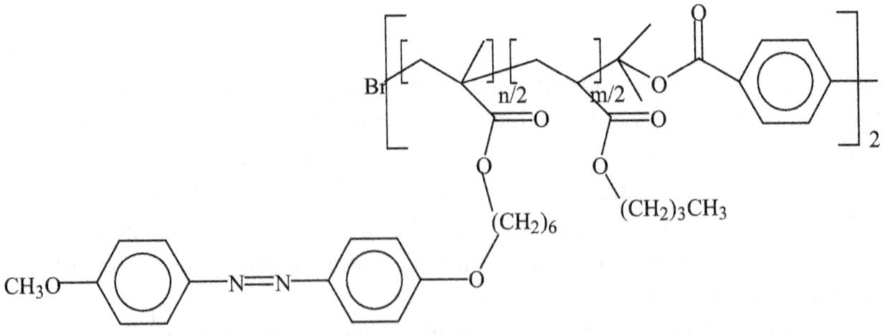

It was found that when the solution-cast films are stretched at temperature greater than T_g, liquid crystalline microdomains can support part of the elastic extension and, at the same time, deform to result in a long-range orientation of azobenzene mesogens. The liquid crystal orientation is retained in the relaxed film at $T < Tg$, which creates a thermoplastic elastomer whose glassy microdomains contain oriented azobenzene mesogens. The reversible *trans-cis* photoisomerization of the azobenzene chromophore can be used to modulate the mechanically induced orientation. Also, Zhao and coworkers [100] reported recently preparation of liquid crystalline ionomers with azobenzene mesogens and three metal ions, copper ++, zinc ++, and manganese++. They found that the metal ions exert a significant effect on the photoinduced orientation of azobenzene mesogens. Ionomers with zinc and manganese ions showed increased photoinduced birefringence and higher diffraction efficiency of birefringence gratings obtained by excitation with spatially modulated light polarization. On the other hand, ionomers with copper showed the opposite effect. The copolymers can be illustrated as follows [100]:

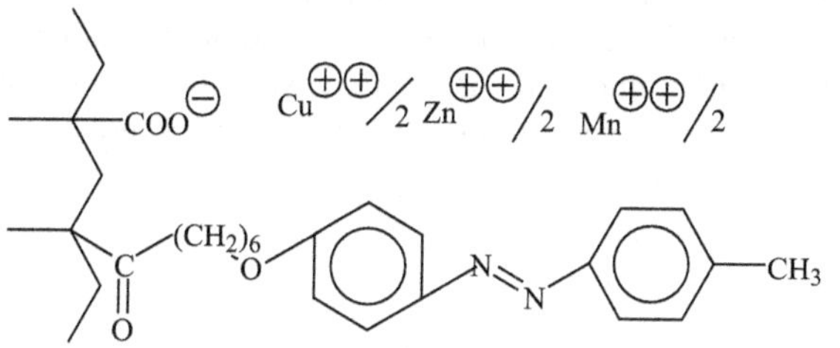

Kawatsuki *et al.*, [101] described two different techniques for thermally enhanced photoinduced reorientations of liquid crystalline polymethacrylate films that contain 4-methoxyazobenzene side groups with various alkylene spacer lengths:

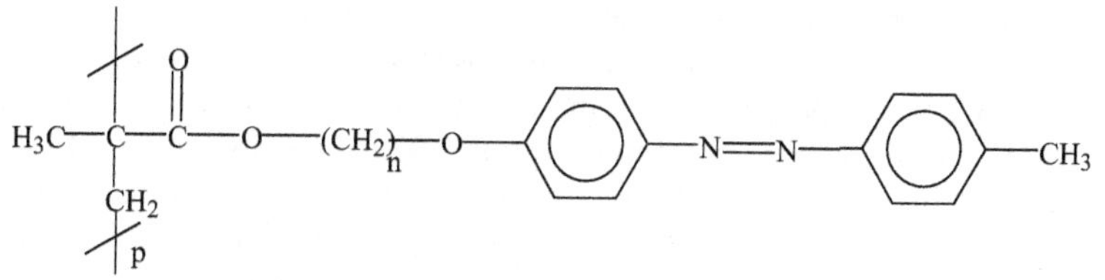

One is the classical method based on photoselection and rotational diffusion in the steady state of the *cis-trans* photoisomerization, which includes irradiating with linearly polarized 365 nm light and subsequent annealing in the liquid crystalline temperature range of the film. The spacer length and exposure dose significantly influence the reorientation direction. In-plane reorientation perpendicular to the electric vector of the linearly polarized 365 nm light generates polymer liquid crystals with short spacers, but out-of-plane reorientation with a small degree of in-plane reorientation is predominant for polymer liquid crystals with long spacers. The other is an alternative technique based on an axis-selective *trans* to *cis* photoisomerization using linearly polarized 633 nm light and thermally enhanced molecular reorientation. Regardless of the spacer length, a high order of in-plane reorientation parallel to *cis* of linearly polarized 633 nm light can be accomplished. [101]

Zettsu and Seki reported [102] preparation of a group of azobenzene containing polymers that can be used in photoinduced surface relief formations. These are soft liquid crystalline azobenzene containing copolymers of acrylate with methacrylate monomers bearing oligo(ethylene oxide) chains. The copolymers display a smectic liquid crystal phase at room temperature. After preexposure to ultraviolet light, thin films of the liquid crystalline polymers show highly sensitive photoinduced material transfer to generate the surface relief structures. The typical exposure dose required for full polymer migration is as low as 50 mJ cm^{-2}

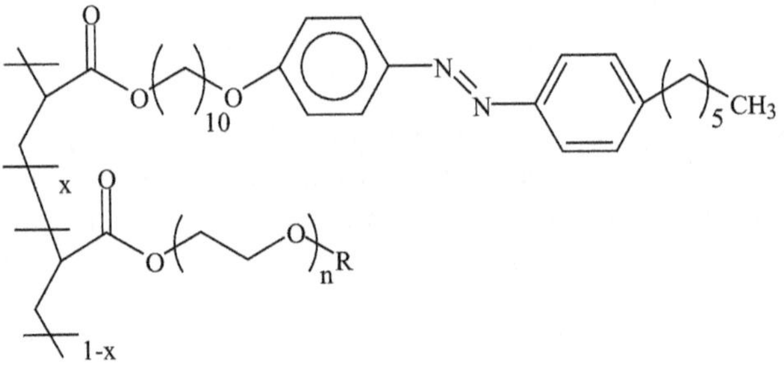

The inscribed surface relief structures can be rapidly and fully erased either by irradiation with incoherent nonpolarized ultraviolet light or by heating close to the clear point of the soft liquid crystalline polymers. It is also possible to chemically crosslink the polymers with mixed vapors of hydrogen chloride and formaldehyde after surface relief inscription. This results in a drastic improvement of the shape stability, maintaining the structure at high temperatures up to 250 °C. After cross-linking, the *trans*-to-*cis* photo-isomerization readily proceeds without any modification of the surface morphology and can, therefore, be applied to the photoswitchable alignment of nematic liquid crystals. [103]

5.3. Photonic Crystals

There is interest in photonic band gap crystals, due to their potential ability to increase light wave guiding efficiency, to increase the efficiency of stimulated emission processes, and to localize light. Original chemical approach to fabricating large face-centered-cubic photonic band gap crystals was based on self-assembly of highly charged, monodisperse colloidal particles into crystalline colloidal arrays. The arrays subsequently self-assemble due to long-range electrostatic repulsions between particles. [103,104]

These crystalline colloidal arrays are complex fluids that consist of colloidal particles give Bragg diffraction pattern in ultraviolet, visible, or near-infrared light, depending on the spacings of the colloidal particle array. More recently, robust semisolid photonic crystal materials were formed by polymerizing a hydrogel network around the self-assembled crystalline colloidal arrays. [105]

This new photonic crystal material can be made environmentally responsive so thermal or chemical environmental alterations as a result of a material volume changes, thereby altering the arrays of the photonic crystal plane spacings and diffraction wavelengths. [105]

It is claimed [106] that these photonic crystals can be useful in novel recordable and erasable memories and/or display devices. Information is recorded and erased by exciting the photonic crystal with ~ 360 nm UV light or ~ 480 nm visible light. The information recorded is read out by measuring the photonic crystal diffraction wavelength. Ultraviolet light excitation forms *cis*-azobenzene cross-links while visible excitation forms *trans*-azobenzene cross-links. The less favorable free energy of mixing of *cis*-azobenzene cross-linked species causes the hydrogel to shrink and blue-shift the photonic crystal diffraction. [106]

5.4. Photoconducting Polymers

Unless polymers contain long sequences of double bonds, they are fairly good insulators, particularly in the dark. Nevertheless, a number of common polymers show measurable increase in conductivity, when irradiated with light. When polymeric materials, like poly(vinyl fluoride), poly(vinyl acetate), poly(vinyl alcohol), or poly(N-vinyl carbazole) are exposed to light they develop charged species. The species can migrate under an electric field and thus conduct electricity. When poly(N-vinyl carbazole) is doped with photosensitizers or compounds that form charge-transfer complexes, the photosensitivity can be increased and even extended into the visible region of the spectrum. Since discovery in 1957 that poly(N-vinyl carbazole) has photoconductive properties, there has been increasing interest in the synthesis and study of this and other polymeric materials with similar properties that allow various photonic applications. Related polymers are presently utilized in photocopiers, laser printers, and electrophotographic printing plates.

Photoconductive polymers can be *p*-type (hole-transporting), *n*-type (electron-transporting) or bipolar (capable of transporting both holes and electrons). To date, all practical photoconductive charge-transporting polymers used commercially are *p*-type.

Poly(vinyl carbazole) and other vinyl derivatives of polynuclear aromatic polymers , such as poly(2-vinyl carbazole) or poly(vinyl pyrene) have high photoconductive efficiencies. These materials may take up a helical conformation with successive aromatic side chains arranged parallel to each other in a stack. In such an arrangement the transfer of electrons is facilitated. Also, it is believed that the primary mechanism for poly(vinylcarbazole) charge carrier generation is due to excitation of the carbazole rings to the first excited singlet state. This polymer absorbs ultraviolet light in the 360-nm region and forms an exciton that ionizes in the electric field. The excited state, by itself is not a conductive species. The addition of an equivalent amount of an electron acceptor, like 2,4,7-trinitrofluorenone shifts the absorption of this polymer into the visible region by virtue of formation of charge transfer states. The material becomes conductive at 550-nm. This associated electron-positive hole pair can migrate through the solid polymeric material. Upon dissociation of this pair into charged species, an electron and a positively charged hole, the electron becomes a conductive state. To achieve this additional energy is required and can be a result of singlet-senglet interaction, [107] singlet-triplet interaction, [108] singlet-photon interaction, [107] triplet-photon interaction, [109] and two-photon interaction. [109] Kepler carried out fluorescence quenching studies and concluded that the migration of the exiton is the most probable energy transfer mechanism of poly(vinyl carbazole). [109] He, furthermore, suggested that the exiton can visit one thousand monomer units during its lifetime. [109] This is a distance of about 200 Å .

Kang and coworkers [111] also explored steady-state and pulsed photo-also *trans*-polyphenylacetylene films doped with inorganic and organic electron acceptors, particularly iodine and 2,3-dichloro-5,6-dicyano-*p*-benzoquinone. They concluded that charge transport occurs by a band like mechanism that is modulated by shallow electron traps in the undoped polymer and by trapping the charge-transfer complex in the doped polymer. [111]

Guillet [112] states that photoconductivity σ is equal to the current density J divided by the applied field strength ε, where J is amperture/unit electrode area. This is related to the number of negative-charge carriers (usually electrons) per unit volume, and p, is the number or positive charge carriers (or positive holes) per unit. [112]

$$\sigma = J/\varepsilon = n\,e\mu_n + p\,e\mu_p$$

where e is the charge on the electron, and μ_n and μ_p are the mobilities of the negative and positive carriers, respectively. Photoconductivity and mobility of the charge carrying species can be determined from a relationship [112]:

$$\mu = d^2/Vt$$

where d is the thickness of the film, V is the applied voltage and t is the carrier drift time. The photoeffect is evaluated in terms of the effective gain, G. It represents the number of generated carriers reaching the external circuit per unit time, compared with the number of photons absorbed at the same time [112]:

$$G = J_p\,/\,eI_0(1 - T)A$$

where J_p is the photocurrent, e is the electric charge, I_0 is the number of incident photons per $cm^2\,s^{-1}$, T is the optical trasmittance of the film, and A is the area of the sample that is being illuminated.

5.4.1. Photoconductive Polymers Based on Carbazole

As stated above, the primary mechanism for charge-carrier generation in poly(vinyl carbazole) appears to be due to the excitation of the carbazole rings to their excited singlet states. [112] While the singlet excited state is not a conductive species, the conductivity is believed to be the result of an associated electron-positive hole pair migrating through the solid polymeric material. Dissociation of the electron pair produces a separate electron and a positive hole in such a way that the electron ends up in the conducting state. [112] This requires acquisition of more energy. One way that can be accomplished is by excitron-surface interaction. [113] Regensburger published an absorption spectrum, fluorescence spectrum and photocurrent spectrum for a 7.6 μ films of poly(N-vinyl carbazole). [114] The shape of the response of the photoconductor to the wavelength of the light flash is very close to the shape of the absorption spectra. Bauser and Klopffer explain this as a result of interaction of singlet excitons with trapped holes. [115]

Lyoo used a low-temperature initiator, 2,2'-azobis(2,4-dimethyl-valeronitrile) to polymerize N-vinyl carbazole in a heterogeneous solution in a mixture of methyl and *t*-butyl alcohols. [116] The polymer that formed has the M_n molecular weights $>3 \times 10^6$. The author emphasized that this method provides ultrahigh-molecular-weight polymer and conversions greater than 80%.

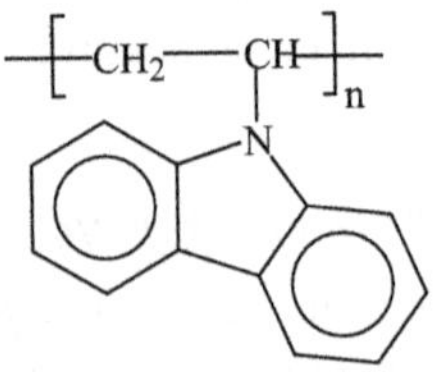

The optical transparency of poly(vinyl carbazole) films produced by this room temperature process appears to be quite high, although transparency decreases at high conversions. In film form, this material is useful for photoconductors, charge-transfer complexes, and electroluminescent devices. The higher polymer molecular weight typically enhances film mechanical properties. [116]

Harhold and Rathe [127] reported that they have prepared poly(9-methylcarbazole-3,6-diyl-1,2–dipenylvinylene). The polymer ($M_n = 10,000$) was formed by dehalogenating polycondensation of 3,6-bis(α,α-dichlorobenzyl)-10,9-methylcarbazole with chromiun (II) acetate. This polymer was found to be also highly photoconductive. It's dark conductivity increases by doping it with arsenic pentafluoride. [117]

Photoconductive polymers and copolymers were also synthesized by Haque and coworkers [118] from 2-(9-carbazolyl)-1-propenyl vinyl ether:

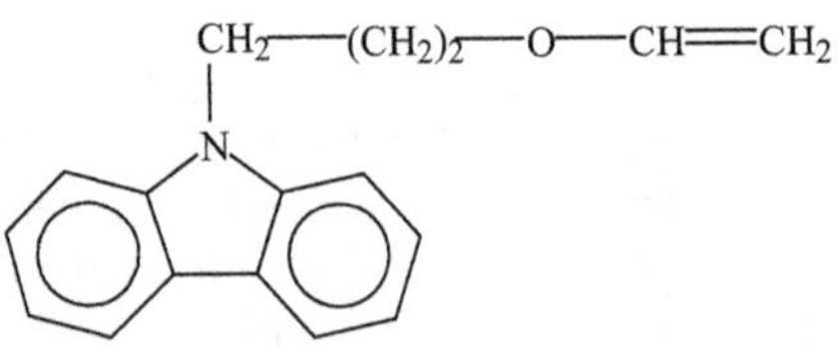

The polymers and copolymers form by a cationic polymerization mechanism, using boron trifluoride or ethylalumionum dichloride as the catalysts. [118]

Charge transfer complexes also form from poly(vinyl carbazole), that acts as the donor, and poly[2-(methacroyloxy)ethyl-4,5,7-trinitro-9-oxo-2-fluorene-carboxylate], that acts as the acceptor [145]:

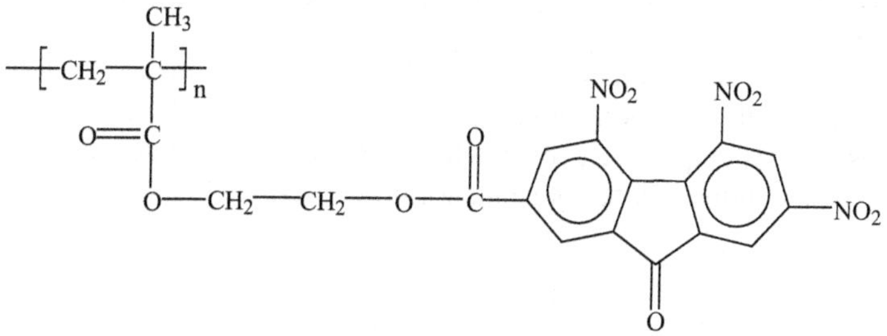

The required mole ratio of components in the complex is 1:1. [145]

Charge transfer complexation occurs in a similar manner in poly(2-carbazolylethyl acrylate) molecularly doped with 2,4,7-trinitrofluorene. Quantum efficiency of the hole propagation of the copolymer with the 0.05 to 1.0 molar ratio of trinitrofluorene to carbazole chromophores is higher than in the corresponding trinitrofluorene and ethyltrinitrofluorene doped homopolymer of poly(2-carbazolylethyl acrylate). [119]

Kim and Webber studied delayed emission spectra of poly(vinyl carbazole) that was doped with dimethylterephthalate and pyrene. [120] On the basis of their results they concluded that at room temperature dimethylterephthalate does not completely quench the triplet excitation state of poly (vinyl carbazole). They also concluded that phosphorescent states of poly(vinyl carbazole)-dimethylterephthalate are similar, implying a significant charge-transfer character in the former.

In 1985 Mitra [55] prepared polymeric triphenylmethane dyes based on condensation polymers such as polyesters and polyurethanes:

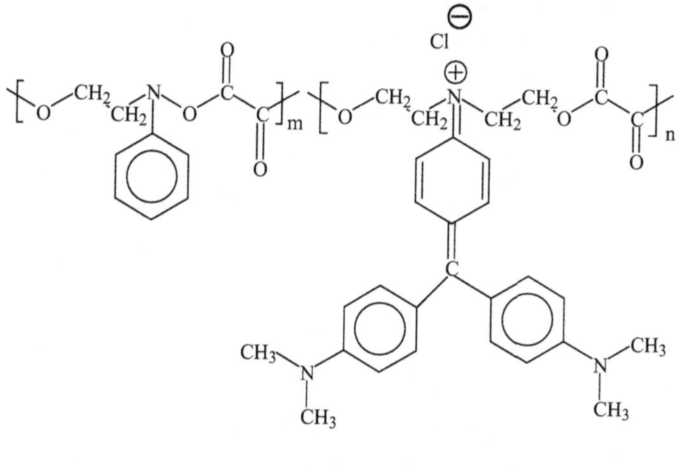

These dyes were then shown to sensitize photoconductivity of poly(N-vinyl carbazole).

Polyacetylene derivatives exhibit unique characteristics such as semi conductivity, high gas permeability, helix inversion, and nonlinear optical properties. [121] Attempts were made, therefore, to incorporate carbazole into polyacetylene in hope of attaining enhance properties. [122]

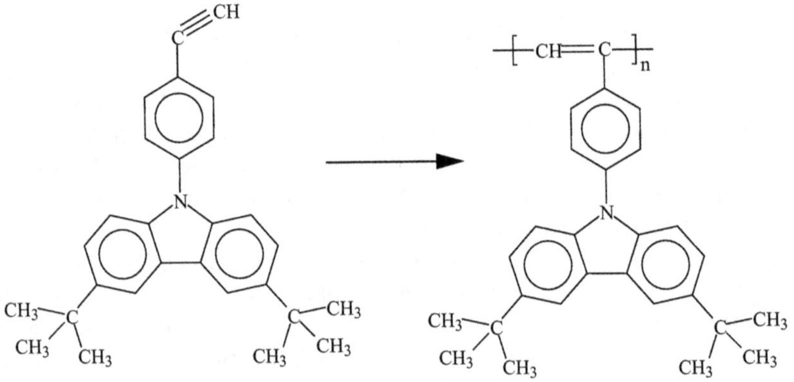

It was found [122] that the current conducted by this polymer during irradiation is 40 to 50 time higher than it is in the dark On the other hand, the electron mobility of the di-*t*-butylcarbazolepolyacetylene (shown above) is lower than in poly(vinyl carbazole). This was attributed to the bulkiness of the butyl groups.[122]

Siloxanes with pendant carbazole groups were synthesized by Strohriegl [73] by the following technique:

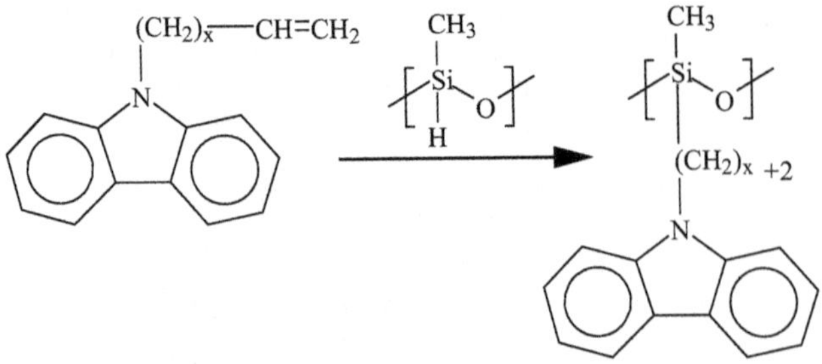

The material, however, did not turn out to be photoconductive. The photoconductivity of copolymers was also investigated. Copolyacrylates with pendant donor and acceptor chromophores, such as 2-N-carozolylethyl acrylate

4,5,7-trinitrofluorenone-2-ethyl acrylate do exhibit photogeneration properties.
[73] These copolymers:

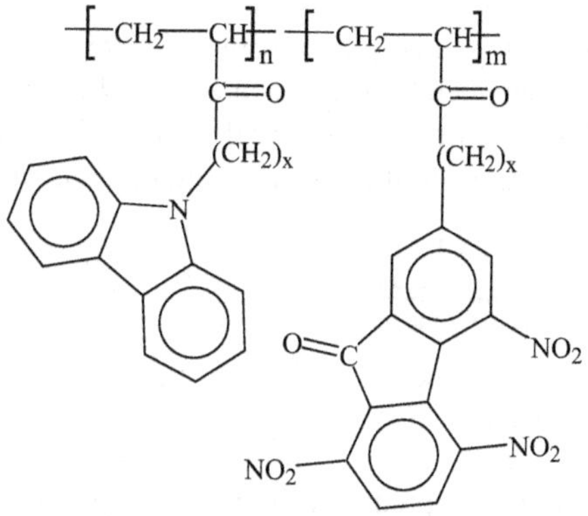

can be prepared by free-radical copolymerization of the appropriate monomers.
Photoconductivity in the visible is obtained by charge transfer complexation.

Similar work was done earlier by Natansohn, [119] who copolymerized N-
methyl, 1,3-hydroxymethyl carbazolyl acrylate with acryloyl-3'-hyroxypropyl-
3,5-dinitrobenzoate:

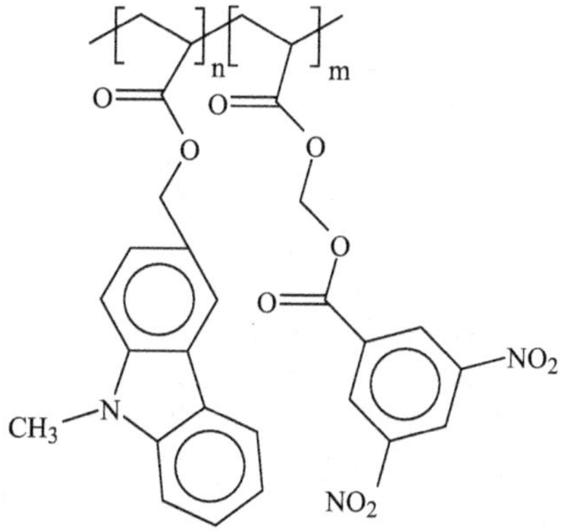

Illumination of the copolymer samples induces a certain degree of ionization
accompanied by proton transfer. There is a permanent increase in the quantity of
radicals generated by light. Another number of radicals apparently forms during
illumination but disappears in the dark. [119]

It was reported [123] that a polymer formed by condensation of N-(3-
isopentyl)-3,6-diformyl-carbazole   and   4-14-bis(-aminoethyl)aminophenyl-
azo]nitrobenzene yields a new polyazomethine, carbazole-azo polymer. The
product is soluble in organic solvents. The polymer possesses carbazole

moieties and azo type nonlinear optical chromophores in the backbone. It shows high photoconductivity and nonlinear optical properties.

5.4.2. Photoconductive Polymers that Are Not Based on Carbazole

When spiropyran is incorporated into plasticized poly(vinyl chloride) membranes and placed between two identical NaCl solutions and irradiated with ultraviolet light for long periods the potential decreases. [125] This was shown by Ryba and Petranek to be a result of the spiran ring opening up [125]:

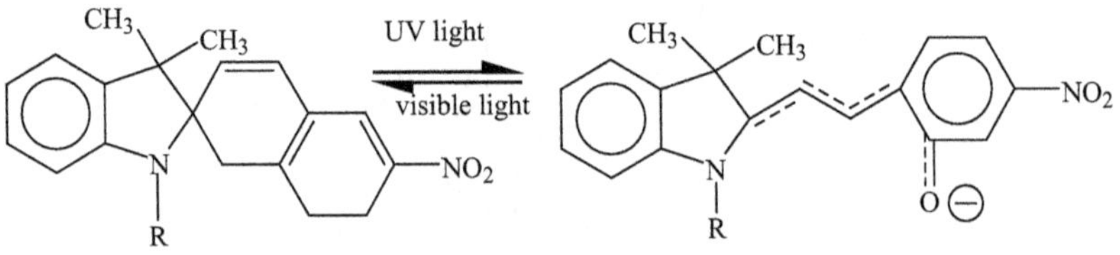

When the irradiation is interrupted and the membrane is irradiated with visible light, the potential returns to its original value, because the ring closes back to the spiran structure. [125]

The addition of electron donors, like dialkyl aniline, to Kapton polyimide film,

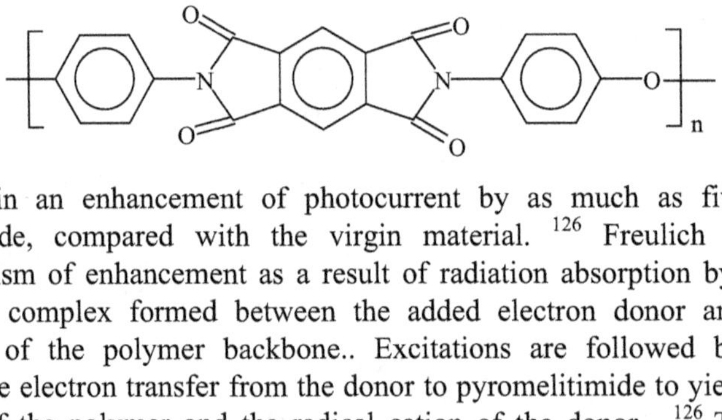

results in an enhancement of photocurrent by as much as five orders of magnitude, compared with the virgin material. [126] Freulich explains the mechanism of enhancement as a result of radiation absorption by the charge-transfer complex formed between the added electron donor and the imide portion of the polymer backbone.. Excitations are followed by rapid and complete electron transfer from the donor to pyromelitimide to yield the radical anion of the polymer and the radical cation of the donor . [126] These species undergo rapid back electron transfer. In other words, the dialkyl aniline donates one of the unpaired electrons in a typical photoreduction reaction to the carbonyl group. The reaction is reversible and the photoconduction is by a mechanism of the ion radical returning to the ground state.

Hoofman et al., [127] studied the thermochromic, solvatochromic, and photoconductive properties of 3-, 4-, 6-, and 9-poly[(butoxycarbonylinethyl urethane)-diacetylenes] that result from changes from rod (red or blue) to coil (yellow) conformations of thepolymer backbone. Photoexcitations of the solutions of these polymers in the rod state result in a large transient photoconductivity, while only very small conductivity signals are observed in the coil state. The thermochromic shift that occurs in going from the rod state at room temperature to the coil state at 65 °C is accompanied by a decrease in the photoconductivity. The large conductivity signal in the rod state is attributed to the formation of mobile charge carriers possibly via interchain charge transfer within aggregates. The decay of the photoconductivity is nonexponential and extends to microseconds.

Wong *et al.*, [128] reported that they prepared a soluble rigid-rod organometallic polymer containing electron-donating and electron-withdrawing *trans*-[-Pt-(tibutylphosphine)$_2$-acetylene-R-acetylene-]$_n$ (where R = bithiazolediyl) groups. The polymer was formed by cuprous iodide-catalyzed dehydrohalogenation reaction. The electron-donating and electron-withdrawing properties of the thiazole ring confer solubility to the polymer. This polyacetylene is luminescent with a singlet emission peak at 539 nm and photoconducting. The glass transition temp. of the polymer is 215° and it shows relatively good thermal stability. The π-conjugation of the ligands extends into and through the metal core and the absorption peaks show a significant red-shift of 17-26 nm compared to the bithienyl counterparts due to the presence of the electron-withdrawing imine nitrogen atoms. [128]

Kimura et al., [129] reported applying organic photochromic compounds to photochemical switching of metal-ion complexation and ionic conduction by combining photochromism with metal-ion binding property of crown ether derivatives. They synthesized vinyl polymers, incorporating a crown spirobenzopyran moiety at the side chain:

The crown spiropyran in the electrically neutral form can bind an alkali metal ion with the crown ether moiety. At the same time the spirobenzopyran portion isomerizes to the corresponding merocyanine form photochemically. The zwitterionic merocyanine form of crowned spirobenzopyran moiety brings about a significant change in the metal-ion binding ability. This prompted the authors to apply the compound to photoresponsive ion-conductive materials. They observed that the ion-conductivity was increased by ultraviolet light and decreased by visible light. [129]

Chan and coworkers [130] prepared polystyrenes and poly(methyl methacrylate)s that contain metal complex cores:

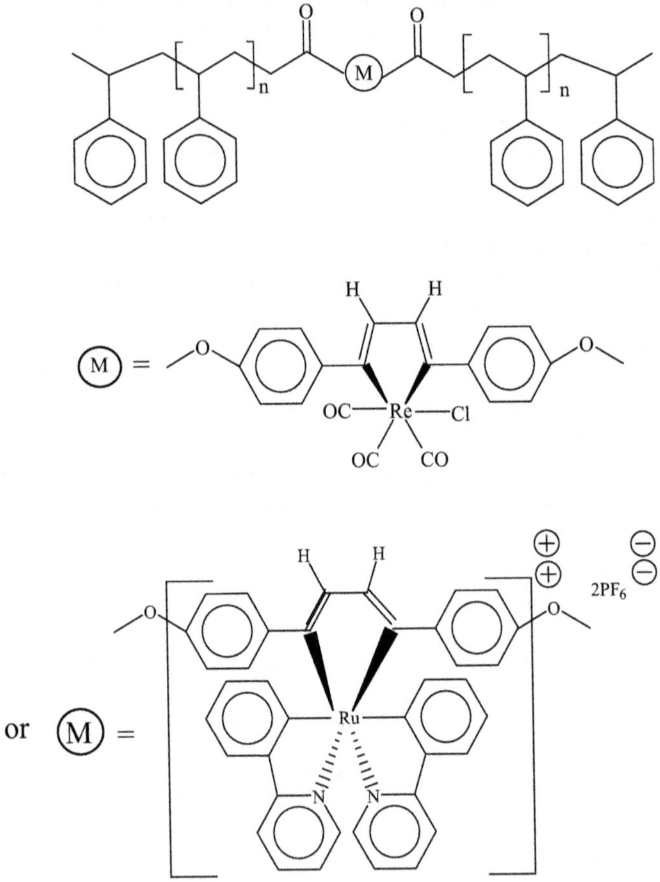

When the polymers are doped with a hole transporting triphenylamine, an enhancement in photoconductivity in the visible region is observed. This suggests that the metal complexes serve as photosensitizers instead of charge carriers. [150] Chan and coworkers [130] observed an electric field dependent charge separation process in these polymers. It is described well by the Onsager's theory of charge germinate recombination. This theory assumes that some fraction of absorbed photons produce bound thermalized electron-hole pairs that either recombine or dissociate under the combined effects of the Coulombic attraction and the electric field. The photogeneration efficiency is given as the product of the quantum yield of thermalized pair formation and the pair dissociation probability:

$$\Phi(r_0, E) =$$

$$\Phi_0\left[1 - \left(\frac{eEr_0}{kT}\right)^{-1} \sum_{g=0}^{\infty} I_g\left(\frac{e^2}{4\pi\epsilon_0\epsilon_r kTr_0}\right) I_g\left(\frac{eEr_0}{kT}\right)\right]$$

where I_g is a recursive formula given by

$$I_{g+1}(x) = I_g(x) - x^{g+1}\exp(-x)/(g + 1)$$

where $I_0(x) = 1 - \exp(-x)$, Φ is the primary yield of thermalized bound pairs, r_0 is the initial thermalization separation between the bound charges, ϵ_r is the relative permeability, and E is the applied electric field strength.

Jenekhe et al.,[132] reported exciplex formation and photoelectron transfer between several n-type {electron accepting) π-conjugated rigid-rod polymers and donor triarylamine molecules. In particular they reported an investigation of an n-type conjugated polymer poly(benzimido azobenzophe- nanthroline ladder). [132] No evidence was observed by them of a ground state charge transfer or any strong interactions between the conjugated polymer pairs Transient absorption spectra of a blend of thin films in the 420-730 nm region were obtained at various time delays following photoexcitation at 532 nm. Dramatically enhanced photoinduced bleaching in the 430-480 nm region was observed. Jenekhe and de Paor propose that they observed enhanced photobleaching in the blends and that it is a consequence of photoinduced electron transfer. [132] The electron transfer was illustrated as follows:

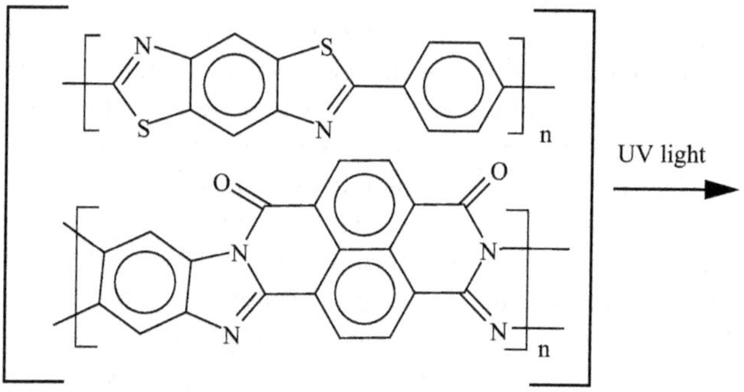

UV light

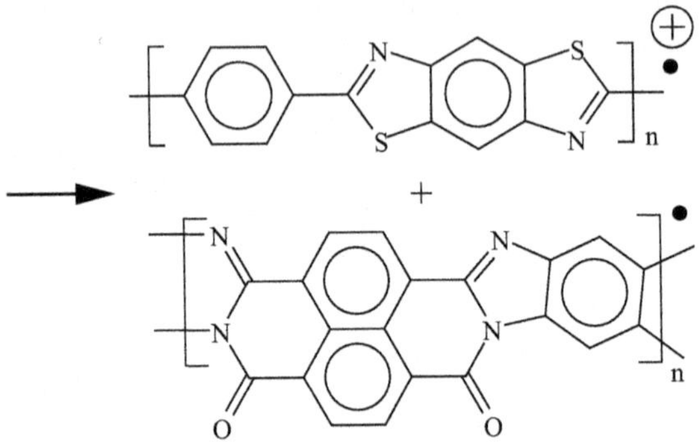

Molecular materials, such as 2-(2-hydroxyphenyl)benzoxazole and 2-(2-hydroxyphenyl)-benzothiazole, which contain intramolecular hydrogen bonds are known to undergo excited state(charge transfer) intramolecular proton transfer upon photoexcitation.

5.5. Proton or Electron Transfer in the Excited State

In 1966 Conwll and Mizes [131] observed that exposure to light frequency beyond the absorption edge in polymers such as poly(p-phenylene vinylene), and polythiophene is expected to create excitons, some of which decay radiactively. Based on the similarity of the emission from thin films and oligomers of poly(p-phenylene vinylene), it was concluded that the exciton is a single chain. Polyimides like Kapton owe their resistance to degradation at least in part to charge transfer character. This generated considerable interest in the role of various types of charge transfer complex states in the photostabilization of polyimides The photoinduced electron transfer processes of π-conjugated polymers are currently of wide interest in view of their importance to the fundamental understanding of the electronic structure and properties of the materials and applications such as photodiodes, photovoltaic cells, electro-photographic photoreceptors, and molecular electronic devices.

Jenckhe at al., [133] explored the effects of molecular size, extent of conjugation, concentration quenching, and competition with excimer formation on intramolecular proton transfer as well as on the electro-luminescent device potential of polymeric materials. Materials like 2-(2-hydroxyphenyl)-benzoxazole and 2-(2-hydroxyphenyl)benzothiazole contain intramolecular hydrogen bonds and are known to undergo excited state charge transfer reactions and intramolecular proton transfer upon excitation. They reported, however, that the results of their studies with model compounds and several polymers, whose structure:

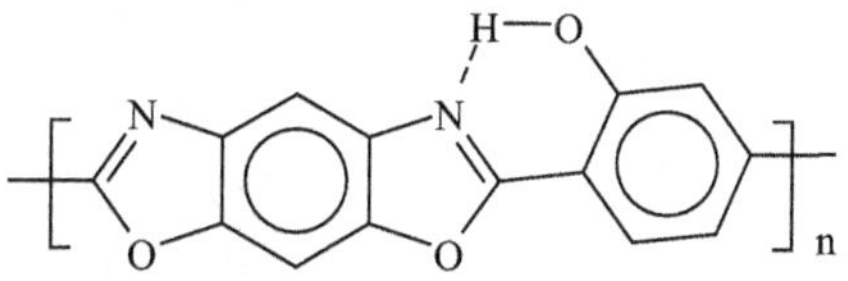

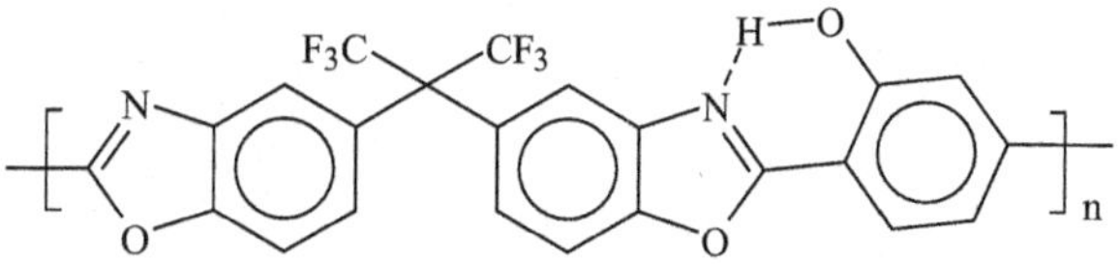

suggest that incorporation of such compounds into the backbones of stiff-chain conjugated polymers does not ensure the occurrence of intamolecular proton transfer upon photoexcitation. [133]

Chang *et al.,* [134] noted from the literature that the excited-state intramolecular proton transfer is a form of phototautomerization occurring in the excited states of the molecules possessing cyclic intramolecular or solvent-bridged hydrogen bonds. Since this process is fast enough to accompany the short-lived first excited singlet state, the electronic excitation of a normal form (N) in the ground state yields a proton-transferred excited-state tautomer (T*) of the much lower energy. This T* relaxes radiatively or nonradiatively to the metastable ground state T, which reverts to the normal form (N) via reverse proton transfer. Therefore, the different absorbing and emitting species result in a large Stokes shift. Chang *et al.,* [134] pointed out that until now, the phenomenon has been studied mainly in the solution systems for small organic molecules like pyridine derivatives, etc. In most cases, however, phototautomeric fluorescence is rather weak for practical applications at ambient temperature due to more effective internal conversion or broken hydrogen bond by torsional motion. To address these limitations, Chang *et al.,* [134] have considered a polymeric system with semirigid prototype polyquinoline groups that contains intramolecular H bonds between the enol and imine groups. The polymer was illustrated as follows:

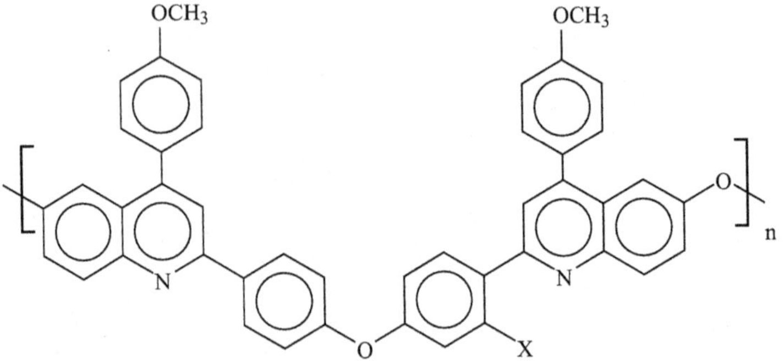

Where the substituent X on the polymer is an OH group with an intramolecular H bond, the polymer in solution shows dual blue and red fluorescent bands. The authors claim that a semiempirical molecular orbital calculation of this emission supports the phenomenon of excited-state intramolecular proton transfer.

Earlier, Nishijima and coworkers were able to generate hydrogen by visible light irradiation with the aid of ruthenium complexes and colloidal platinum stabilized by viologen bearing polymers in aqueous media. [124] The polymer was prepared as follows:

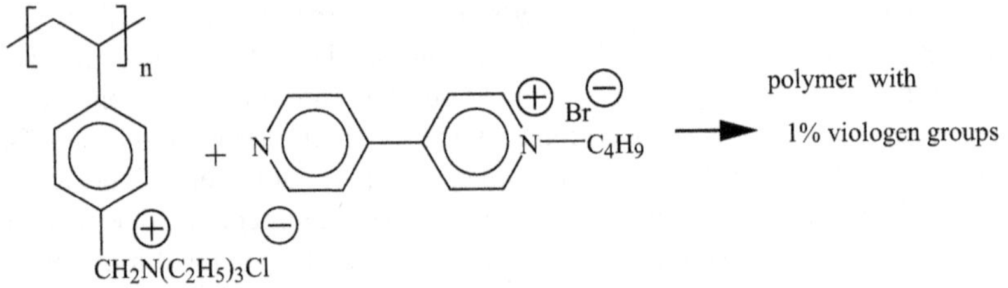

Hydrogen generation requires a photosensitizer, an electron donor, an electron mediator and a multi-electron redox catalyst. To obtain the polymer the reaction mixture must be refluxed in methanol for 20 hours. This yields a polymer with 1% viologen groups. In the system, water donates the protons and electrons are supplied by ethylenediamine tetraacetic acid disodium salt. [124] The viologen groups transport the electrons to the multi-electron redox catalyst (2,2'-bipyridine)ruthenium(II), $Ru(bpy)^{++}$.

5.6. Reversible β-Sheet Transitions

A convenient, fully reversible method for ligation of deoxyo-ligonucleotides was reported by Saito and co-authors. [135] 5- vinyldeoxy -

uridine at the 5'-end of the DNA strand and a thymine residue at the 3'-end were used to form a *cis*-syn [2+2] adduct between the vinyl group and the 2,3-double bond by simple irradiation at 366 nm. This adduct formation is completely reversible by irradiation at 302 nm to restore the original compounds

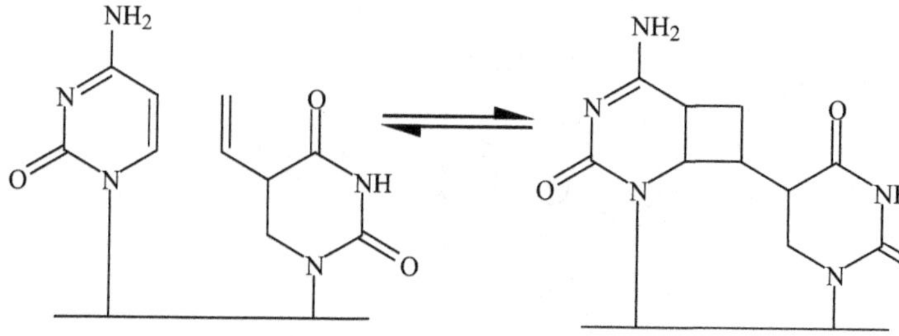

The authors demonstrated this coupling principle by simultaneous ligation of several single-stranded deoxyoligonucleotides on a DNA template to produce a longer DNA. The process is then quantitatively reversed to give the original oligonucleotides.

5.7. Polymers that Shed Small Molecules, Ions, or Free-Radicals Upon Irradiation

Into this category of materials belong polymers that shed molecules, ionic species, or free-radicals when irradiated with light of the appropriate wave length. One such material is poly[*p*-(formyloxy)styrene]. It can be prepared from poly(*p*-hydroxy styrene). The polymer was reported to decarbonylate smoothly in ultraviolet light [136]:

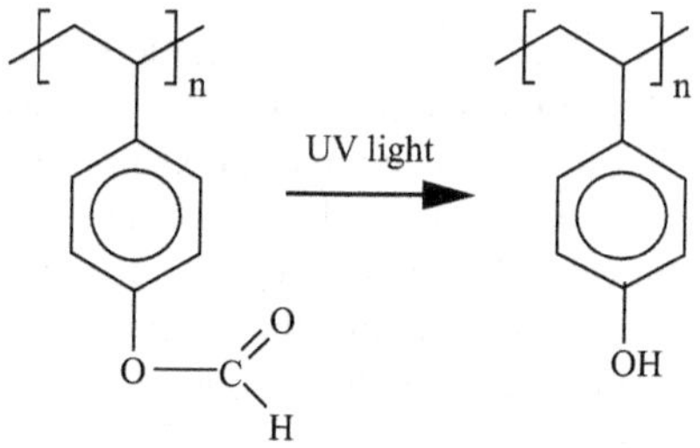

The reaction can take place in solid state or in solution. It was reported that the material can be used in photolithographic processes and that it produces sharp images. [136] Such images can be developed positively or negatively.

Crivello and coworkers [137] prepared aromatic polyimides bearing sulfide linkages in the main chain by condensing sulfur containing dianhydride with aromatic diamines. They also condensed sulfur containing aromatic diamines with dianhydrides. Phenylation with diphenyliodonium salts converts the diarylsulfide groups into triarylsulfonium salts. The resultant photosensitive polyamides undergo main chain cleavage during photolysis with ultraviolet light. These polymers are offered as positive high temperature photoresists. [137] The decomposition upon UV light illumination can be illustrated as follows:

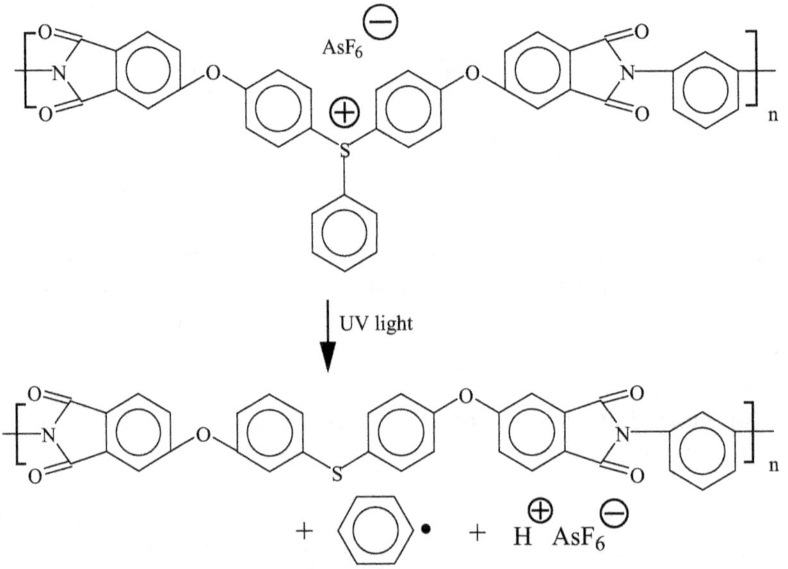

The details of the decomposition mechanism of sulfonium salts upon irradiation with ultra-violet light are described in Chapter 2. The illustration above is a simplified version.

Preparations of multifunctional photopolymers with both pendant phenacyl ester and vinyl ether groups were reported. [138] These materials were synthesized by one-pot method for the reaction of poly(methacrylic acid) with 2-chloroethyl vinyl ether, This was followed by the reaction with phenacyl bromides using 1,8-diazobicyclo-[5.4.6]-undecene-7. It was found that the pendant phenacyl ester groups cleave upon photoirradiation to give pendant carboxylate groups. The produced carboxylate groups react with pendant vinyl ether groups to form acetal linkages. [138]

5.8. Color Changes

Two distinct triplet excimetric emissions were observed in the solid films of poly[N-(vinyloxy)-carbonyl)-carbazole]. [139] Phosphorescence spectra of this polymer and its monomers in solutions are essentially the same as those observed for poly(N-vinylcarbazole). The fluorescene spectrum, however, shows a significant shift to the blue. [139]

Copolymers of vinyl viologens, such as N-vinylbenzyl-N'-alkyl and benzyl and N-(χ-methacroyloxy)propyl-N'-propyl-4,4'-bipyridinium dihalides and polar aprotic comonomers, such as N- vinyl-2-pyrrolydone or N.N'dimethylacrylamide exhibit a strong color formation upon irradiation. [140] One such copolymer can be illustrated as follows:

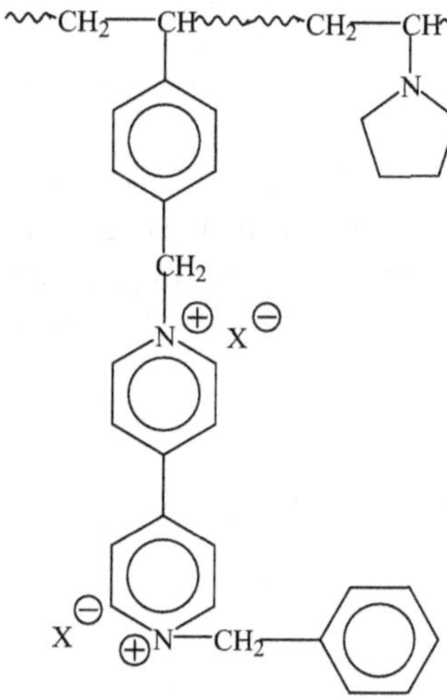

where, x - Br, $BF_4(SO_4)_{1/2}$. The photocolor developments of these aprotic copolymers in the film state are completely reversible and faster than in corresponding copolymers with protic comonomers, such as acrylamide. A characteristic absorption spectra of the copolymer is attributable to single radical cations. The mechanism of reversible color development was shown to be as follows [140]:

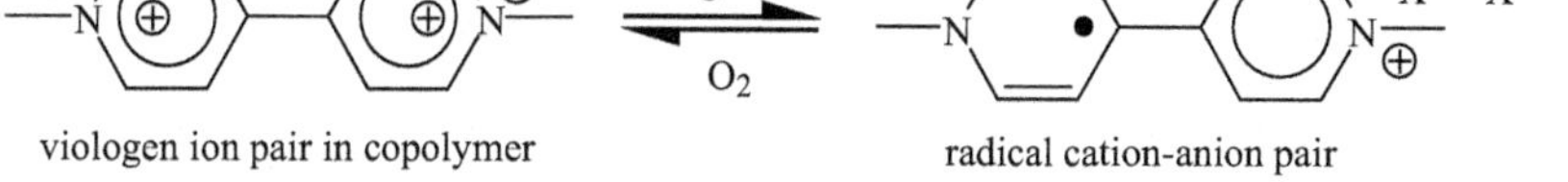

viologen ion pair in copolymer
(colorless)

radical cation-anion pair
(highly colored)

Photochromic compounds based on 1,2-bis(3-thienyl)-cyclopentene derivatives undergo reversible photocyclizations between their colorless ring-open and colored ring-closed forms when irradiated with light of appropriate wavelengths [141]:

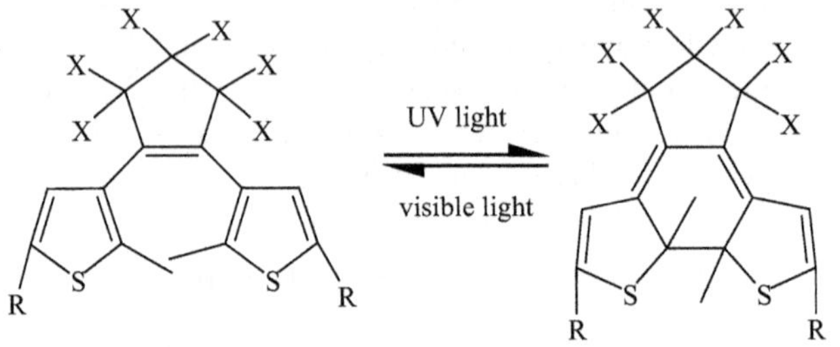

where X = F or H

This was utilized to form several novel homopolymers with 1,2-bis(3-thienyl)cyclopentene derivatives using ring-opening metathesis polymerization techniques. [141] The color forming cyclization reaction in the polymers can be illustrated as follow:

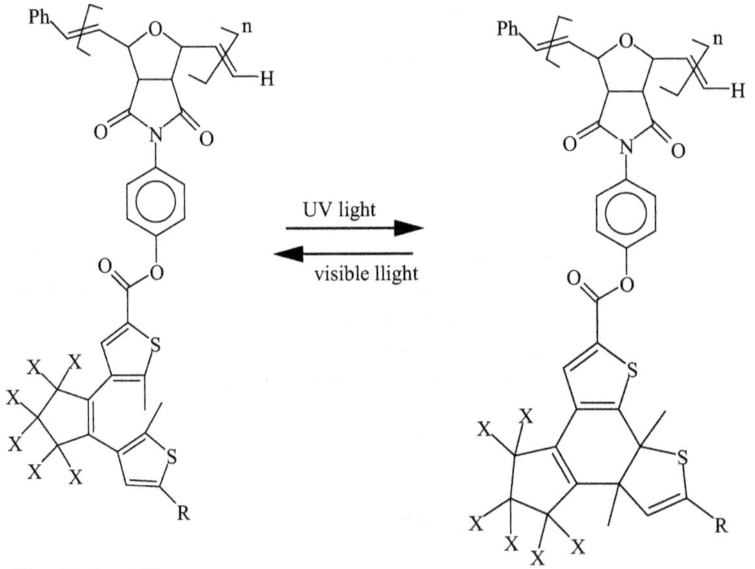

X = H R = Cl

colorless   colored

Lim, An and Park [166] synthesized polymers with similar photochromic groups for use in erasable optical memory and fast optical switching. The strong fluorescent of the polymer in a neat polymeric film shows that it can be photoswitched through highly efficient bistable photochromisms. They also demonstrated erasable fluorescence photoimaging on a spin coated film. The photoreaction of the polymer was illustrated as follows [142]:

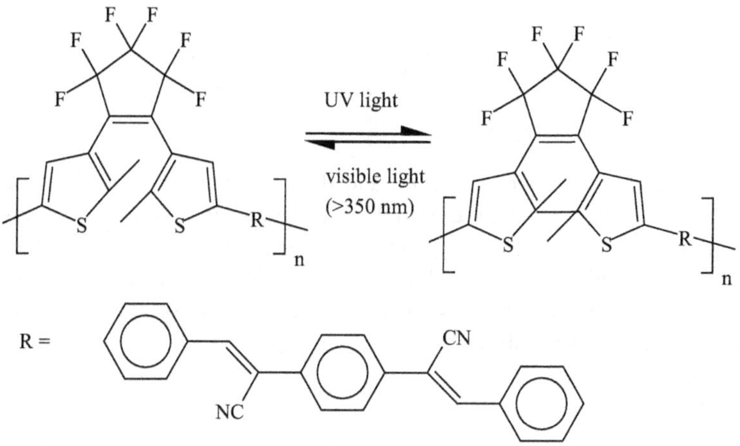

R =

5.9. Photoresponsive Gas Barrier Films

By incorporating photolabile groups into the polymeric backbone Eisner and Ritter reported that they were able to form photoresponsive membranes. [167] The polymer was illustrated as follows [143]:

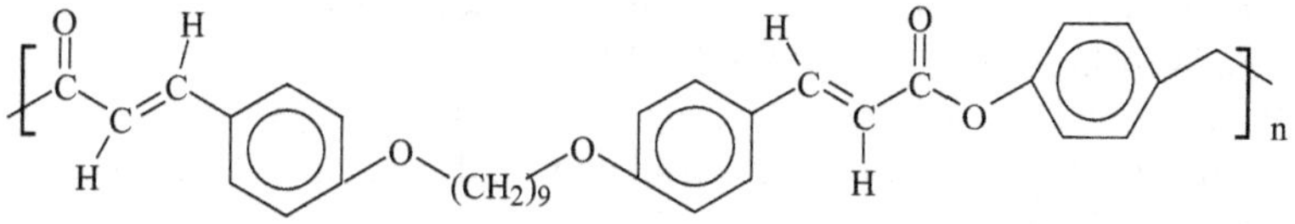

When this polymer as well as its precursor are blended with poly(vinyl alcohol) and poly(m-phenylene-isophthalamide) and formed into membranes, the permeation of gases through the membranes can be controlled by ultraviolet light irradiation. [143]

5.10. Photoisomerizable Rotaxanes

Okada and Herada reported formation of photoresponsive poly-ethylene glycol rotaxanes with anthracene derivatives stopper groups and cyclodextrin cyclic components. [99] Light-associated reversible changes of these polyrotaxanes were observed. [99] Thus, light irradiation with ultraviolet light or heat causes a rearrangement to a mixture of polyrotaxane together with a catenane.

5.11. Photoresponsive Shape Changes

Landlein, Langer, and coworkers, [144] reported grafting photoresponsive groups to copolymer films. The photoresponsive groups can be illustrated as follows:

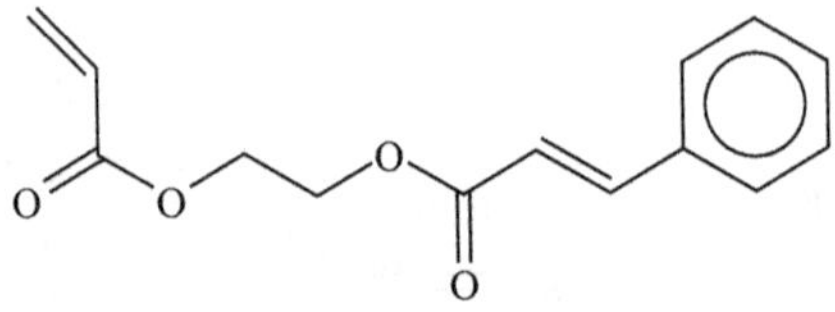

When the copolymer films are stretched and then illuminated with ultraviolet light, the light responsive groups crosslink, maintaining the elongated state of the films after the removal of the external stress. Subsequent exposure of these elongated films to ultraviolet light of a different wave length causes the cross-links to cleave and the film returns to its original shape. Formation of other temporary shapes is also possible. [144]

References

1. J.C. Craneand R.J. Guglielmetti, *Organic Photochromic and Thermochromic Compounds*, Plenum Press, New York, 1999
2. S.E. Webber, *Chem. Rev.*, **1990**, *90*, 1469
3. R.B. Fox and R.F. Cozzens, *Macromolecules*, **1969**, *2*, 181
4. A. Ikeda, A. Kameyama, T. Nishikubo, and T. Nagai *Macromolecules* **2001**, *34*, 2728-2734
5. T. Nagai, M. Shiimada, Y. Ono, and T. Nishikubo, *Macromolecules*, **2003**, *36*, 1786
6. N. Kawatsuki, H. Ono, H. Takatsuka, T. Yamanioto and O. Sangen, *Macromolecules,* **1997**, *30,* 6680
7. M. Sampei, K. Hiramatu, A. Kameyama, and T. Nishikubo, *Kobunshi Ronbunshu* **2000**, 57(9), 569-576
8. N. Kawashima, A. Kameyama, T. Nishikubo, and T. Nagai, *Reactive & Functional Polymers* **2003**, 55(1), 75-88
9. H. Ritter, R. Sperber, and C.M. Weisshuhn, *Macromol. Chem. Phys.*, **1994**, *195*, 3823; T. Deutschmann, and T. Ritter, *Chem. Phys.*, **2000**, *201*, 1200
10. A. Theis, B. Menges, S. Mittler, M. Mierawa, T. Pakula and H. Ritter, *Macromolecules,* **2003**, *36*, 7520; A. Theis and H. Ritter, *Macromolecules,* **2003**, *36*, 7552

11. W. Kuhlbrandt, and D.N. Wang, *Nature* **1991**, *350*, 130; W. Kuhlbrandt, D.N. Wang, and Y. Fujiyoshi, *Nature* **1994**, *367*, 614; G. McDermott, S.M. Prince, A.A. Freer, A.M. Hawthornthwaite-Lawless, M.Z. Papiz, R.J. Cogdell, and N.W. Isaacs, *Nature* **1995**, *374*, 517; W. Kühlbrandt, *Nature* **1995**, *374*, 497.

12. G.R. Fleming and G.D. Scholes, *Nature 2004*, *431*, 256; S. Jang, M.D. Newton, and R.J. Silbey, *J. Phys. Rev. Lett.* **2004**, *92*, 218301

13. C. Devadoss, P. Bharathi, and J.S. Moore, *J. Am. Chem. Soc.* **1996**, *118*, 9635; J.M. Serin, D.W. Brousmiche, and J.M. Fréchet, *J. Chem. Commun.* **2002**, 2605; T. Weil, E. Reuther, and K. Mullen, *Angew. Chem., Int. Ed.* **2002**, *41*, 1900; I. Yamazaki, N.Tamai, T.Yamazaki, A. Murakami, M. Mimuro, and Y.J. Fujita, *Phys. Chem.* **1988**, *92*, 5035; M.N. Berberan-Santos and J.M.G. Martinho, *J. Phys. Chem.* **1990**, *94*, 5847.

14. S. Kawahara, T.Uchimaru, and Murata, *Chem. Commun.* **1999**, 563; A.K. Tong, S, Jockusch, Z. Li, H.-R.Zhu, D.L. Akins, H.J. Turro, and J. Ju, *J. Am. Chem. Soc.* 2001,123, 12923

15. R.S. Deshpande, V. Bulovic, S.R. Forrest, *Appl. Phys. Lett.* **1999**, *75*, 888.; K.-Y. Peng, S.-A. Chen, W.S. Fann, *J. Am. Chem. Soc.* **2001**, *123*, 11388; J. Kim, D.T. McQuade, A. Rose, Z. Zhu, T.M. Swager, *J. Am. Chem. Soc.* **2001**, *123*, 11488.

16. P.L. Burn, A.B. Holmes, A. Kraft. D.D.C. Bradley, A.R. Brown, R.H. Friend, R.W. Gymer, *Nature* **1992**, *356*, 47; Z Yang; I. Sokolik, F.E. Karasz, *Macromolecules* **1993**, *26*, 1188.

17. K.T. Nielsen, H. Spangard, and F.C. Krebs, *Macromolecules*, **2005**, *38*, 1180

18. Y.-J. Cheng and T.-Y. Luh, *Macromolecules* **2005**, *38*, 456-4568

19. M.-C. Fang, A. Watanabe, M. Matsuda, *Macromolecules* **1996**, *29*, 6807

20. C.H. Hu, R. Oshima, and M. Seno, *J. Polymer Sci., Chem. Ed.,* **1988**, *26*, 1239

21. J. March, "Advanced Organic Chemistry", 3-ed., Wiley, New York, 1985

22. M. Onciu, C. Chirlac, S. Vlad, G. Stoica. G. Grigoriu, and D. Daniel, *Rev. Roum. Chim.* **1998**, 43(7), 641-648.

23. T. Nishijima, T. Nagamura, and T. Matsuo, *J. Polymer Sci., Chem. Ed.,* **1981**, *19*, 65

24. S. Mayer and R. Znetel, *Macromol. Chem. Phys.,* **1998**, *199*, 1675

25. R. Mruk and R. Zentel, *Macromolecules*, **2002**, *35*, 185

26. O. Pieroni, A. Fissi, N. Angelini, and F. Lenci, *Acc. Chem. Res.,* **2001**, *34*, 9

27. Zimmerman, G.; Chow, L.-Y.; Paik, U.-J. *J. Am. Chem. Soc.* **1958**, *80*, 3528-3531

28. Kumar, G.S.; Neckers, D.C. *Chem. Rev.* **1989**, *89,* 1915-1925
29. Delang, J.J.; Robertson, J. M.; Woodward, 1. *Proc. R. Soc. London, Ser. A* **1939**, *171,* 398.
30. F. Agolini and F.P. Gay, *Macromolecules,* **1970**, *3*, 249
31. C. Gilles, I. Faysal, and V.M. Rotello, *Macromolecules* **2000**, *33*, 9173
32. Y. Nishikata, J. Minabe, K. Kono, and K. Baba, Kazuo **Jpn. Kokai Tokkyo Koho JP 2000 109,719** (April 18, 2000)
33. A. Izumi, R. Nomura, and T. Masuda, *Macromolecules* **2001**, *34,* 4342-4347
34. A. lzumi, R. Nomura, and T. Masuda, *Chem. Lett.* **2000** 728-729.
35. A. lzumi, M. Teraguchi, R. Nomura, and T. Masuda, *Macromolecules* **2000**, 33, 5347-5352.
36. A. lzumi, M. Teraguchi, R. Nomura, and T. Masuda, *J. Polym. Sci., Polym. Chem.* **2000**, *38,* 1057-1063.
37. A. Izumi, M. Teraguchi, R. Nomura, and T. Masuda, *Macromolecules* **2000**, *33,* 5347-5352
38. D. Grebel-Koehler, D. Liu, S. De Feyter, V. Enkelmann, T.Weil, C. Engels, C. Sainyn, K. Muelleii, and F.C. De Schryver, *Macromolecules* **2003**, 36(3), 578-590
39. D. Grobel-Koehler, D. Liu, S. De Feyter, V. Enkelmann, T. Weil, C. Engels, C. Samyn, K. Mullen, and F.C. De Schryver, *Macromolecules,* **2003**, *36,* 579
40. D.M. Junge, and D.V. McGrath, *J. Am. Chem. Soc.* **1999**. 121(20), 4912-4913
41. N. Kosaka, T. Oda, T. Hiyama and K. Nozaki, *Macromolecules,* **2004**, *37,* 3159
42. A. Rachkov, N. Minoura, and T. Shimizu, *Optical Materials (Amsterdam, Netherlands)* **2003**, *21*(1-3), 307-314.
43. R. Lovrien, *Paroc. Natl. Acad. Sci. U.S.A.,* **1967**, *57*, 236
44. L Matejka and K. Dusek, *Makromol. Chem.,* **1981**, *182*, 3223
45. M.L. Hallensleben and H. Menzel, *Brit. Polymer J.* **1990**, *23*, 199
46. M. Irie, Y. Hirano, S. Hashimoto, and K. Hayashi, *Macromolecules,* *1981,* *14*, 1912
47. G.S. Kamar, P. DePra, and D.C. Neckers, *Macromolecules,* **1984**, *17*, 1912; G.S. Kamar, P. DePra, K. Zhang, and D.C. Neckers, *Macromolecules,* **1984**, *17*, 2463
48. M. Irie and T. Sizuki, *Makromol. Chem. Rapid Commun.* **1987**, *8*, 607; M. Irie and M. Hsoda, *Makromol. Chem. Rapid Commun.* **1985**, *6*, 533

49. M. Moniruzzaman, C.J. Sabey and G.F. Fernando, *Macromolecules*, **2004**, *37*, 2572

50. P. Strohriegl, *Makromol. Chem., Rapid Commun.,* **1986**, *7*, 771

51. R.N. Thomas, **J. Polymer Sci., Chem. Ed.**, *32*, 2727 (1994)

52. O. Ryba and J. Petranek, **Makromol. Chem., Rapid Commun.**, *9*, 125 (1988)

53. A. Natansohn, P. Rochon, J. Gosselin, and S. Xie, *Macromolecules*, **1992**, *25*, 2268

54. M. Irie and R. Iga, *Macromolecules*, **1986**, *19*, 2480

55. S. Mitra, **Am. Chem. Soc. Polymer Preprints**, *26* (1), 78 (1985)

56. C.T. Lee, Jr., K. A. Smith, and T.A. Hatton *Macromolecules* **2004**, 37, 5397

57. I. Tomatsu, A. Hashidzume, and A. Harada, *Macromolecules* **2005**, *38*, 5223–5227

58. T.J. Ikeda, *J. Mater. Chem.* **2003**, 13, 2037

59. A. Natansohn and P. Rochon, *Chem. Rev.*, **2002**, *102*, 4139

60. H. Ishitobi, Z. Sekkat, M. Irie, and S. Kawaata, *J. Am. Chem. Soc.,* **2000**, *122*, 12802

61. A. Brown, A. Natansohn, and P. Rochon, *Macromolecules*, **1995**, *28*, 6116-6123

62. D. Hore, Y. Wu, A. Natansohn, P.J. Rochon, *J. Appl. Phys.,* **2003**, *94*, 2162

63. P. Rochon, J. Gosselin, A. Natansohn, S. Xie, **Appi,** Phys. Lett., **1992**, *60*, 4

64. M. Eich, J.H. Wendorff, B. Reck, and H. Ringsdorf, *Makromol. Chem., Rapid Comman.* **1987**, *8*, 59.

65. A. Natansohn, and P. Rochon, *Adv. Mater.* **1999**, *II*, 1387

66. Y. Wu, A. Natansohn, and P. Rochon,. *Macromolecules*, **2004**, *37*, 6801

67. C. Li, T. Xia, *Y.* Xiaohu, L. Guojun, and Z. Yue, *Macromolecules*, **2003**, *36*, 9292

68. W. Liu, S. Bian, L. Li, L. Samuelson, J. Kumar, and S. Tripathy, *Chem. Mater.* **2000**, 12(6), 1577-1584

69. W. Chen, Y. He, and X. Wang, *Gaofenzi Xuebao*, **2003** (2), 225 (from *Chem. Abstr.*)

70. K. Meerholz, B.L. Volodin, Sandalphon, B. Kippelen, N. Peyghambarian, *Nature*, **1944**, *371*, 497

71. B. Kippelen, B. Domercq, J.A. Herlocker, R.D. Hrera, J.N. Haddock, C. Fuentes-Hernandez, G. Ramos-Ortiz, P.A. Blanche, N. Peyghambarian, A. Schulzgen, Y. Thang and S.K. Marder, *Am. Chem. Soc. Polymer Preprints,* **2002**, *43*(I), 158

72. Z. Sekka, J. Wood, W. Knoll, Willi Volksen, V.Y. Lee, R.D. Miller, A. Knoesen, Am. *Chem. Soc. Polymer Preprints,* **1997**, *38* (1), 977

73. P. Strohriegl, *Makromol. Chem., Rapid Commun.,* 1986, *7,* 771

74. K.D. Belfield, K.J. Schafer and S.J. Andrasik, *Am. Chem. Soc. Polymer Preprints*, **2002**, *43*(I), 83

75. K.D. Belfield, D.J. Hagan, Y. Liu. R.A. Negres, M. Fan, F.E. Hernandez, in *Organic Photorefratlives, Photoreceptors. and Nanocomposites*, Proc. SPIE Vol. 4104, K.L. Lewis, and K. Meerholz, eds., SPIE-The International Society for Optical Engineering. Bellingham, WA. **2000**. *15,* 22

76. L. Wu, X. Tuo, H. Cheng, Z. Chen, and X. Wang, *Macromolecules* **2001**, 34, 8005-8013; K. Tamura, A.B. Padias, J.H.K. Hall, and N. Payghambarian, *Appl. Phys. Lett.*, **1992**, *60*, 108

77. G.D. Jaycox, *Journal of Polymer Science, Part A: Polymer Chemistry* **2004**, *42*(3), 566-577

78. A. Yu. Bobrovskii, N.I. Boiko, and V.P. Shibaev, *Vysokomol. Soedin., Ser. A Ser. B* **1998**, *40*(3), 410

79. P.J. Collings and J.S. Patel, eds., *Handbook of Liquid Crystal Research*, Oxford University Press, Oxford, 1997

80. M. O'Neill and S.M. Kelly, *J. Phys D: Appl Phys.*, **2000**, *33*, R67; K. Ichimura, *Chem. Rev.*, **2000**, *100*, 1847

81. M. Schadt, H. Seiberle and A. Schuster, *Nature (London),* **1996**, *381,* 212

82. R.A. Hickmet, *Adv. Mater.*, **1995**, *7*, 300

83. C.A. Guymon, L.A. Dougan, P.J. Martens, N.A. Clark, D.M. Walba, and C.N. Bowman, *Chem Mat.*, **1998**, *10*, 2378

84. S.W. Lee, S.I. Kim. B. Lee, W. Choi, B. Chae, S.B. Kim and M. Ree, *Macromolecules*, **2003**, *36*, 6527; H. Kamogawa and M. Yamada, *Macromolecules*, **1988**, *21*, 918

85. Shannon. P.J.; Gibbons, W.M.; Sun, S. T. *Nature* **1994**, 368,532

86. Schadt, M.; Schmitt, K.; Kozinkov, V.; Chigrinov, V. *Jpn. J. Appl. Phys.* **1992**, *31*, 2155

87. Gibbons, W.M.; Shannon, P.J.; Sun, S.-T. *Mol. Cryst. Liq. Cryst.* **1994**, *251*, 191

88. L. Corvazier and Y. Zhao, *Macromolecules*, **1999**, *32*, 3195

89. Y. Zhao Y. Chenard, and N. Paiement, *Macromolecules*, **2000**, *33*, 5891

90. Y. Zhao and Y. Chenard, *Macromolecules*, **2000**, *33*, 5891

91. M. Han, S. Morino, and K. Ichimura, *Macromolecules* **2000**, *33*, 6360-6371

92. R. Rosenhauer, Th. Fischer J. Stumpe, R. Gimenez, M. Pinole, J.L. Serrano, A. Vinuales, and D. Broer, *Macromolecules* **2005**, 38, 2213-2222

93. Y. Wu, J. Mamiya, A. Kanazawa, T. Shiono, T. Ikeda, and Q. Zhang, *Macromolecules* **1999**, *32,* 8829-8835

94. I. Zebger, M. Rutloh, U. Hoffmann, J. Stumpe, H.W. Siesler, and S. Hvilsted, *Macromolecules,* **2003**, *36,* 9373

95. Y. Zhao and N. Paiement, Adv. Mat., 2001, 13, 1891

96. Y. Zhao, N. Paiement, S. Sevigny, S. Leclair, S. Motallebi, M. Gigueri, and M. Bouchard, *SPSI Proc.*, **2003**, *5003,* 150

97. S. Sevigny, L. Bouchard, S. Motallebi, and Y. Zhao, *Macromolecules,* **2000**, *33,* 9033

98. X. Tong, L. Cui, and Y. Zhao, *Macromolecules,* **2004**, *37,* 3101

99. M. Okada and A. Harada, Macromolecules, 2003, 36, 9701

100. Q. Bo, A. Yavrian, T. Galstian, and Y. Zhao, *Macromolecules,* **2005**, *38,* 3079

101. E. Uchida, T. Shiraku, Hi. Ono, and N. Kawatsuki, *Macromolecules* **2004**, *37,* 5282-5291

102. N. Zettsu and T. Seki *Macromolecules* **2004**, *37,* 8692-8698

103. P.A. Hiltner and I.M. Krieger, *J. Phys. Chem.* **1969**, 73, 2386; J.W. Goodwin, R.H. Ottewill, and A. Parentich, A. *J. Phys. Chem.* **1980**, *84,* 1580

104. S.A. Asher, P.L. Flaugh, and G. Washinger, G. *Spectroscopy* **1986**, *1,* 26; R.J. Carisonand and S.A. Asher, *Appl. Spectrosc.* **1984**, *38,* 297; C. Reese and S.A. Asher, *J. Colloid Interface Sci.* **2002**, *248,* 41.

105. S.A. Asher, J. Holtz, L. Liu, and Z. Wu, *J. Am. Chem. Soc.*, **1994**, *116,* 4997

106. M. Kamenjicki and S. A. Asher *Macromolecules* **2004**, 37, 8293-8296

107. C.L. Braun, *Phys. Rev. Lett.*, **1968**, *21,* 215

108. L. Fourny, G. Dalacote, and M. Scott, *Phys. Rev. Lett.*, **1968**, *21,* 1085

109. R.G. Kepler, *Phys. Rev. Lett.*, **1967**, *18,* 951; P. Holtzman, R. Morris, R.C. Jarnagin, and M. Silver, *Phys. Rev. Lett.*, **1967**, *19,* 506

110. F.C. Strome, Jr., *Phys. Rev. Lett.*, **1968**, *20,* 1; W. Klopffer, *J. Chem Phys.*, **1969**, *50,* 2337

111. E.T. Kang, P. Ehrlich, A.P. Bhatt, and W.A. Anderson, **Macromolecules**, *17,* 1020 (1984)

112. J. Gullet, *Polymer Photophysics and Photochemistry*, Cambridge University Press, Cambridge, 1985

113. F. Goodwin and I.E. Lyons, *Organic Semiconductors*, Wiley, New York (1967)

114. P.J. Regensburger, *Photochem. Photobiol.*, **1968**, *8*, 429

115. H. Bauser and W. Kloepfer, *Chem. Phys. Lett.*, **1970**, *7*, 137

116. W.S. Lyoo, *J. Polym. Sci. A: Polym. Chem.* **2001**, 39, 539-54555

117. H.H. Horhold and H. Rathe, *Makromol. Chem.*, **1987**, *188*, 2083

118. S.A. Haque, T. Uryu, and H. Ohkawa, *Makromol. Chem.*, 1987, *188.* 2521

119. C.H. Hu, R. Oshima, and M. Seno, *J. Polymer Sci., Chem. Ed.*, **1988**, *26.* 123922. A. Natansohn, *J. Polymer Sci., Chem. Ed.*, **1984**, *22,* 3161

120. N. Kim and S.E. Weber, **Macromolecules**, *18*, 741 (1085)

121. T. Masuda, *Acetylenic Polymers* in *Polymeric Materials Encyclopedia*, J.S. Salamone, ed., CRC Press, New York, 1996

122. F. Sanda, T. Nakai, N. Kobayashi and T. Masuda, *Macromolecules*, **2004**, *37*, 2703

123. S.C. Suh, and S.C. Shim, *Synth. Met.* **2000**, 114(1), 91-95

124. T. Nishijima, T. Nagamura, and T. Matsuo, *J. Polymer Sci., Chem. Ed.*, **1981**, *19*, 65

125. O. Ryba and J. Petranek, *Makromol. Chem., Rapid Commun.,* **1988**, *9,* 125

126. S.C. Freilich, *Macromolecules*, **1987**, *20,* 973

127. R.J.O.M. Hoofman, G.H. Gelinck, L.D.A. Siebbeles, M.P. de Haas, J.M. Warman, and D. Bloor, *Macromolecules* **2000**, *33*, 9289-9297

128. W.-Y. Wong, S.-M. Chan, K.-H. Choi, K.-W. Cheah, and W.-K, *Mucromol. Rapid Commun.* **2000**, *21*(8), 453-457

129. K. Kimura, H. Sakamoto, and R.M. Uda, *Macromolecules*, **2004**, *37*, 1871

130. S.H. Chan, L.S.M. Lam, C.W. Tse, K.Y.K. Man, W.T. Wong, A.B. Djurisic, and W.K. Chan, *Macromolecules*, **2003**, *36*, 5482

131. E.M. Conwell and H.A. Mizes, *Am. Chem. Soc. Polymer Preprints*, **1966**, *37* (1), 96

132. S.A. Jenekhe and L.R. de Paor, *Am. Chem. Soc. Polymer Preprints*, **1966**, 37 (1), 94 (1996)

133. S.A. Jenekhe, X. Xhang, and M. Tikka, *Am. Chem. Soc. Polymer Preprints*, **1997**, *38* (1), 347

134. D. W. Chang, S. Kim, S.Y. Park, H. Yu and D.-J. Jang, *Macromolecules* **2000**, *33*, 7223-7225

135. I. Saito, *J. Am. Chem. Soc.* **2000**, *122,* 5646-5647

136. S.A. Haque, T. uryu, and H. Ohkawa, **Makromol. Chem.**, *188*, 2521 (1987)

137. J.V. Crivello, J.L. Lee, and D.A. Canlon, *J. Polymer Sci., Chem. Ed.*, **1987**, *25*, 3293

138. K. Inomata, S. Kawasaki, A. Kameyama, and T. Nishikubo, *Kobunshi Ronbunshu* **2000**, 57(7), 457-466 (from *Chem. Abstr.*)

139. S.C. Freilich, **Macromolecules**, *20*, 973 (1987)
140. H.H. Horhold and H. Rathe, **Makromol. Chem.**, *188*,. 2083 (1987)
141. A. J. Myles and N. R. Branda, *Macromolecules* **2003**, *36,* 298-303
142. S.-J. Lim, B.-K. An, and S.Y. Park, *Macromolecules*, **2005**, *38*, 6236
143. E. Elsner and H. Ritter, *Makromol. Che., Rapid. Commun.*, **1987**, *8,* 595,
144. A. Landlein, R. Langer, *et al.*, *Nature*, **2005**, *434*, 879
145. K. Abe, S. Haibara, and Y. Itoh, *Makromol. Chem.*, **1985**, *186*, 1505

Chapter 6

Photorefractive Polymers for Nonlinear Optics

Photorefractivity is defined ¹as a modulation of the index of refraction in an electrooptic material by internal electric fields produced by optical redistribution of charge carriers. It is based on the combined effects of photoconductivity and electrooptic property. [2,3] Under the illumination of nonuniform light formed by the interference of two coherent laser beams, a spatially oscillating space-charge field is formed. This field forms from generation and redistribution of photoinduced charges. To be more specific, in photorefractive materials, the absorption of light is followed by diffusion and drift of the free charges. This generates a space-charge field that modulates the refractive index. [1] The refractive index of material is subsequently modulated *via* an electrooptic effect. Photorefractive materials exhibit unique features, such as photosensitivity, reversibility, and beam amplification. This provides important advantages in their applications in the fields of optical data storage and information processing. [4]

6.1. Photorefractive Materials

Photorefractive materials are unique in their ability to generate large changes in the indices of refraction in response to relatively low power light. This photorefractive effect, a persistent but reversible change in the refractive index of an electrooptic material is caused by nonuniform illumination. It was first observed as a detrimental "optical damage" in lithium niobate and in other inorganic crystals used for second harmonic generation. Since the phenomenon became understood, such materials found application in integrated optical devices. Presently used inorganic materials are hard to process, however. Extensive research is, therefore, being carried out to develop polymeric materials that could replace inorganic ones, such as $LiNbO_3$ or $BaTiO_3$.[5] As a result, it was first reported in 1991 that it is possible to form polymers that exhibit the photorefractive effect .[6] Such polymers must have the ability to convert electrical signals into optical output. In other words, they must function electrooptically when changes in applied voltage result in changes in their refractive index. Such changes are produced by optical redistribution of charge carriers. In other words, in order for the molecules to exhibit second-order optical effects, they must be "noncentrosymmetric" or not centrally symmetrical [7,8,9]. These effects include the generation of second harmonics. This generation of the second harmonics doubles the frequency of light while the electro-optical effects enable the electric current to control and modulate light. To put this in other words, the electric field can alter the index of refraction and thereby change the light propagating through polymeric materials. This in turn modulates the phase of light.

Within a relatively short time, many polymeric materials were developed that do exhibit the photorefractive effects. [10-14] These materials consist of polymers with charge transport agents and photosensitizing dyes. The quantity of change in a material's refractive index that is altered by alterations in an applied electric field is expressed as the electrooptic coefficient r. Lithium niobate has an r value of 30 pm / volt. Several polymers are already known that possess coefficients (at 1.3 μm) greater than 30 pm/V and the aim now is to develop materials with r values of 50 pm / volt or even higher. It is also known that the photorefractive effect can also be observed in many organic materials, such as polymer films, [15] liquid crystals, [16] and organic glasses. [17]

Device-quality materials must possess: 1) large optical nonlinearity, 2) high temporal stability of the dipole orientation, 3) low optical loss, and 4) good processibility. The first requirement was addressed by a number of research groups and various high μβ chromophores have been developed. Some of these exhibit large bulk nonlinearity in composite systems. Unfortunately, however, the bulk nonlinearity of composite samples tend to decays rapidly, due to fast relaxation of chromophore orientation. Long-term stability of dipole orientation is a problem associated with this type of materials. To realize a stable dipole alignment, chromophores have been linked to high-Tg polymers such as polyimides, to lock the dipole orientation into rigid matrix after poling. This too is a problem, because many nonlinear optics chromophores cannot survive the high processing temperatures (ca. 220-250 °C) and the harsh acidic environment of polyamic acid during imidization. In addition, to be industrially useful, the device-quality nonlinear optics polymeric materials must retain high optical quality in thin films, high optical damage thresholds, and exhibit low optical propagation loss. They should also lend themselves to device fabrication without considerable difficulty and maintain sufficiently large and stable nonlinear optics susceptibilities. [18] To be of practical use the developed materials must also exhibit reasonable thermal stability (greater than 95% retention of optical nonlinearity for 1000 hours at 100 °C). Although it is very difficult to solve all the problems, very impressive progress has been made to date. [19] This chapter tries to illustrate what has been achieved so far.

In summary, the photorefractive polymers are required to do the following:

1. Transport the charges through the material.

2. Trap the charges.

3. Provide nonlinear optical properties or changes in the index of refraction of the material in response to changes in an applied electrical field (Exhibit an electrooptic effect).

4. Respond to light by generating electrons and positively charged vacancies. *Nonlinear optics effects* are generally described by a *polarization equation* for the optical response *(P)* of a material to an optical electric field *(E)*[20]:

$$P = \chi^{(1)}E \; + \; \chi^{(2)}E^2 \; + \; \chi^{(3)}E^3 \; + \; ...$$

In this equation the $\chi^{(1)}$ term describes linear optics of such phenomena as light reflection and refraction. The other terms, beyond the $\chi^{(1)}$ namely, $\chi^{(2)}$ and $\chi^{(3)}$ describe the second- and third-order optical effects, respectively.

The above equation is called the *polarization equation*. In order for the materials to exhibit second-order effects, the π electrons must de-localize in response to applied electric fields. This results in spatial asymmetry, as already stated above.[8] Also, for the second-order effects to be large, there must be large differences in the dipole moments between the ground state and the excited states.

There are some essential differences between photorefractive polymers and inorganic crystals. In the organic materials the quantum efficiency for generation of mobile charge, Φ, is highly field dependent due to Onsager germinate recombination.[21] High fields are, therefore, generally required for facile charge generation. Also, the mobility is highly field-dependent.[21] It can vary as $\log \sqrt{E}$. This has been widely observed in many molecularly doped polymers.[22] This means that higher fields are required to increase the speed of charge transport.

It was pointed out[23] that the performance of photorefractive polymer composites is too large to be explained by the simple electrooptic photorefractive effect alone. A theoretical model was offered[10] where both the birefringence and electrooptic coefficient are periodically modulated by the space-charge field due to the orientational mobility of the nonlinear chromophores at ambient temperatures.

The methods that are used in preparation of the polymeric materials with nonlinear optical properties can be as follows.[24] They include (1) doping or simply dissolving in the polymeric matrix chromophores, (2) covalent attachment of such chromophores[25] to amorphous and liquid-crystalline polymers, and (3) assembling nonlinear optics chromophores into Langmuir-Brodgett layers. The statistical center of symmetry in polymers is removed via dipolar alignment by application of an external electric field (called *poling*). A strong electric field is applied while the polymer is heated at, or preferably below, T_g, or glass transition temperature, of the material. The active molecules become mobile and the dipoles are forced to line up. This is done, because the chromophores must be aligned in such a way that the material has no center of symmetry. While the electric field is still applied, the films are then cooled. This locks in the asymmetry. The electrooptic response in these materials is, therefore, a result of the electric field induced alignment of the chromophores, which are second order nonlinear. Due to strong interactions among the electric dipoles, the poled structures are in a metastable state. This means that the poled nonlinear optics chromophores that possess large dipole moments will tend to relax back to the randomly oriented state. [26] The stability of the poled structures, therefore, strongly depends on the rigidity of the overall material. To maintain this rigidity, polymers with very high T_g and thermal stability have been used with a fair amount of success. Another approach is to crosslink the polymeric matrix. This has also yielded fruitful results. To obtain large nonlinearities the quantity of chromophores in the system must also be large. The problem is, however, that chromophores tend to have high dipole moments, and above a critical loading density, electrostatic interactions occur. Such interaction favor centro-symmetrical ordering of the chromophores in the polymer matrix. As a result, $\chi^{(2)}$ will not increase linearly with chromophore loading N, but it will show a maximum at relatively low loading levels. This maximum is approximately 15 wt % for dipole moments and decreases at higher chromophore concentrations.

Several approaches to diminish the dipolar interactions were investigated. Dalton et al., [27,28] studied the influence of the shape of the chromophore on their dipolar interaction. The conclusion that was reached was that a spherical shape inhibits dipolar interactions. Another approach was to use chromophore-functionalized dendrimers, where dendrons effectively decrease the interactions among the chromophores due to steric hindrance.[29]

6.2. Chromophores for Nonlinear Optics

A typical nonlinear optics material consists of an electron-accepting group, a bridge, and an electron-donating group. The donor-acceptor charge transfer groups are connected with and separated by a π-electron segment (consisting typically of polyene, azo, or heteroaromatic moieties). Thus, chromophores that are active in nonlinear optics are conjugated molecules that contain electron-donating group on one end and electron accepting groups at the other end. Applications of external electric fields cause the electric charges in the molecules to separate. The resultant asymmetric distributions of charges markedly affect the interactions of the chromophores with light. The function of the conjugate bridge is to allow electronic communication between the donor and the acceptor. Examples of the electron donors can be amino or methoxy groups and of the acceptors can be nitro or cyano groups.[30,31] It was shown, that large molecular nonlinearities can be achieved by using extended polyene bridge systems.[32] For instance, one compound , a derivative of isoxazolone exhibits $\mu\beta$ value of 3156 x 10^{-48} esu at 1.907 μm and a low absorption maximum of 562 nm in chloroform.[32] (where μ is the dipole moment and β is the molecular first hyper-polarizability). More precisely, the term β was defined by Jen, Marder, and Shu[33] as:

$$\beta \propto (\mu_{ee} - \mu_{gg})(\mu^2_{ge} / E^2_{ge})$$

where g is the index of the ground state, e is the index of the charge-transfer excited state, and μ is the dipole matrix element between the two states. [33] Unfortunately, the unsubstituted polyene segments of the above mentioned chromophore make that compound thermally, photochemically, and chemically unstable. [34] On the other hand, it was shown that thermal stability of many chromophores can be significantly improved by using aromatic amino functional groups in place of aliphatic amino functional groups as the electron donors. [35,36] The same is true when a simple polyene bridge is replaced with a thiophene moiety.

Some examples of chromophores that were incorporated into side chains and main chains of polymers are donor/acceptor substituted stilbenes and azobenzenes. [37] Another example are the two compounds shown below:

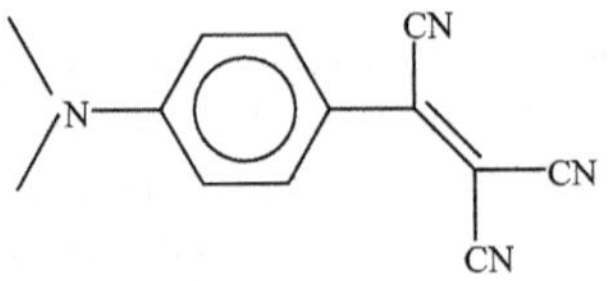

In some of the early preparations polymers and copolymers of poly(n-vinylcarbazole) were used as the polymeric materials [38]:

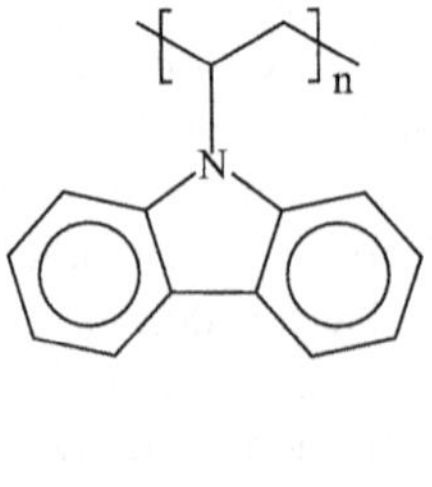

The polymers were combined (or doped) with such diverse chromophores as (diethylamino)benzaldehyde diphenylhydrazone:

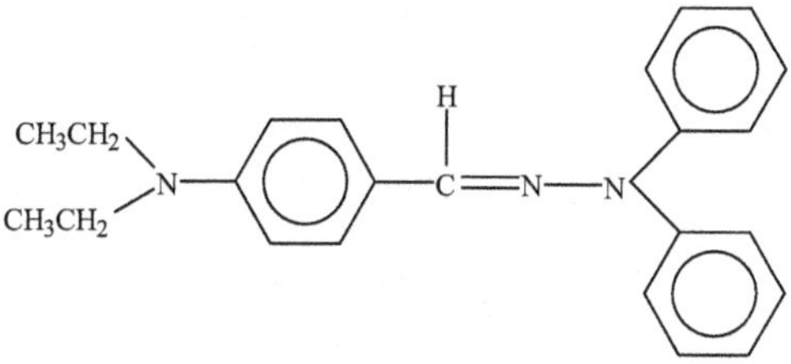

or 2,5 dimethyl-4-(*p*-nitrophenylazo)anisole:

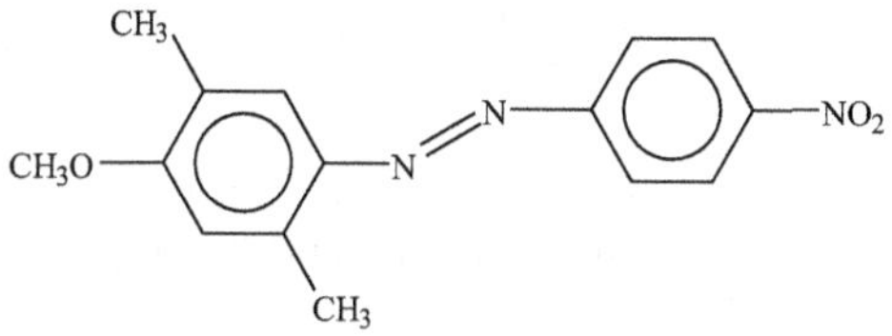

For instance, one material was prepared from photoconducting poly(N-vinyl carbazole) that was doped with a blue-shifting optically nonlinear chromophore, 3-fluoro-4-N,N-diethylamino-*p*-nitrostyrene. This material was sensitized for charge generation with 2,3,7-trinitro-9-fluorenone.[39]

A material was reported recently that is claimed to be very efficient. It is a plasticized guest-host polymer with a low glass transition temperature. The T_g. is below room temperature. [40] The material is prepared from a photoconducting poly(9-vinylcarbazole), a plasticizer, 9-ethylcarbazole, a nonlinear optical chromophore, 2, 5-dimethyl-4-(p-nitrophenyl) azoanisole, and a sensitizer, 2,4,7-trinitro-9-fluorenone (2% by weight). The sensitizer forms charge-transfer complexes with the carbazole moieties. If a periodic light-intensity pattern is incident to the material, charges are generated in the bright regions due to the optical excitation of the charge-transfer complexes. The carbazole units act as hole-transport agents. This enables the holes to move to the dark regions of the grating by diffusion or drift under the influence of an applied external field. In most photorefractive materials, the space charge field is produced by inhomogeneous charge distribution that modulates the refractive index through the electro-optic effect. Because, the material has a T below room$_g$temperature, the index modulation amplitude is enhanced further by birefringence.[41] This material was reported to combine close to 100% diffraction efficiency in a 105 µm thick film with sub-second response time and high resolution. Unfortunately, however, the polar dye in the nonpolar matrix tends to crystallize slowly and undergoes space separation separation, thus reducing the dielectric strength of the material. [42]

Another chromophore, 4-(4'-nitropnenyl)azo-1,3-bis[(3'- or - 4'-vinylbenzyl)-oxy]benzene:

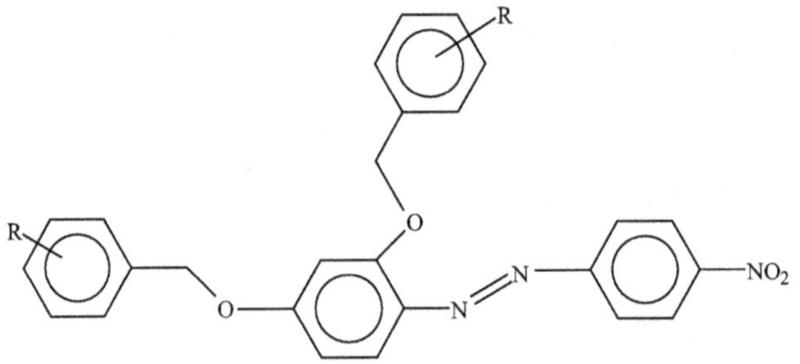

is claimed to exhibit very good phase stability and a nearly complete diffraction at 633 nm. [42]

Also, polymers that contain carbazole moieties in the backbone with two acceptor groups as multifunctional chromophores were synthesized. [43] The sytheses were carried out by a Knoevenagel polycondensation of divinylcarbazoles with bis(cyanoacetate)s using 4-(N,N'-dimethyl)pyridine as a base. [43] Second-order nonlinear optical properties of the polymers were confirmed by second harmonic generation. The photorefractive optical gain of these polymers was demonstrated to be high.

In 1996, work that focused upon the exploration of alkyl and aryl amine and ketene dithioacetal donor groups together with a variety of electron acceptor groups was reported. [44] The acceptor groups included thiobarbituric acids, isoxazolones, cyanovinyls, sulfoximines, tetracyanoindanes, and a cyanosulfone. Improvements in $\mu\beta$ values of the order of 3-10, relative to the values observed for commonly used stilbenes and azobenzene chromophores were readily realized. More than one hundred such chromophores were incorporated into processable polymer lattices and evaluated for stability in the polymers under conditions of corona poling (explained above) and optical illumination (at 1.3 μm). This led to a conclusion [44] that promising chromophore-containing polymeric materials that can be poled to yield electrooptic coefficients greater that 30 pm/V, are among those that contain isoxazole and tetracyanoindane acceptors. They appear to exhibit the good stability at condition of corona poling and optical irradiation (in the presence of atmospheric oxygen). A polymeric material of this type can be illustrated as follows:

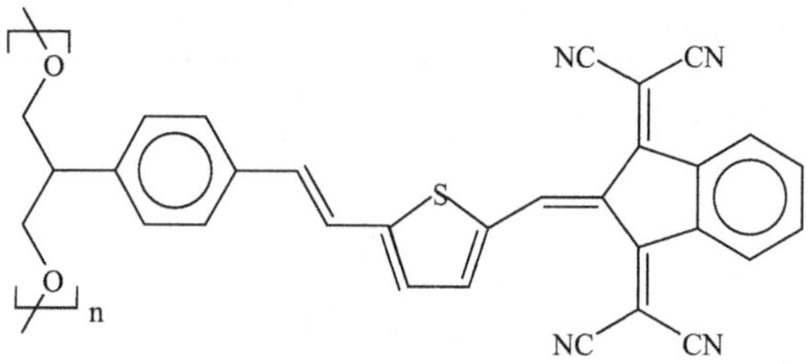

Table 6.1. gives examples of some nonlinear chromophores. These nonlinear optics chromophores are intended for incorporation into polyimides, polyurethanes, polyureas, and polyamides.[45] Among them are some promising chromophores that were reported by Jen and coworkers[33].

Additionally, new thermally, photochemically, and chemically stable chromophores were reported recently.[46] These were formed by incorporating isophorone moieties into chemical structure, such as:

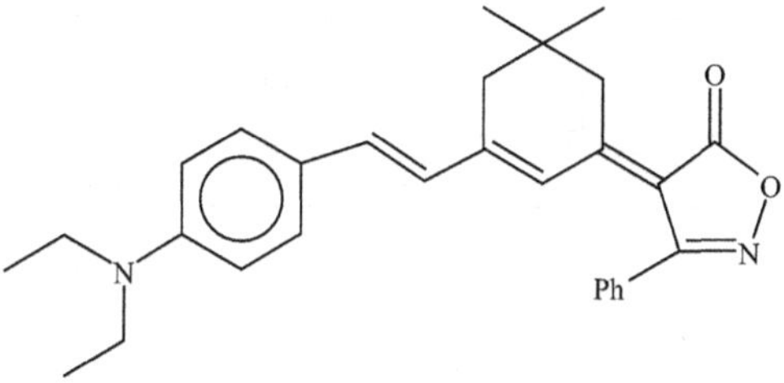

The above shown chromophore, was reported to yield up to 30 pm/V at an optical loss of 0.96 dB/cm with no loss attributable to intrinsic absorption of the chromophore.[47]

Table 6.1. Some Examples of Chromophores for Nonlinear Optics [a]

NLO Chromophores	$\mu\beta(\text{x } 10^{-48}\text{esu})$
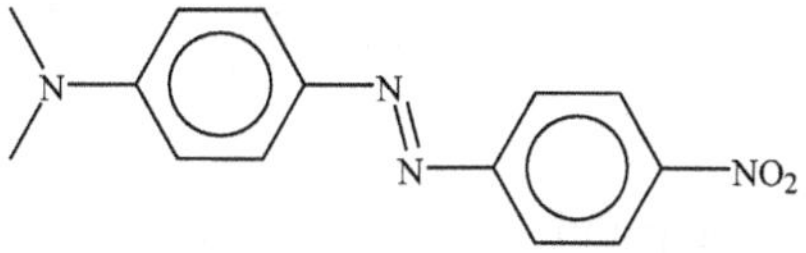	800 (1.6μm)
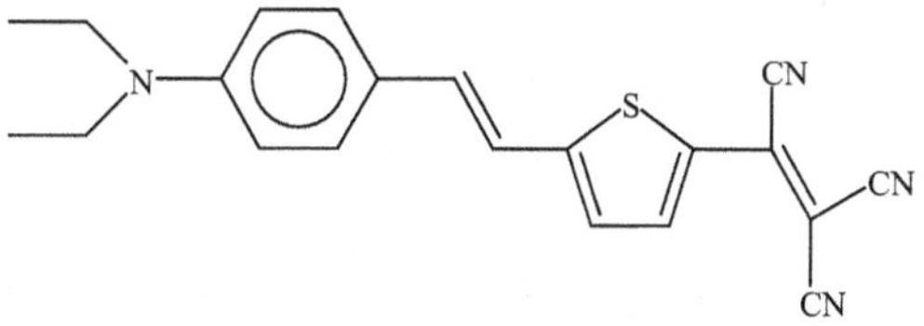	6200 (1.9 μm)

Table 6.1. (Continued)

NLO Chromophores	$\mu\beta(x\ 10^{-48}esu)$
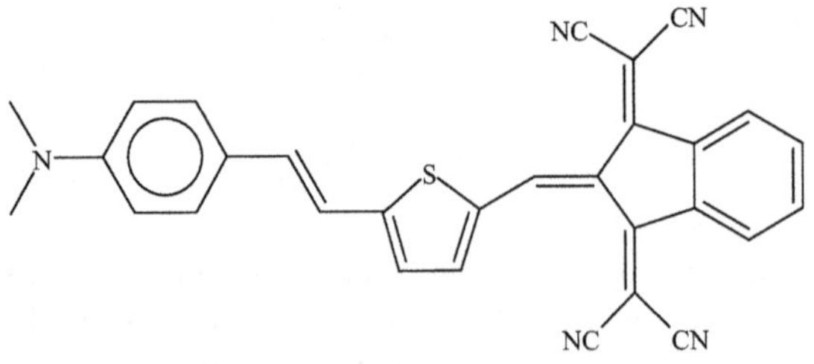	6144(1.9 μm)
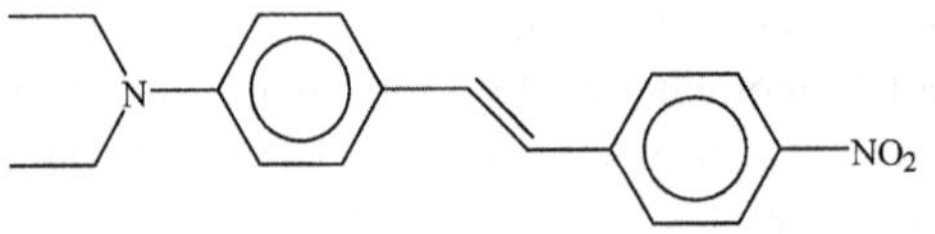	580
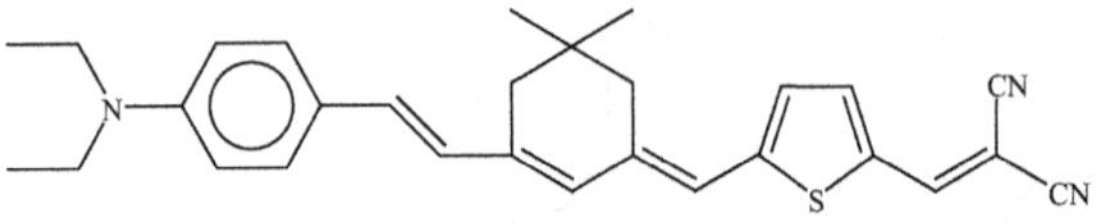	3450
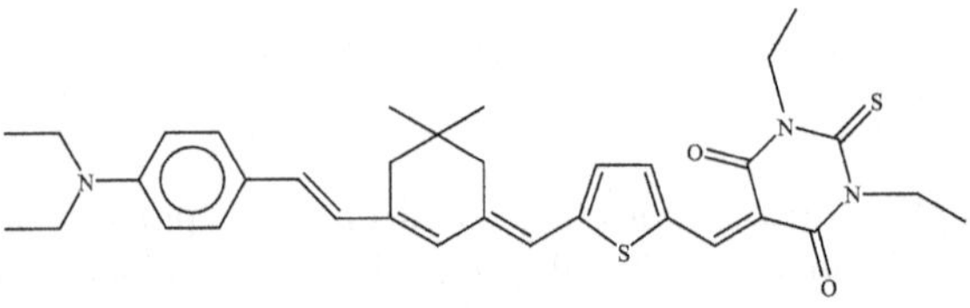	5380
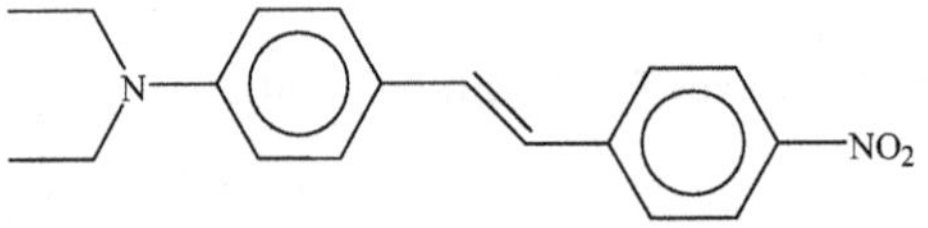	580

[a] from various literature sources

Another recent study reports [46] a preparation of two thermally stable chromophores that have the following structure:

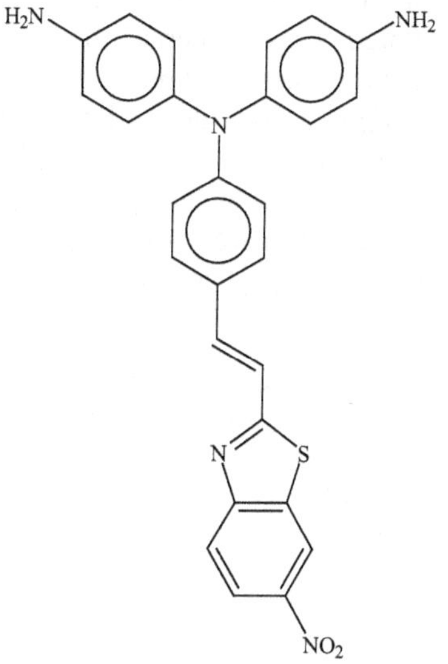

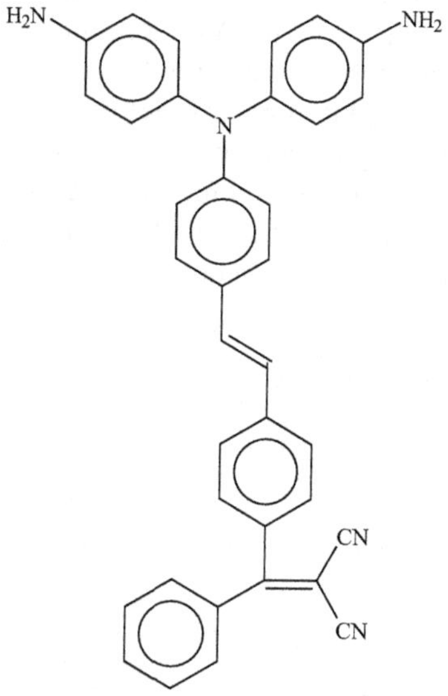

The trapping sites may be provided by additional doping materials. It is also possible to have the nonlinear optics groups perform charge-generating or trapping functions. This, for instance, was accomplished by reacting bisphenol A diglycidyl ether with 4-nitro-1,2-phenelenediamine, a chromophore.[10]

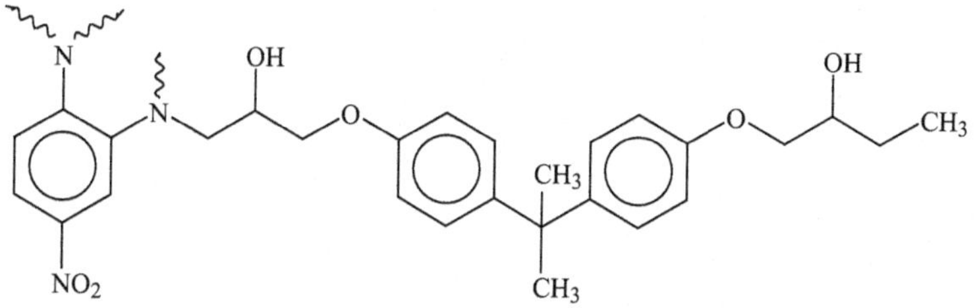

The crosslinked, optically nonlinear material was doped with a charge-transporting agent, (dimethylamino) benzaldehyde diphenylhydrazone. Subsequent dramatic improvement in performance was achieved in a similar polymer by using nitroaminostylbene as the charge transport agent[48].

You and coworkers,[49] reported that they developed a series of dendrized chromophores and fluorinated chromophore-containing dendrimers that possess nanostructures and electro-optic activities. By controlling their size, shape, and molecular architecture, they claim to have achieved dramatically enhanced poling efficiency in these materials. They incorporated these chromophores that contain fluorinated dendrons onto a hydroxystyrene photoresist polymer to form side-chain dendrized nonlinear optics polymers. A large electrooptic coefficient (97 pm/V at 1.3, um) was demonstrated. These materials, however, do not possess the thermal stability that is needed for practical applications. The poling-induced polar order decays quite rapidly upon heating due to low glass-transition temperature (Tg) of the polystyrene backbone.

To overcome this, You and coworkers,[49] synthesized a high-Tg aromatic polyimide with pendant dendronized nonlinear optics chromophore with a cardo-bisphenol linkage to a rigid aromatic polyimide backbone:

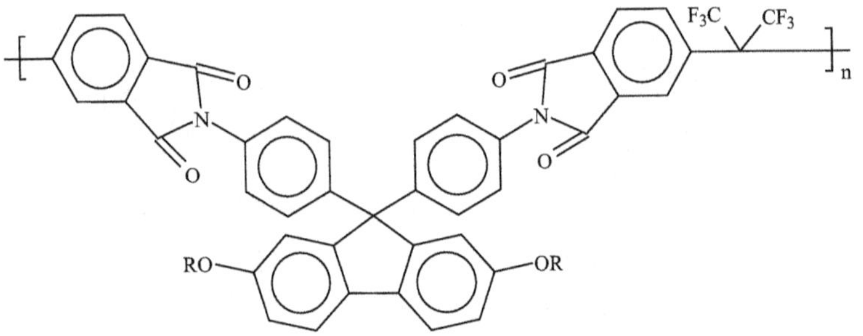

where R =

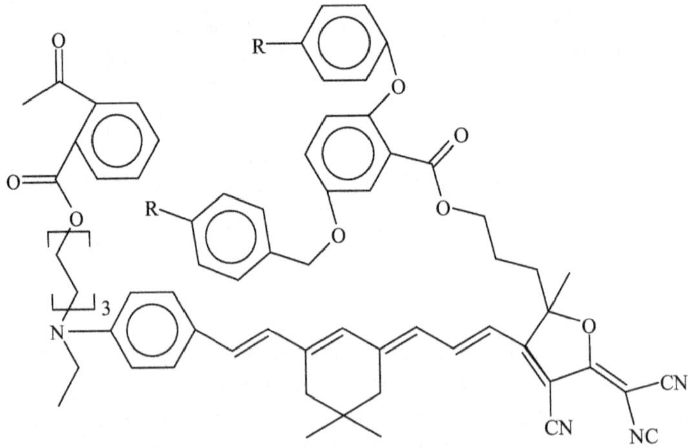

where R = O—

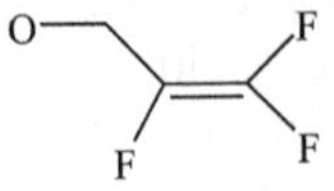

The authors claim to have successfully applied the site isolation principle to a rigid 3-F cardo-type polyimide with very high T_g. At the same time high poling efficiency is also claimed to yield very large electrooptic coefficient (71pm/V at 1.3 μm) More than 90% of this value is claimed to be retained at 85 °C for more than 650 hours.

Marder *et al.*,[50] describe the synthesis of several new photorefractive polymers, which are composed of a new type of nonlinear optical chromophore attached to conjugated polymer, poly(*p*-phenylene-thiophene). Since the NLO chromophore is labile in many reaction conditions, the Stille coupling reaction was used to prepare these polymers. The resulting polymers exhibit high PR performances. An optical gain coefficient of 158 cm $^\wedge$ at a field of 50 V/$^\wedge$m and a diffraction efficiency of 68% at a field of 46 V/$^\wedge$m for polymer PI were obtained, which are among the best values for fully functionalized PR polymers to date.

6.3. Polymeric Materials for Nonlinear Optics

Poling, as stated earlier, is applied to lock in asymmetry. This, however, is very difficult to do permanently. With time and subsequent heating the orientations of the polymers tend to become disordered. A slow decay of the optical nonlinearity results. In trying to overcome the draw-back various approaches were tried. One concept is to synthesize polymeric materials with high T_g , such as aromatic polyimides and polyamides. Other approaches to preserve chromophore alignment in poled polymers involve restricting freedom of movement of the chromophores by various crosslinking schemes as well as by using liquid crystalline materials.

6.3.1. Crosslinked Polymeric Materials

The crosslinking process in some systems must be carried out at 200 °C. [51] Also, the high T_g approach required high poling temperatures, like 225 to 250 °C and for integrated optic applications, poling temperatures above 350 °C might be necessary due to the high temperatures that are required during the fabrication of the devices. These temperatures are extremely demanding for organic molecules, particularly for chromophores with extended conjugation. Many nonlinear optics chromophores just cannot survive the high processing temperatures (ca. 220-250 °C) or the harsh acidic environment of polyamic acid (in preparation of polyimides) during imidization.[50] Polymer lattice hardening under mild conditions is much desired for processing chromophores into useful device-quality materials.[51]

6.3.1.1. Polymers Crosslinked by the Diels-Alder Reaction

Dolton et al., [52] pointed out that typical crosslinking processes to achieve stability possess additional deficiencies. These are entanglements of the polymers during the lattice-hardening process that significantly increase the difficulty in aligning dipolar chromophores. The result is lower electrooptic activity of these poled thermoset materials. To overcome this nonlinearity-stability trade off, Dolton et al., [52] suggested that a new poling process is required to increase rotational freedom of nonlinear optics chromophores. Ideally, this process should be prior to and separate from any crosslinking process. Moreover, to achieve and maintain polar order in the resulting poled material, the reaction for subsequent crosslinking should be triggered only by very mild conditions, preferably without further temperature elevation. By taking into account all these requirements a lattice-hardening approach based on a Diels-Alder [4 + 2] cycloaddition reaction was developed. [52] This is claimed to provide significant advantages over the conventional nonlinear optics thermosets. It consists of preparing linear polymers first with pendant protected dienophile groups, These groups become unprotected under mild conditions and subsequently undergo Diels Alter addition reactions with other polymer molecules containing pendant diene groups. The process can be illustrated as follows [52] :

6.3.1.2. Polyurethanes

Formation of urethane crosslinking linkage from terminal hydroxy groups and isocyanates occurs at relatively low temperatures and essentially neutral conditions. Polyurethanes are, therefore, suitable for the above described purposes and received some attention. Linear polyurethanes with covalently attached chromophores have lower T_g (ca. 130 °C) than polymers like polyimides (ca. 250 °C), and, consequently are less stable at higher temperature. Many tried to overcome this by forming crosslinked three dimensional structure.

Beecher et al. reported that they carried out photocrosslinking of polyurethane polymers with the aid of free-radical photoinitiators. [53] By placing trans-stilbene groups in the chromophore moieties, the photocrosslinking steps are restricted to the methacryloyl double bonds, and proceed without concurrent trans-cis isomerization of the stilbene. This polymer is illustrated as follows [53] :

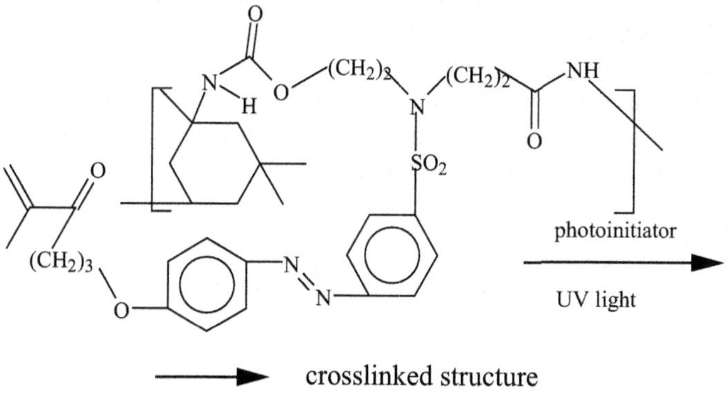

For this reaction an organometallic photoinitiator was used that absorbs at 532 nm. Photocrosslinking of the oriented polymer can be used to produce a patterned material that exhibits a stable second-order nonlinear optics signal that lasts for several weeks at 100 °C . Addition of phenol is necessary to prevent thermal crosslinking.

The following scheme was developed for forming polyurethane thermosets [54] :

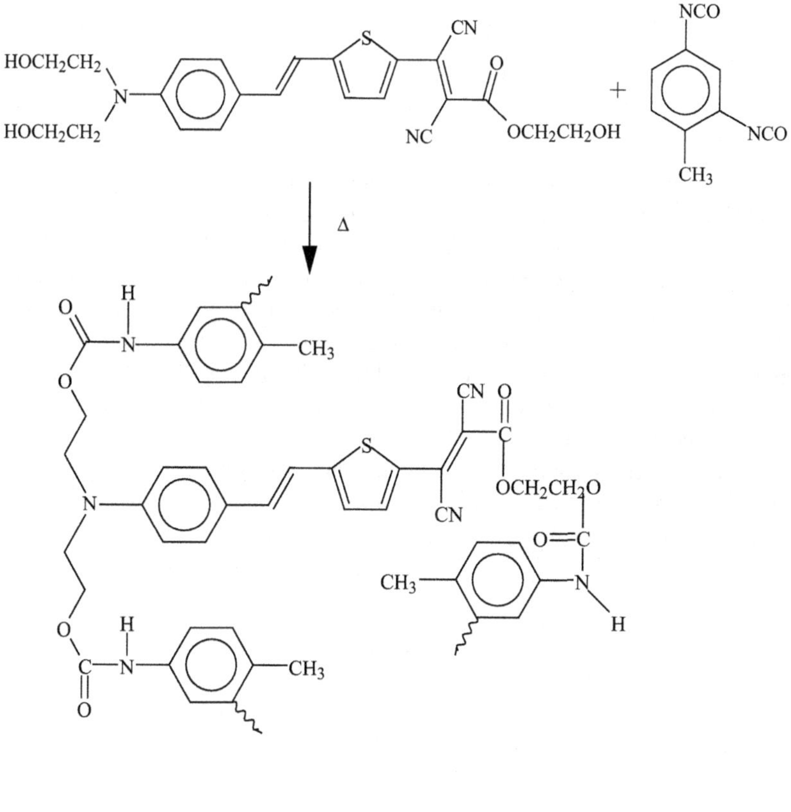

Based on the absorption spectrum of the above material there is a strong charge-transfer (CT) band at 626.6 nm. This is an indication of an enhanced charge separation and greater nonlinearity.

One can compare the above with another polyurethane preparation where a structure with a hemicyanine dye is attached as a pendant group. [55] Following preparation procedure was used:

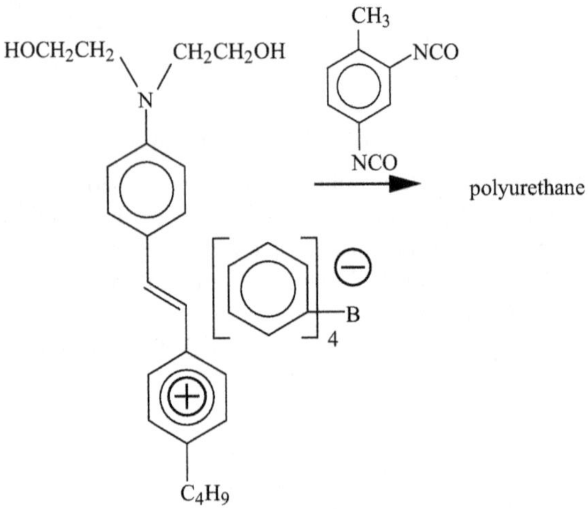

The polymer, with an $M_n = 12000$, is soluble in dimethylformamide and can be processed into optical quality films by spin casting. The $T_g = 121\ °C$. The $\chi^{(2)}$ of this polyurethane was reported to be in the range of 1.8×10^{-7} to 5.0×10^{-7} esu, depending upon poling conditions. This high second-order activity is attributed to the strong electrons accepting nature of the pyridinium groups. The alignment of the nonlinear chromophore moieties induced by electric poling exhibits temporal stability. This is believed to be due to the hydrogen bridges that form between neighboring polyurethane chains. These bonds prevent relaxation of oriented molecular dipoles. [55]

In a different approach, a polyurethane was formed from a terpolymer of methyl methacrylate that was functionalized with pendant hydroxy groups [51]: and then reacted with a diisocyanate.

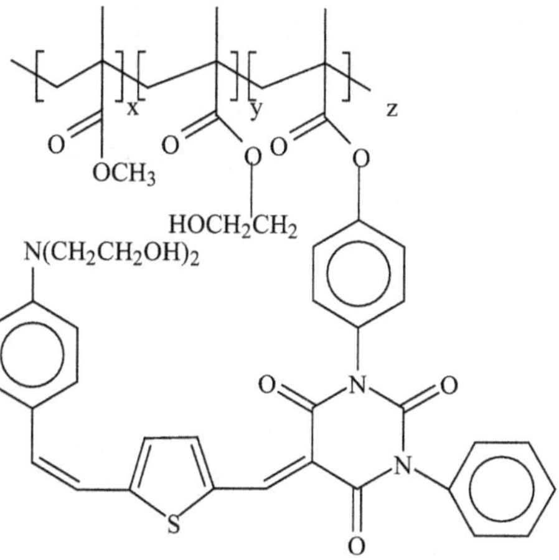

The technology involves combining the acrylic polymer with the diisocyanate and then spin casting the material into films that are electrically poled at T_g to align the chromophores. The films are then heated to higher temperature for crosslinking and "hardening" the lattices and locking in the dipole alignment. The products exhibit electrooptic coefficients as high as 20 pm per volt and maintain their optical nonlinearity for several thousand hours at high temperature.

A synthesis and characterization of an active styrylthiophene monomer and formation of polyurethane with second-order optical nonlinearity was reported by Wang et al.[56] This was carried out on (trans)-7-[4-N,N-(dihydroxyethyl)aminobenzene-ethenyl-3,5-dinitrothiophene to form the corresponding prepolymer, polyurethane.. The prepolymer and polyurethane are claimed to exhibit good thermal stability and good solubility. in common organic solvents. The d_{33} coefficient of the dielectric polarized films was reported. to be 40.3 pm/V. [56]

Dalton et al.,[51] described preparation of new trifunctionalized Disperse-Red type chromophore, Disperse-Red Triol compound by reaction of this chromophore with diisocyanate at 80 °C. It resulted in formation of soluble oligomers that were spin cast to form high optical quality films. Concurrent corona poling and thermosetting of these films resulted in polyurethane networks with electrooptic coefficients (r_{33} at 1.06 μm) of 14.5 pm/V and good dynamic thermal stability. This system is claimed to offers some prominent advantages in terms of simpler material processing, wider processing window and higher temporal stability of dipole alignment [51].

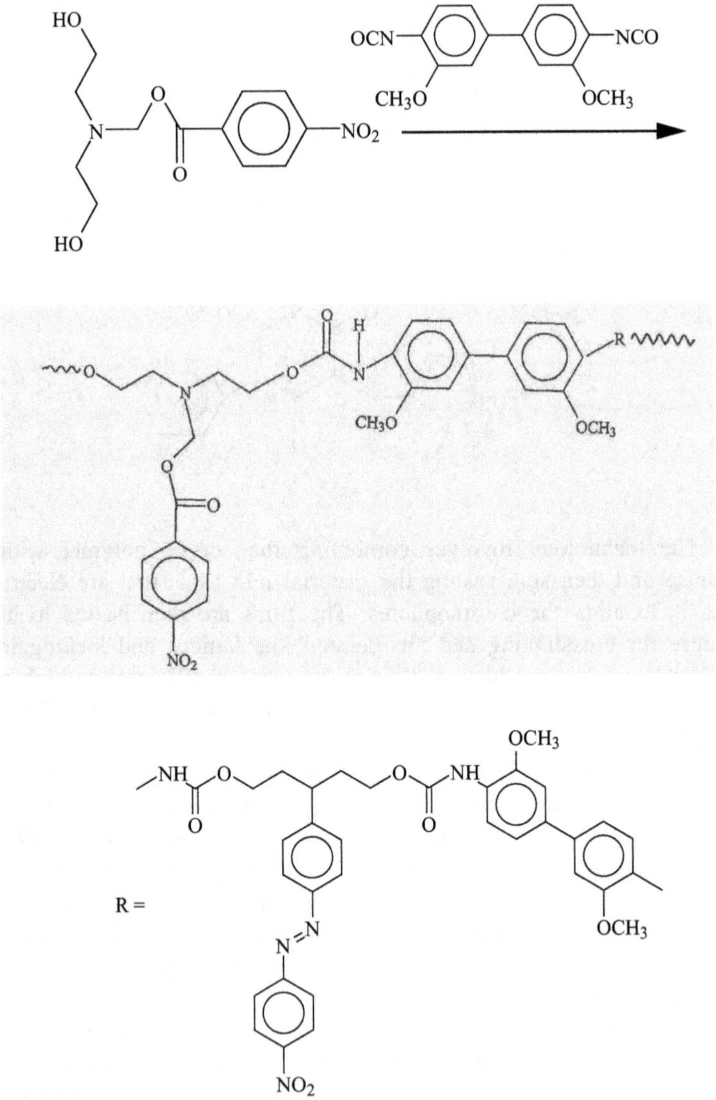

Another, somewhat different preparation of a polyurethane was also reported by Dalton et al. [51] It is shown as follows:

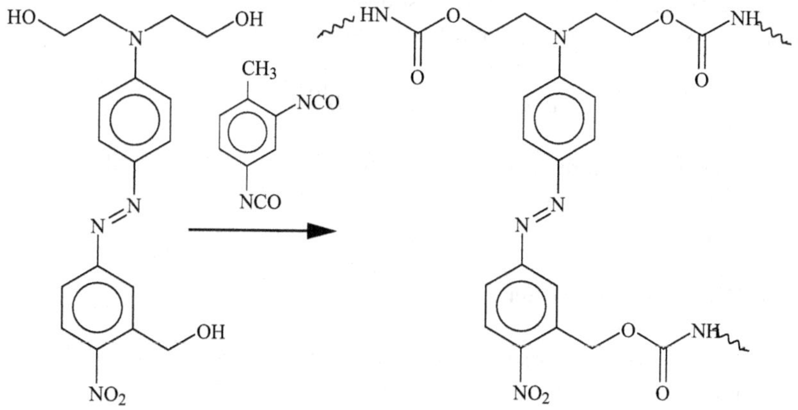

6.3.1.3. Epoxy Crosslinked Polymers

Thermal epoxy crosslinking, using relatively rigid reagents at optimized stoichiometry substantially enhances the orientational stability of the aligned chromophore substituents by forming ether linkages between polymeric chains and reestablishing hydroxy groups for hydrogen bonding. This produces a dense, three-dimensional network of covalent bonds that restrict the freedom of chromophore motion, thereby enhancing temporal stability characteristics. This was shown on orientational relaxation of the nonlinear optics chromophores of polyamides with the following structure [57]:

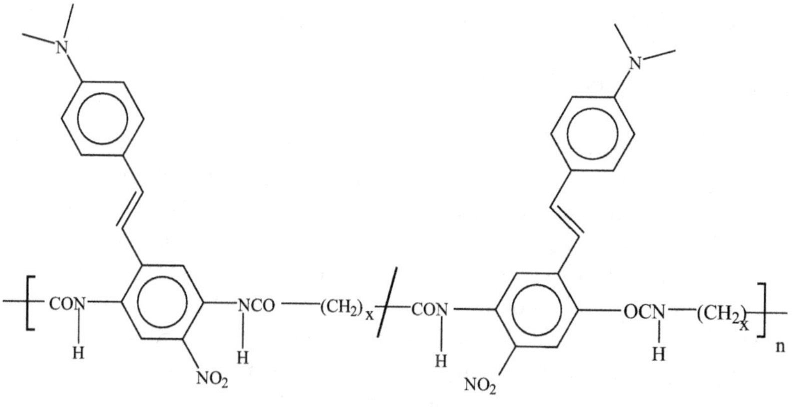

where x = 4,6,8,10,12.
These polymers were studied at different temperatures below the glass transition by the decay of the nonlinear optical susceptibilities of corona-poled films. The

time dependence of the decay was shown to be well represented by the Kohlrausch-Williams-Watts stretched exponential function:

$$d(t) = d_0 \exp[-(t/\tau)^\beta]$$

The temperature dependence of the decay correlates with the glass transition temperature, T_g, using normalized relaxation law with $(T_g - T)/T$ as the relevant scaling parameter.

6.3.1.4. Crosslinked Polyesters

Trollsas and coworkers [58] prepared a crosslinked polyester by in situ photocrosslinking of the ferroelectric monomer, which was obtained via an 11-step synthesis. The resulting polymer has a large electro-optical coefficient (15-35 pm/V; in the range of organic single crystals), which means the material may be suitable for use in electro-optic devices. [58]

6.3.2. Polymeric Materials with High Second Order Transition Temperature

Generally, above T_g, thermoplastic polymers undergo cooperative localized motions. Below the second order transition temperature there is not enough energy in the system to enable whole segments of the polymeric chains to move. In the attempt to lock in asymmetry, as stated earlier, one approach is to synthesize polymeric materials with a high T_g. Following is a description of some these materials.

6.3.2.1. Polyimides and Polyamides

It was shown that the thermal stability of conventionally formed polyimide polymers and copolymers (bearing nonlinear optical chromophores) is adequate for numerous device applications. There is a problems, however, with doped systems in that they tend to undergo phase separation. This limits the amount of the nonlinear optics chromophore that can be incorporated into the system. To try and make doping unnecessary, multifunctional polymers were synthesized that contain all the necessary components. One limitation of this technology, however, is the small nonlinear optical response (r_{33}) values for many such polyimides. This is not true of all of them.

One photorefractive polyimide [59] was formed by linking thee units, a substituted porphyrin (charge generator), 4-amino-4'-nitrostilbene derivative (the nonlinear optics molecule), and a charge transporting agent, pyromellitic dianhydride. This polyimide retains its nonlinear optical properties even after heating at 170 °C for extended periods. It works in the absence of an external electric field. The optical gain coefficient is approximately 22 cm^{-1} in zero field. The absorption coefficient, however, is much higher. The polymer consists of two sections —A—B— :

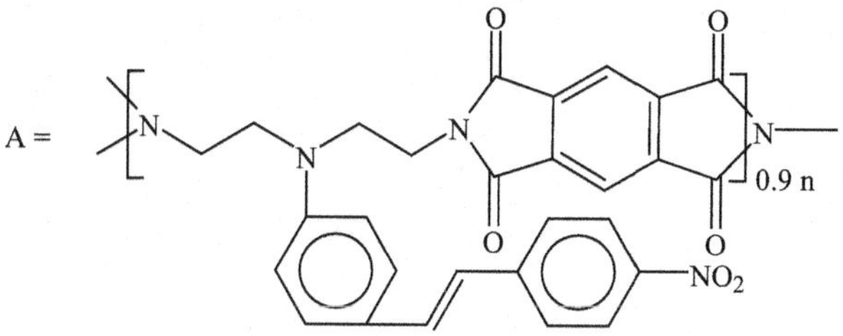

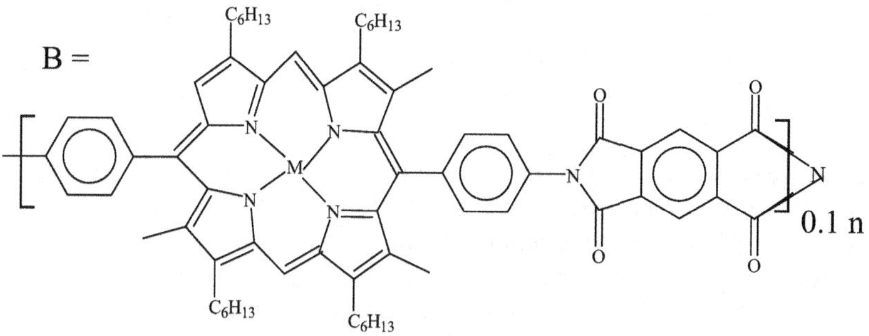

A =

B =

The approach by Drost, Jen, and Rao was to incorporate into typical nonlinear optics materials [60] electron-rich five-membered heteroaromatic groups, like thiophene as part of an all π conjugation between the donor and acceptor substituents. The function of the conjugating unit is to transfer electrons. In addition, they also incorporated a very active acceptor, a tricyanovinyl group into the thiophene-based moiety [60]:

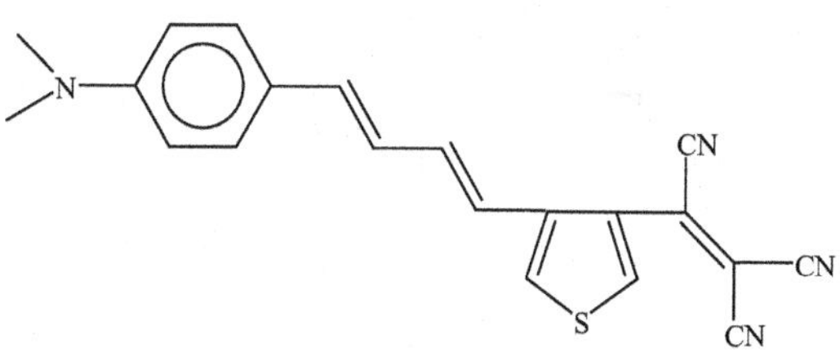

Drost and coworkers [60] also chose materials for nonlinear optics applications with high T_g. The considerations was that if these materials are poled at the glassy temperatures (>250 °C), then at operating temperatures of 80 to 120 °C the dipoles of the chromophores would not randomize, and the alignment would be preserved. An example of such a polyimide can be represented as follows [60]:

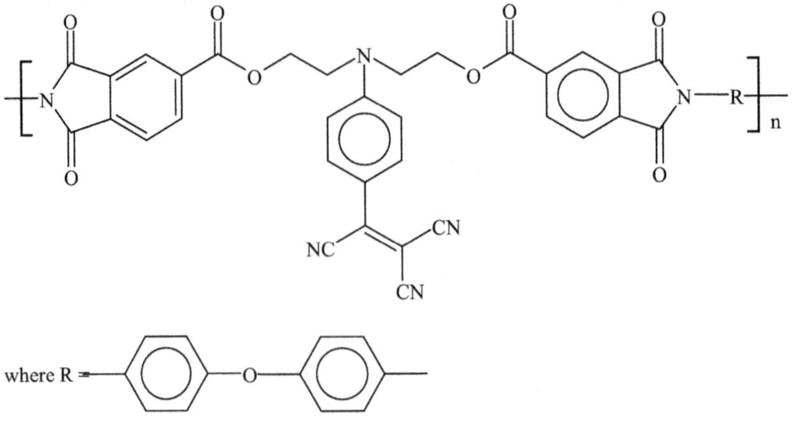

where R =

The above polyimide, one that contains 32 mol% of nonlinear optics chromophore, was reported to have an r_{33} of 16 pm/V (at 0.83 μm) with the electro-optical signal showing almost no change when the polymer is heated to 90 °C for greater than 1000 hours [60].

Polyimides with diamino chromophores were prepared with the aid of the Misunobi reaction [61]:

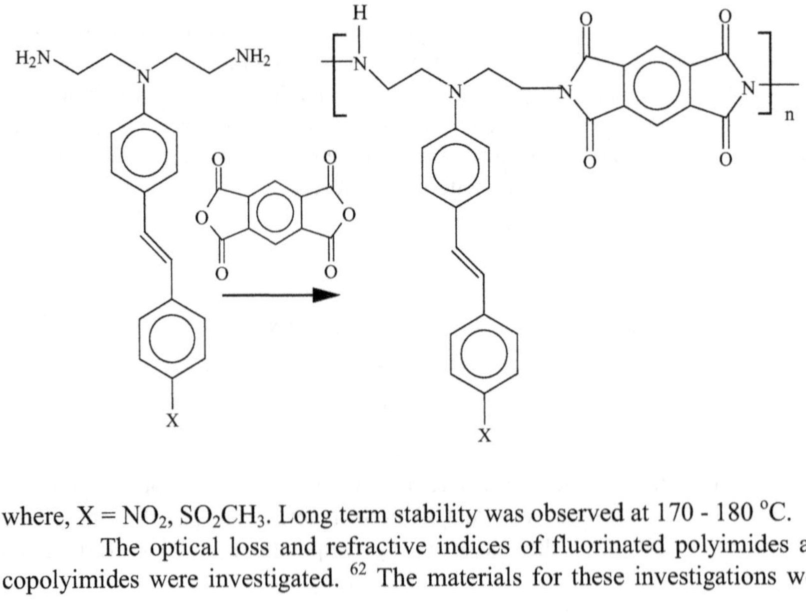

where, X = NO$_2$, SO$_2$CH$_3$. Long term stability was observed at 170 - 180 °C.

The optical loss and refractive indices of fluorinated polyimides and copolyimides were investigated. [62] The materials for these investigations were prepared from 2,2'-bis(3,4-dicarboxyphenyl)hexafluoro-propane dianhydride, pyromellitic dianhydride, and 2,2'-bis(trifluoromethyl)-4,4'-diaminobiphenyl. They can be illustrated as follows:

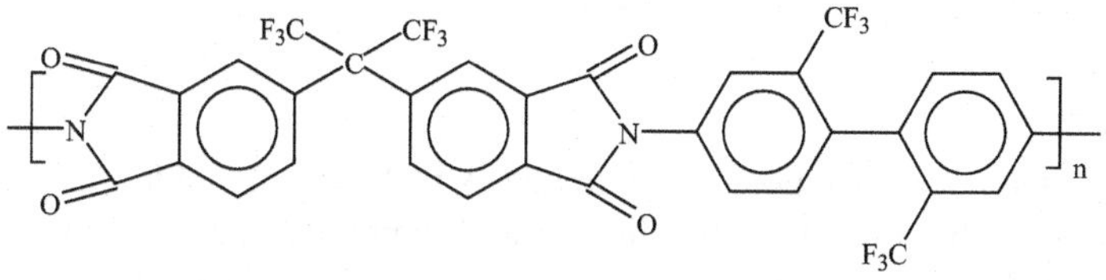

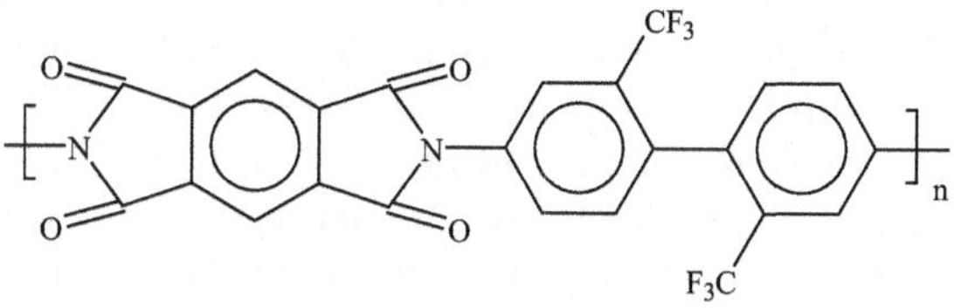

A two step, generally applicable synthetic approach for polymers with non-linear optics side-chains, aromatic polyimides was developed Chen, *et al.* Following synthesis was used in the preparation of the polymers [63]

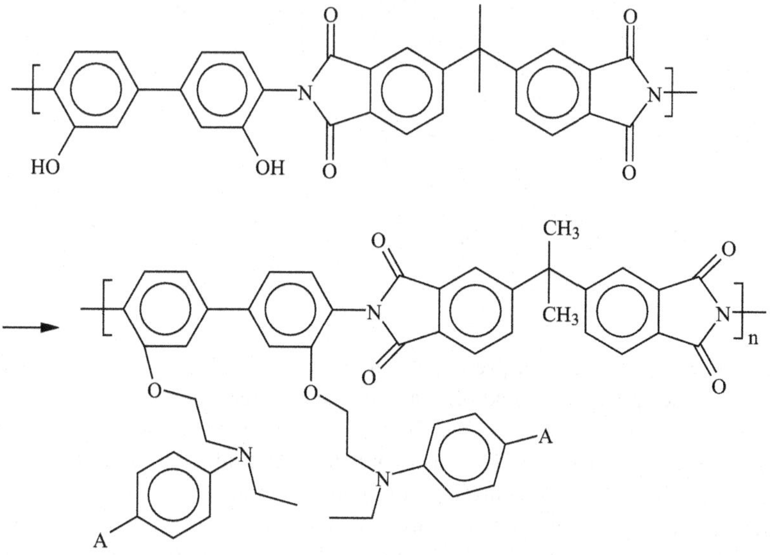

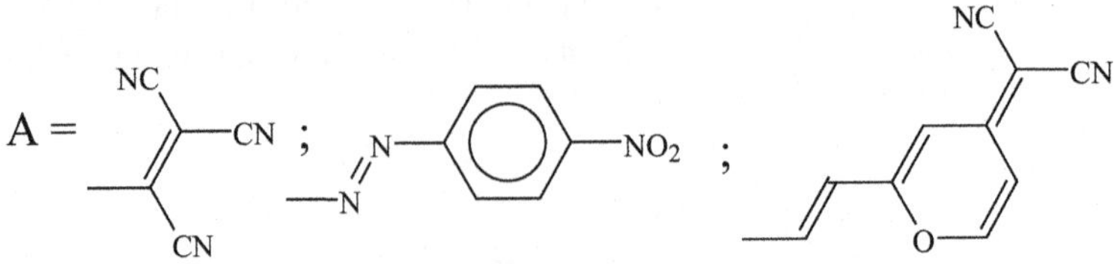

$$A =$$

One-pot preparation of a preimidized, hydroxy polyimide was reported.[64] It is followed by a covalent bonding of a chromophore onto the backbone of the polyimide via a mild postMisunobi reaction. Using this technique, aromatic polyimides with different polymer backbones and different chromophores were synthesized, and the chromophore loading level in these polyimides was controlled efficiently from 0 to 50% by weight. These polyimides were reported to have high glass transition temperatures ($T_g > 220$ °C) and are claimed to be thermally stable. Large electrooptic (E-O) coefficient (r_{33}) value (11 pm/V measured at 0.83 μm and 34 pm/V at 0.63μm) and long term stability (>500 h at 100 °C) of the dipole alignment were observed. Earlier this group also reported preparation of another polyimide by the same technique [64] :

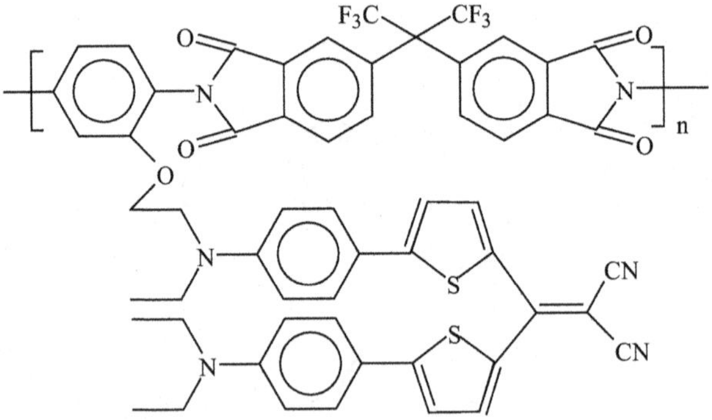

The advantage of this one-pot preparation method is that the chromophores are attached to the polymer backbone at the last stage using a very mild condensation reaction between the hydroxy substituted polyimide and a hydroxy-substituted chromophore. This avoids the harsh imidization process and the necessity of preparing the chromophore-containing diamine monomers. The method also allows a wide variety of polyimide backbones.

Burland and coworkers [65] synthesized somewhat different types of polyimides. They can be illustrated as follows :

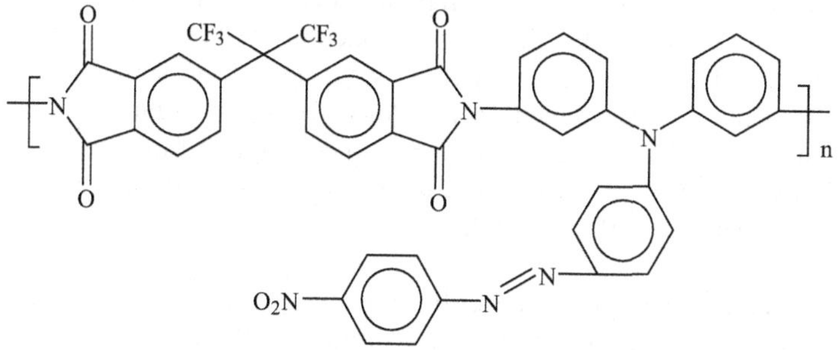

In this compound the electron donor is the diarylamino group and the acceptor is the nitro group. They are connected by an azobenzene bridge. In the above polymer the chromophore is present in every repeat unit and accounts for more that 50 % of the polymer's weight.

Yu and coworkers[66] prepared a similar polyimide that bears phenylene diamine and a diazobenzene-type nonlinear optical chromophore:

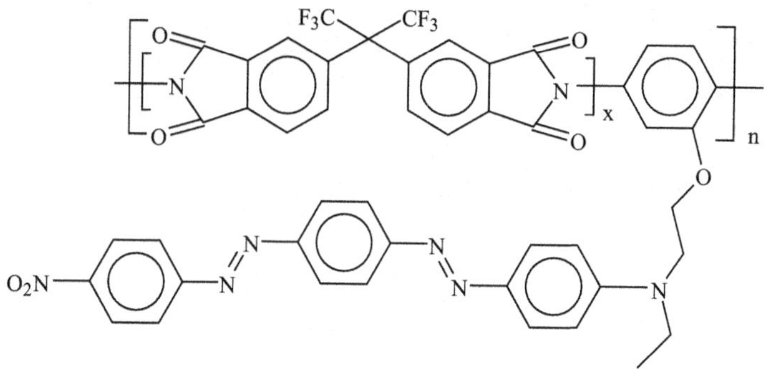

This high glass transition temperature polymer exhibits large electrooptic coefficients, $r_{33} \sim 14 - 35$ pm /V. The second harmonic measurements indicate long-term stability of the dipole orientation (> 800h at 100 °C) .

In still another report in the literature, the syntheses and characterizations of two aromatic polyimides with nonlinear optical chromophore side chains were reported through a two-step synthetic route[103] These two polymers were prepared by polycondensation of 4,4'-(hexafluoroisopropylidene)diphthalic anhydride with 4-(4-amino,2-hydroxy) phenoxyaniline and 2,2-bis(3-amino-4-hydroxyphenyl)hexafluoropropane, respectively. The resulting polyimides bearing hydroxy groups were found to react easily with the terminal hydroxy group of the chromophore via the Mitsunobi reaction. It was claimed that the resulting polyimides possess high glass transition temps. (Tg > 185°C), excellent solubilities and process abilities even though the extent of chromophore grafting is up to 95 mol %. [67]

Also, Zhu and coworkers [68] reported preparing of two high T_g (250° and 258°) polyimides embedded with triphenyl imidazole chromophores. Simultaneous polymerization and poling was carried in situ. The second harmonic generation measurements technology yielded moderate d_{33} values of 15 pm/V and 16 pm/V. The signals of these poled polymer films were reported to be without any decay at < 150°; The relaxation starts rapidly over 195°. The half-decay temperatures, of the samples orientation are as high as 240° and 224° respectively.

Yu and coworkers [69] describe a technique to retain stability and maximize the electrooptic response. They preformed the imide structure with an aryl dihalide monomer and incorporated highly efficient chromophore groups before polymerization. This highly functionalized monomer was subjected to a Pd-catalyzed coupling reaction with 2,5-bis(tributyltin)thiophene to give the polymer [69]:

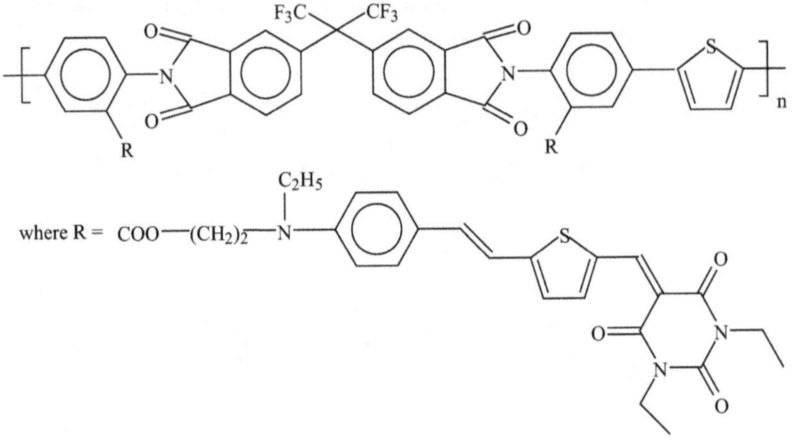

The polymer, shown above is soluble in organic solvents, has a glass-transition temperature (T_g) of 170 °C and its electro-optical response is 33 pm/V. [69]

Polycondensation and imidization of *m,m'*-diaminobenzophenone and pyromellitic dianhydride under microwave radiation was also carried out. [70] The product polyimide was obtained in a two-step process. It is claimed that this product of microwave radiation polymerization compares favorably with a product of conventional thermal polymerization, because it exhibits third-order nonlinear optical coefficient of 1.642×10^{-13} esu and response time of 24 ps. The third-order optical nonlinearity of this polymer is dependent on the chain length and the molecular structure.

6.3.2.2. Polycarbonates

Polycarbonates are also high melting rigid polymers. They were, therefore, also looked at for potential use in nonlinear optics. To achieve high molecular weights, some attempts were made to prepare them via ring opening polymerization. This, however, led to intractable materials. [66] The cyclic structure can be shown as follows:

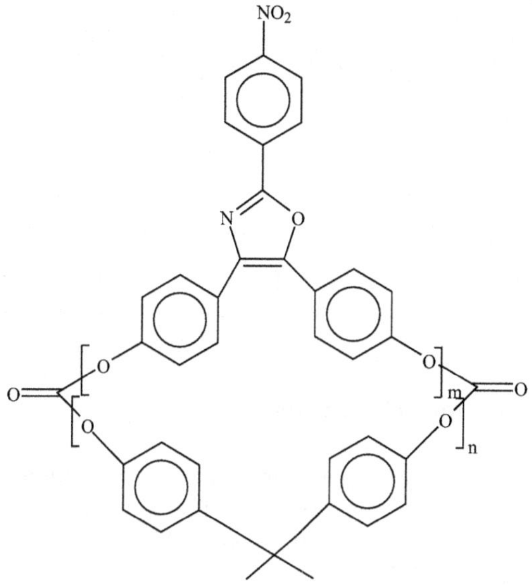

A copolymer, was, therefore, formed in the investigations [66]:

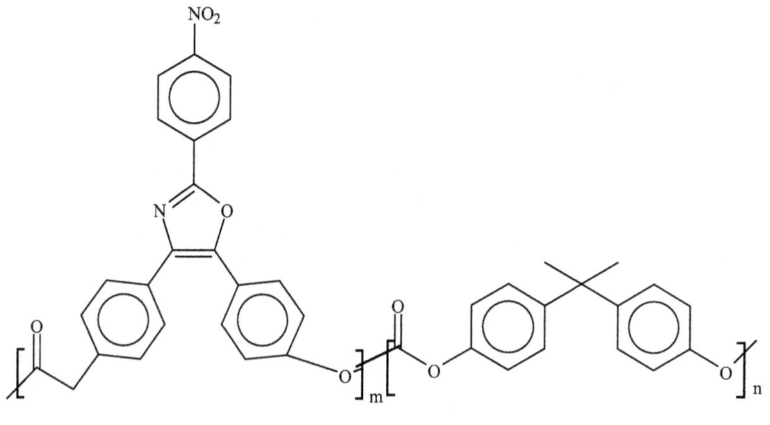

The above shown copolymer, after poling, had an electrooptic coefficient of 55 pm per volt at about 1.3 μm. The chromophore, not attached covalently to the polymer, in this case is based on a heterocyclic acceptor that contains two electron withdrawing groups, dicyanomethilidine and sulfone. The donor is a di-*n*-butylamino moiety. This chromophore can be illustrated as follows.

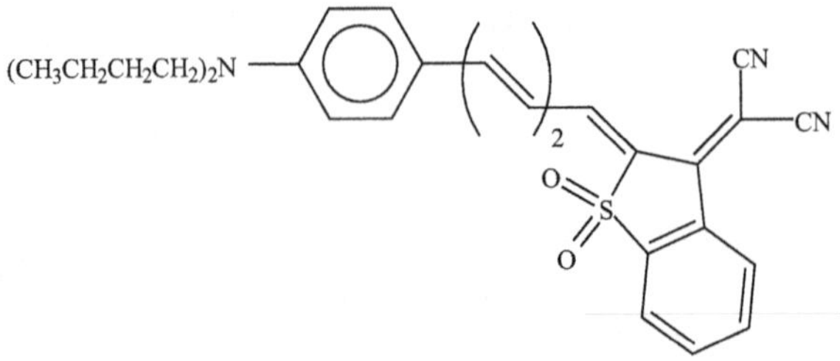

This chromophore was dispersed in a polycarbonate matrix and the composite film poled at 80 °C. The thermal stability of this material, however, does not exceed 150 °C.

6.3.2.3. Highly Aromatic Polyquinolines

Chen *et al.* [71] prepared polyquinolines with both aliphatic and aromatic in-chain structures to provide sites far the side-chain nonlinear optics chromophores. The aromatic ones are also be high melting, rigid molecules. They can be illustrated as follow:

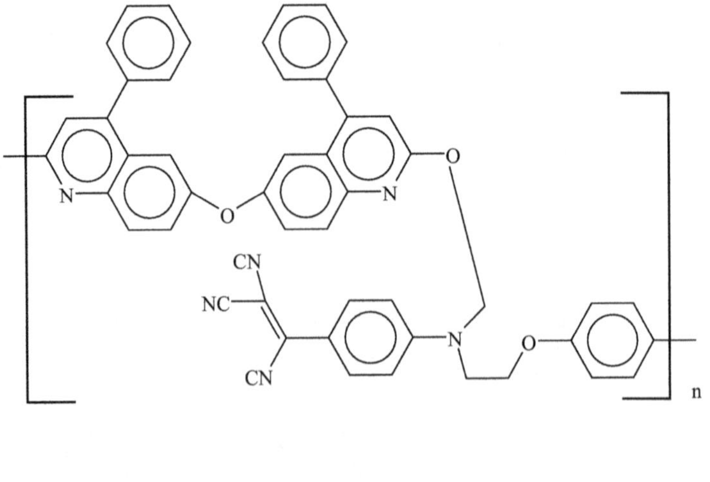

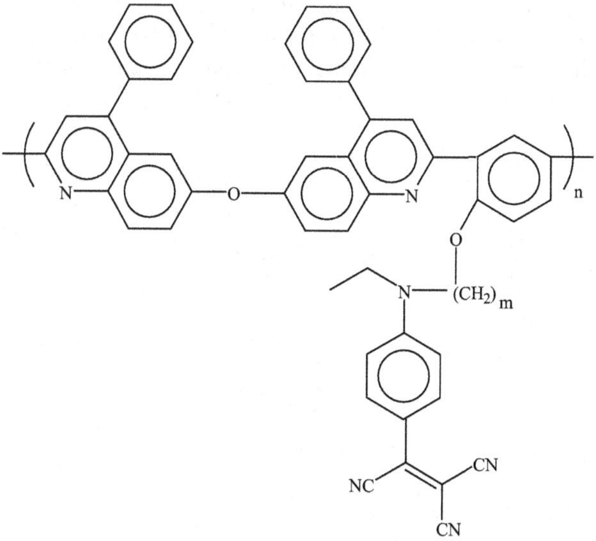

These polymers were processed into optical-quality thin films for potential device fabrication. The authors claim good solubility for several of these materials in common organic solvents with molecular weights as high as 41,000. These materials are reported to be especially advantageous for use in nonlinear optics applications due to combination of film forming ability, good thermal stability, and high glass transition temperature. It is also claimed that when chromophore dipoles are aligned by contact or corona poling, useful levels of second order nonlinearity are induced that appear stable at temperatures necessary for the fabrication of optical devices. [71]

6.3.2.4. Polynaphthalenes

Polymers containing binaphthalene units can adopt a helical configuration. Due to the twisted configuration of the binaphthalene moiety, the polymers that form through rigid groups with triphenylamine derivatives are helices. It is the rigidity of the system that gives rise to rigid, rod like, helical structures. All bonds that form the helix have the same configuration (either *S* or *R)*. As a consequence, if both *R* and S monomers are incorporated in the same backbone, no helical configuration can be formed. The rigidity of the backbone can greatly diminish unwanted dipolar interactions between the chromophores, allowing a very high chromophore density.[72] In addition, when the polymers are chiral, they can increase the nonlinear optical response. [72]

Samyn *et al.,*[72] reported preparation of chiral chromophore-functionalized polybinaphthalenes for nonlinear optics. These polymers were formed by direct polymerization, using Stille coupling between a bis(trimethyltin) compound and dibromo-substituted binaphthalene monomers. The chromophores were attached to the binaphthalene units via alkyl spacers:

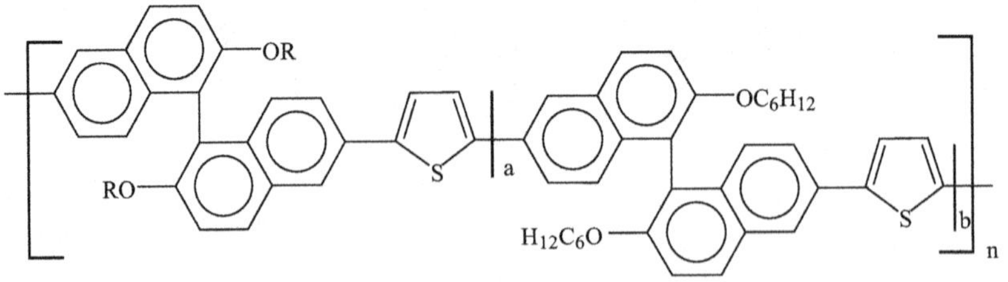

where R is ,

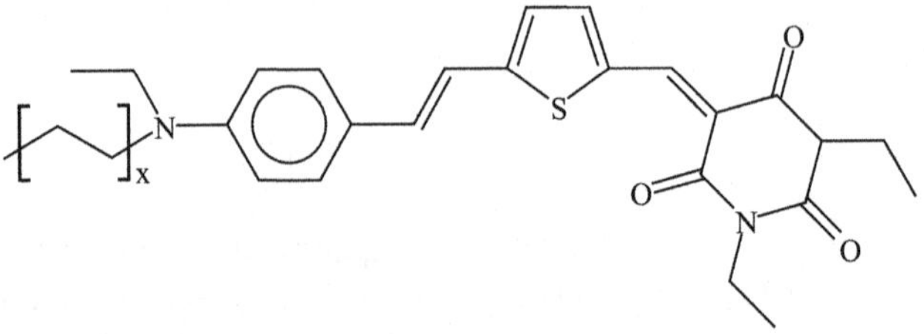

x is hexyl. The authors report that the typical tree-like macromolecular architecture of these molecules gives rise to a unique behavior in the glass transition temperature as well as in nonlinear optical properties. The nonlinear optical response shows a continuous increase in function depending on the chromophore content. In this way, the nonlinear optical properties can be increased in a way that is not possible with other chromophore-functionalized polymer materials.[72] By attaching chromophores as side chains to the rigid non bendable backbone a treelike structure with flexible branches forms.

The same group reported [72] preparation of eight other chiral, chromophore-functionalized donor-embedded polybinaphthalenes. The polymers were prepared by direct polymerization, using Stille coupling reaction between a chiral bis(trimethyltin) binaphthalene derivative and diiodo-functionalized chromophores. They reported that use of diiodo-functionalized instead of dibromo-functionalized chromophores results in a significant increase of molecular weight. The reaction conditions allow use of a great variety of chromophores with different structures. The typical tree-like macromolecular architecture of the polymers is reflected in the behavior of the glass transition temperature and, more clearly, in the nonlinear optical properties. The nonlinear optical response shows a continuous, linear increase as a function of chromophore concentration, indicating that the dipolar interactions between the chromophores are eliminated. This group of polymers can be illustrated as follows [72]:

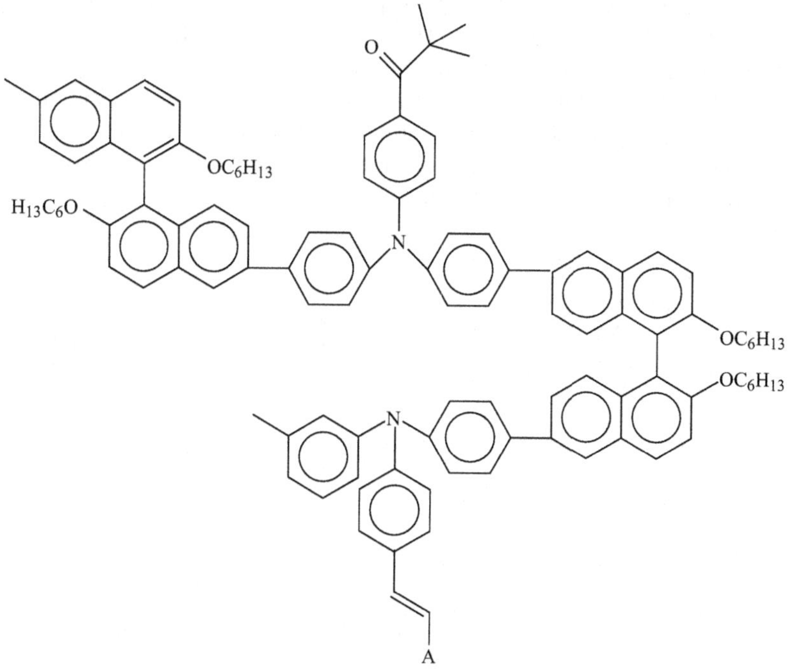

where A ca be:

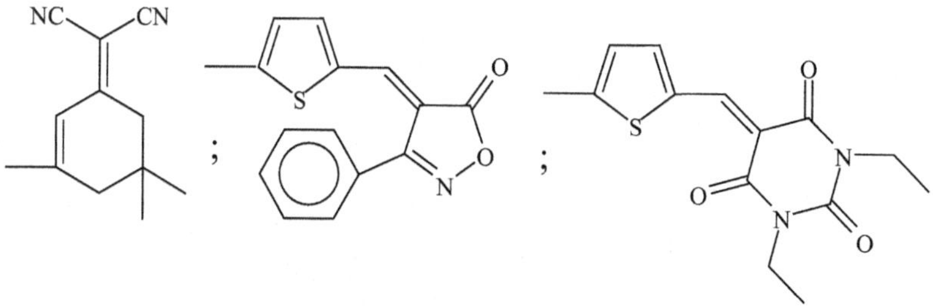

6.3.2.5. Fluorinated Poly(arylene ethers)

Another approach, starts with preparation of two fluorinated poly(arylene ether)s, containing perfluorophenyl moieties in the main chains and second-order nonlinear optical chromophores in the side chains.[73] The polymers were formed by Knoevenagel condensation reactions between perfluorophenyl-containing poly(arylene ether) and cyano acetylated chromophores. The molar percentages of the pendant chromophores were estimated to be 162% (1) and 138% (2) for the two polymers, respectively, and their weight percentages were both 42.3%. Both of the two polymers are thermally stable and readily soluble in common organic solvents. The glass

transition temperatures $(T_g's)$ of the two polymers were determined to be 186
(1) and 192 °C (2), respectively. The in-situ second harmonic generation
measurement revealed the resonant nonlinear optics coefficient (dss) values of
60 (1)and 31 (2) pm/V for the poled films, respectively. [73] The polymer consists
of three sections. In section A both R groups are substituted, in section B only
one R group is substituted and the other one is an aldehyde, while in section C
both R groups are aldehyde.groups.

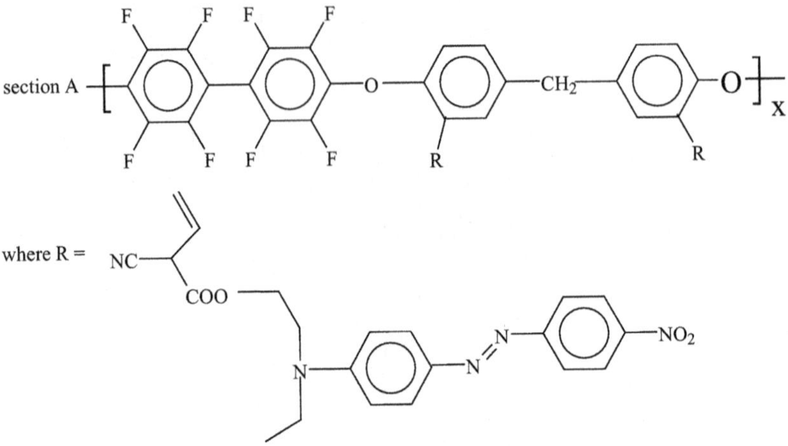

where R =

the second copolymer is very similar, except that the R groups are:

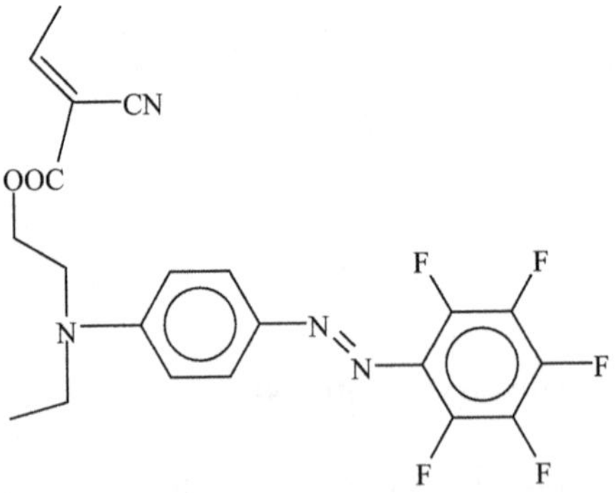

6.4. Polysiloxanes

The polysiloxanes are not noted for high grass transition temperatures.
Actually it is the opposite, and many have very low T_g-s. If rigid moieties are

introduced into the polymeric structures, however, they can form liquid crystals. As such, they can maintain an orderly arrangement.

It was reported that siloxanes containing up to 20 % spiropyran groups are liquid crystalline. [74] At higher compositions they are amorphous. With increased spiropyran content, the selective reflection band characteristic of cholesteric mesophases shifts to lower wavelength:

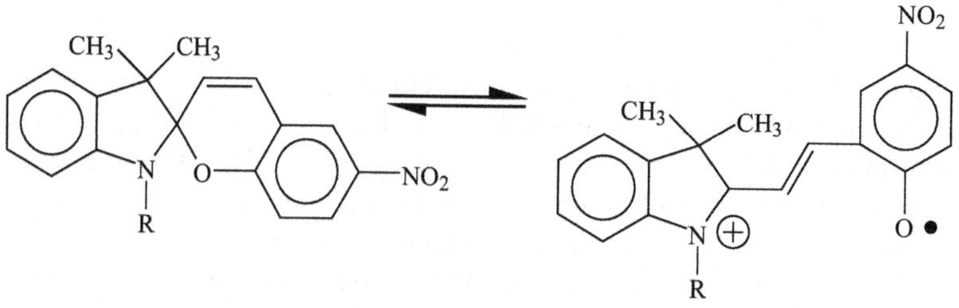

spyropyran   "merocyanine" blue

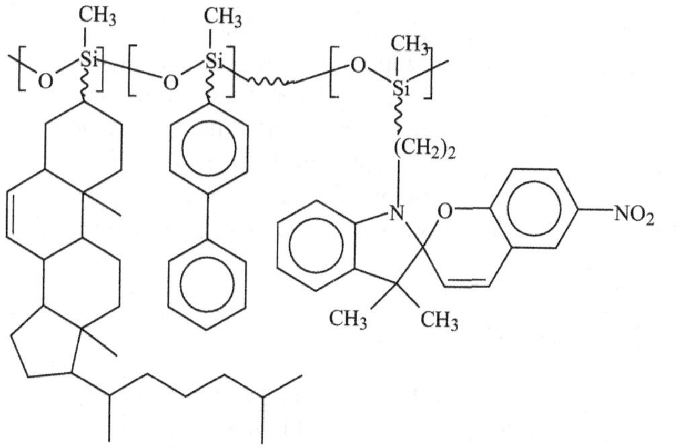

The photochromic reaction leading to the blue merocyanine formation results in a narrowing of the reflection bandwidth.

Sanchez and coworkers [74] obtained very high second order nonlinear optical properties from materials derived from sol-gels and based on organic-inorganic hybrid structures. These materials are formed by reactions of an organic polyfunctional alkoxy silane derived from Red 17 dye with tetramethoxysilane in the sol-gel formation. The Red 17 structure, 4-(amino-N,N-diethanol)-2-methyl-4'-nitroazobenzene was reacted with 3-(isocyanatopropyl)-triethoxysilane via urethane linkages to introduce the alkoxysilyl reacting groups:

This structure, carrying the nonlinear optics chromophore, was co-hydrolyzed with tetramethoxysilane in a tetrahydrofuran and water medium acidified with HCl to form sols, which were subsequently aged 3 days. Transparent coatings were then spin coated onto glass substrates and dried to prepare films for optical property measurements. Chromophore orientation of these films by the standard corona poling technique resulted in second harmonic generation responses as high as 150 pm/V. It was claimed that thermally precuring the samples considerably improves their nonlinear optics response and thermal stability. [75]

On the other hand, formation of photorefractive silicone composite with good performance was reported [76] A carbazole-substituted polysiloxane that was sensitized by 2,4,7-trinitro-9-fluorenone was used as a photoconducting medium and 1-[4-(2-nitrovinyl)phenylpiperidine was added as an optically nonlinear chromophore. The photorefractive property of polymer was determined by diffraction efficiency using a 100 μm-thick film. The maximum diffraction efficiency (η max) of 71% was obtained at the electric field of 70 V/μm. [76]

6.5. Polyacetylenes

Cheong and coworkers,[77] demonstrated that a soluble, asymmetric polydiacetylene can be used to form stable monolayers at the air-water interface of a Langmuir trough. The monolayers of this amphiphilic polymer were then repeatedly deposited onto hydrophobic substrates during every upstroke cycle of the vertical dipping method to form Z-type monolayers. The second harmonic generation increases in these films with the number of layers, and there are indications of a high degree of orientation anisotropy of the backbone. The second order nonlinear coefficient (d_{33}) was estimated to be 1.52 pm/V at 1064 nm of Nd:YAG laser.

In addition, it was reported recently, [78] that a diacetylene monomer with a rigid backbone and capable of forming hydrogen bonds was polymerized in such a way as to form two-dimensional super-molecular assembly. The two-dimensional structure self-assembles when UV light generates polydiacetylene comb polymers, and hydrogen bonds are established within molecular layers. The material forms blue solid thin films which generate third-order nonlinear optical signals and exhibit photochemical stability to 1064 nm radiation from a

Q-switched Nd:YAG laser. Heating the polymer to 62 °C changes the color to bright red. This is reversible. The two-dimensional structure is still maintained. [77] This thermochromic process is accompanied by endothermic and exothermic signatures detected by differential scanning calorimetry. Third-harmonic generation signals retain much of their original intensity through the thermochromic transitions. It was observed, however, that the results do not conform in a consistent manner to both the theory of third-order effects and to a previously suggested connection between intramolecular conjugation and optical absorption of polydiacetylenes. The explanation that was offered is that possibly intermolecular interactions in these highly ordered structures play a role in defining optical properties. [78] The acetylene groups containing monomer was illustrated as follows:

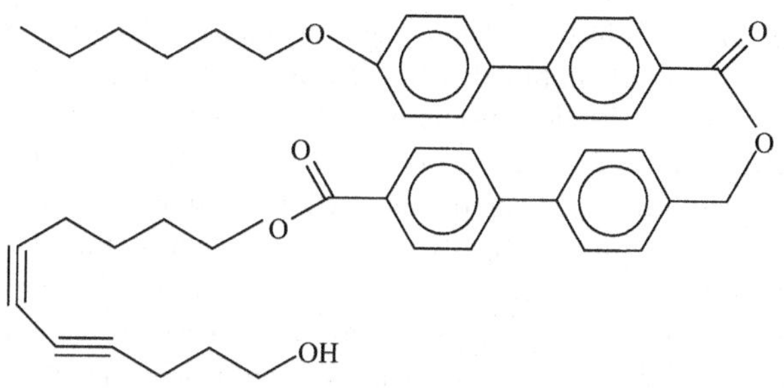

Wang et al, [79] synthesized of a soluble polymer from diacetylene with push-pull azobenzene and pyrimidine rings attached. The polymer was formed from an asymmetric polydiacetylene having a pyrimidine ring and an azobenzene chromophore directly linked to the two ends of a diacetylene moiety. The material was polymerized thermally or under light illumination. The resulting polydiacetylene with push-pull azobenzene chromophores as side chain and a pyrimidine ring covalently bonded to the diacetylene conjugated system was found to have nonlinear refractive index of 1.18×10^{-11} esu. The macro and microscopic susceptibilities $X^{(3)}$ and y for the functionalized polydiacetylene; $X^{(3)}$ is 1.01×10^{-11} esu and y is 1.06×10^{-30} esu.

6.6. Polymers with Carbazole Structures in the Backbone

Zhang and coworkers [80] prepared carbazole group containing polymers by the Knoevenagle reaction. The reaction was carried out in two stages, first in THF and then as a solid state condensation. This yielded high molecular weight polymers. Thin films of the polymers exhibit good optical quality. They can be obtained from their chloroform solutions by spin-coating. After electric poling, the films are reported to show reasonable second-order nonlinear optical responses [80]:

6.7. Liquid Crystal Polymers

Chiral smectic ferroelectric liquid crystals are liquids that possess spontaneous polar order. Combined with their excellent processibility on silicon integrated circuits, these liquid crystals provide an attractive potential approach to synthesis of materials for second order nonlinear optics, provided adequate second order susceptibility $X^{(2)}$ can be obtained. Unfortunately, the second-order nonlinear optical susceptibility ($x^{(2)}$) of the ferroelectric liquid crystals are usually low and their thermal stability is limited. Several, very interesting approaches to utilization of liquid crystals, however, were carried out.

Walba et al. [81] tried to create a "side-by-side" type dimer structure where two conventional liquid crystal core structures that are covalently connected by an azo link, and are appropriately substituted to afford a system where the DR1 unit is essentially forced by the two liquid crystal cores to orient normal to the liquid crystal director. While this strategy did indeed afford the desired supermolecular stereochemistry, the dopants were only soluble up to about 30% by weight in a standard ferroelectric liquid crystal host. Furthermore, it was not clear whether the side-by-side dimer structure was compatible with smectic mesogenicity. They reported instead preparation, structure and properties of two similar side-by side dimers that possess good smectic liquid crystallinity; structures A and B. Both materials possess a broad monotropic smectic A* phase. While crystallization from this phase occurs eventually, both structures can be kept in the A* phase at room temperature for days. The polyunsaturated analog B possesses improved mesogenicity, showing a narrow enantiotropic A* phase. Polarized light spectroscopy showed that the DR1 unit makes an angle of about 60" with the liquid crystal director, in good agreement with the expected super-molecular structure. Furthermore, compound B is shown to mix in all proportions with a standard C phase host, showing the ferroelectric C* phase at concentrations up to 60% by weight of B . The ferroelectric liquid crystal mixtures show a good linear relationship between the maximum observed ferroelectric polarization and concentration of B, and most importantly the expected sign of the polarization, demonstrating that compound B "fits the C* binding site" in the desired manner.

Walba et al.[81] synthesized a low molar mass ferroelectric liquid crystal specially designed for second-order nonlinear optics that showed a second harmonic coefficient (d_{22}) of 0.6 ± 0.3 pm/V in the chiral smectic C (SC*) phase. The nonlinear optics-chromophores aligned in the direction of the polarization, perpendicular to the long axis of the molecules were later successfully varied. [81] This can be illustrated as follows:

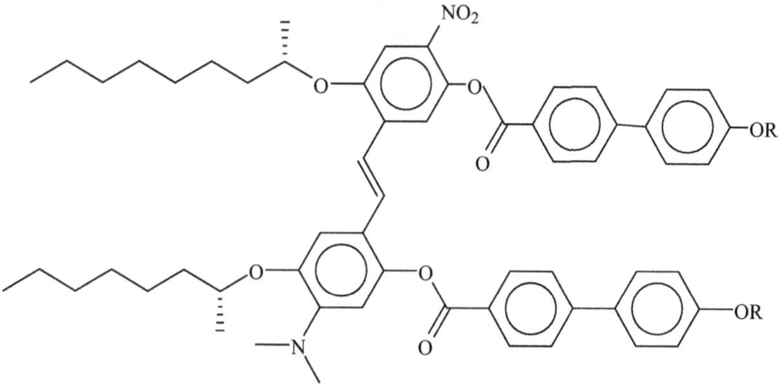

where, R is decyl in A and 9,12-octadecadienyl in B.

The thermal and mechanical stability of nonlinear optics materials were improved [82] when a polymer was synthesized with ferroelectric liquid crystalline side-chain and then further improved by Keller et al [83] when they formed a ferroelectric liquid crystalline main-chain polymer. The thermal and mechanical stability and the long-term properties of nonlinear optics-materials are known to be improved by crosslinking. The idea was therefore to synthesize a crosslinkable monomer or a crosslinkable monomer mixture that possesses a chiral smectic C (S_{C*}) mesomorphism over a wide temperature range. This ferroelectric mixture can then be aligned and photocrosslinked into a pyroelectric polymer. The system, developed by Walba et al., [81] is based on two monomers, where only one is nonlinear optics-active81. The acrylate monomer mixture displayed a ferroelectric SC* phase which could be photocrosslinked into a pyroelectric polymer, shown below. The crosslinked material displays a clear second harmonic signal (d-coefficient = 0.4 pm/V).[81] The molecular structure of this polymer was later modified in order to increase the nonlinear optics activity. [81]The possibility to modify the monomer system are strongly limited due to the demands of ferroelectricity and possibility to crosslinking. However, promising one single monomer system was recently developed. [81] The new chiral monomer displays a large spontaneous polarization (175 nC/cm^2)

in the chiral smectic C phase. And the photocrosslinked pyroelectric polymer shows a considerably enhanced second-order nonlinear optical activity. The modifications of the molecular structure was possible through the use of the enantioselective biocatalyst *Candida antarctica* lipase B.

Kajzar and coworkers [1] reported forming side chain liquid crystalline polymers with high T_g containing maleimide units along the backbone. The choice of maleimide was based on its high dipole moment at right angle to the backbone and its cyclic structure imparting great stiffness to the chains, yielding a high T_g. A cyano-biphenyl-based chromophore was selected as an nonlinear optics mesogenic moiety and an oligomethylene space was used to decouple the chromophores from the main chain and to enhance the possiblity of self organization into anisotropic mesophase [92]:

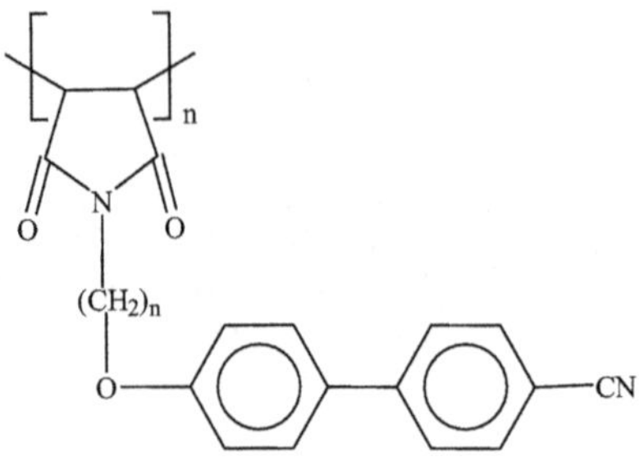

These polymers were also reported to demonstrate good second-order nonlinear optical properties.

You *et al.* [49] describes the synthesis and physical study of several new photorefractive polymers that consist of a nonlinear optical chromophore attached to conjugated poly(p-phenylene-thiophene)s backbones. A Stille coupling reaction was used to prepare these materials.

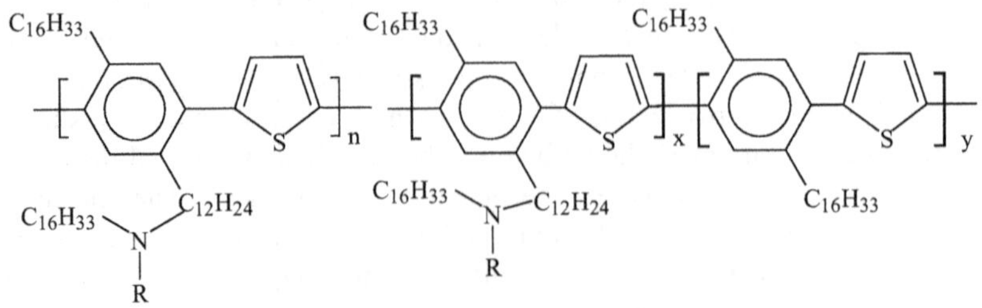

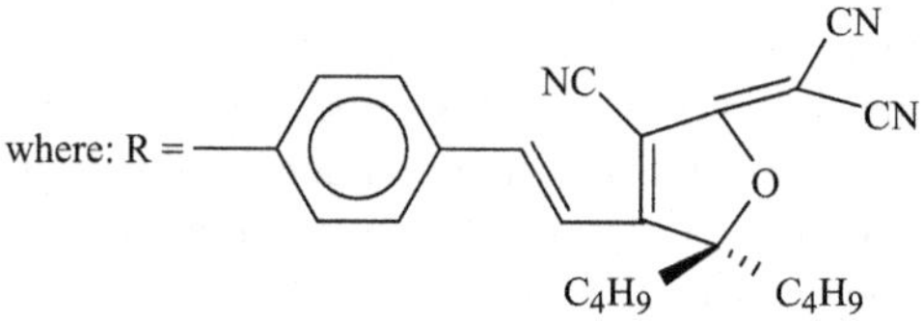

where: R =

The resulting polymers exhibit high photorefractive performances. An optical gain coefficient of 158 cm^{-1} at a field of 50 V/μm and a diffraction efficiency of 68% at a field of 46 V/μm for polymer were obtained. [49]

Barmatov and coworkers [84] took an approach of forming photo-optically active polymers by using hydrogen bonding between functionalized liquid crystalline copolymers and low-molecular-mass dopants. The photochromic dopants used by them contain azobenzene components:

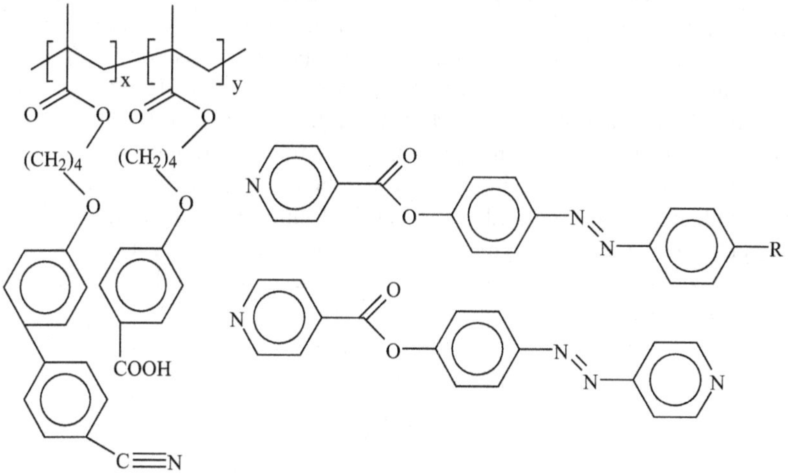

The formation of hydrogen bonds between the carboxylic acid groups of the functionalized liquid crystal copolymers and the pyridine portion of the dopants leads to stable, none separating mixtures. In mixtures containing up to 30 % of the dopants no separation was observed. Induction of a nematic mesophase is observed in the case of a smectic polymer matrix doped with low molecular weight photochromic dyes.

6.8. Acrylic Polymers

Syntheses and characterization of three fully functionalized photorefractive polymethacrylates containing different chromophores were reported[85] Carbazole and nonlinear optics-functionalized methacrylate monomers in a 1:1 ratio were polymerized with 20 mol % of dodecyl methacrylate to obtain photorefractive polymers. The resulting polymers had glass transition temps, of 48, 47, and 52° C, respectively. These polymers, when doped with 1 wt % 2,4,7-trinitro-9-fluorenylidene)malononitrile, showed good photorefractive properties. A net two-beam coupling gain and a diffraction efficiency of 60% were observed. at 58 V/μm. [85]

Hirai et al., [86] reported synthesis and photochromic behavior of methyl methacrylate copolymers that contain anil groups. These groups are pendant 4-(methacryloyloxyalkyl)-N-(4-methoxysalicylidene)-aniline units. The molecules contain 4-12 carbon alkyl chain spacers between the polymer backbone and the pendant sallicylidene aniline units. The photochromic behavior was investigated in the solid state. Photochromism was observed at 445 nm in five copolymer films under UV irradiation. The color thermal decay reactions increased with increasing temperature, and the rate was five to seven times slower than those of monomers blended in with poly(methyl methacrylate). The rate of thermal decay reactions depended on the alkyl chain length of sallicylidene aniline units. It decreases with decreasing alkyl chain spacer because of restriction of the thermal conformational change by the polymer backbone. The difference in the thermal decay reaction is small when the carbon number of the alkyl spacers was greater than 10, and contains more than 0.08 mol azo groups.

Chromophores with indole and nitrobenzene push pull groups were reacted with methacroyl chloride and acryloyl chloride. The resultant acrylic monomers were then copolymerized with methyl methacrylate and butyl acrylate to yield photorefractive acylic copolymers that can be illustrated as follows [87]:

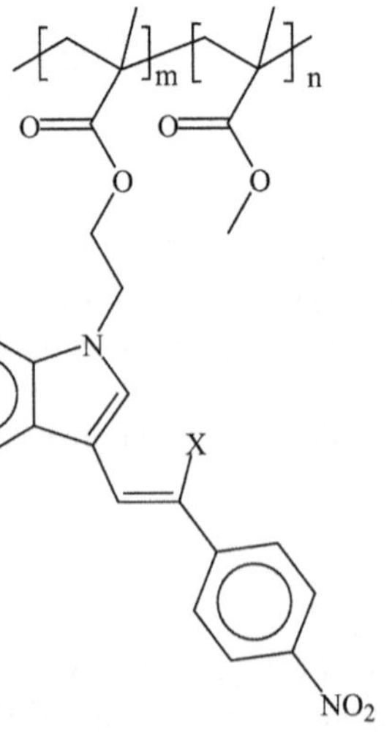

These polymers are very soluble in organic solvents. They fabricate into optically clear polymer films. The glass transition temperature, is 150 °C for the methyl methacrylate copolymer and 8 °C for the butyl acrylate copolymer. Large nonlinear optical coefficients $(d_{33}$ as high as 49 pm/V) and photoconductive sensitivity (on the order of 10^{-9} S-cm $^{-1}$/W-cm $^{-1}$) were observed for both copolymers.

The syntheses and nonlinear optical properties of methacrylate polymers based on 2-[4- (N-methyl-N-hydroxyethylamino)phenylazo]-phenyl-6-nitrobenzoxazole chromophores were reported.[88] Methacrylate polymers containing different molar contents of these nonlinear optical active molecular segments were synthesized. Polymers containing 6-17 mol % pf chromophore segments yielded amorphous and optically clear thin films. Some mesomorphic structural order was exhibited by a polymer with 33 mol % chromophoric units. Values ranging from 40 to 60 pm/V were measured with increasing chromophore molar contents.[88]

6.9. Polyphosphazenes

Qin et al.,[89] reported a new post functional approach to preparation of second-order nonlinear optical polyphosphazenes containing a sulfonyl-based chromophores. The sulfonyl groups act as acceptors. Two polyphosphazenes, one containing aniline and the other containing indole groups as side chains were obtained from poly(dichlorophosphazene), by nucleophilic substitution reaction. Then a postazo coupling of p-ethylsulfonylbenzenediazonium fluoroborate or p-octylsulfonylbenzenediazonium fluoroborate was carried out with the aniline or indole ring. This yielded sulfonyl-based chromophore-functionalized polyphosphazenes. The polymers exhibit good solubility in common organic solvents and are thermally stable. The maximum absorption in one appeared at about 440 nm, while at about 393 nm in the other one. The poled films of both exhibit a resonant d_{33} value of 27 and 18 pm/V, respectively, by second harmonic generation measurements.

These polyphosphazenes can be illustrated as follows [90]:

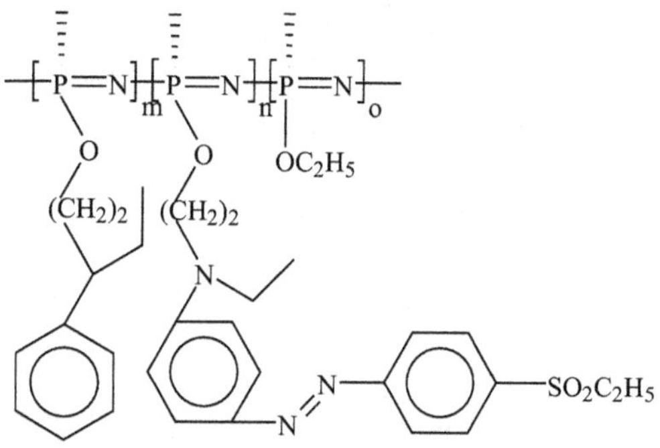

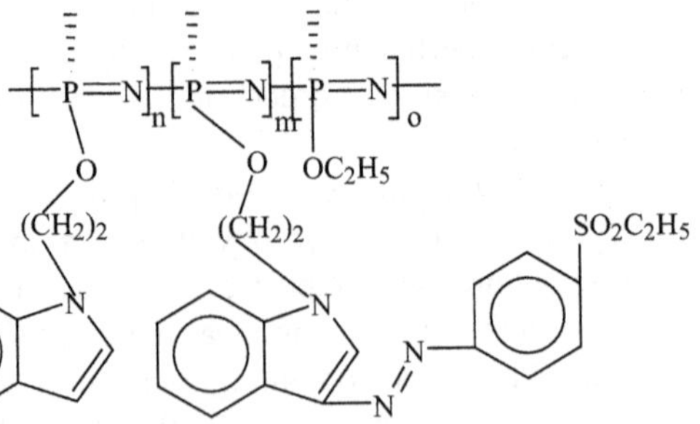

6.10. Miscellaneous Materials

Yu *et. al*,[91] found that molecules containing two 2,6-diacetamido-4-pyridone groups and a nonlinear optical chromophore will self-assemble with diimides or bis(uracil) structures, to yield polymer-like hydrogen-bonded assemblies.

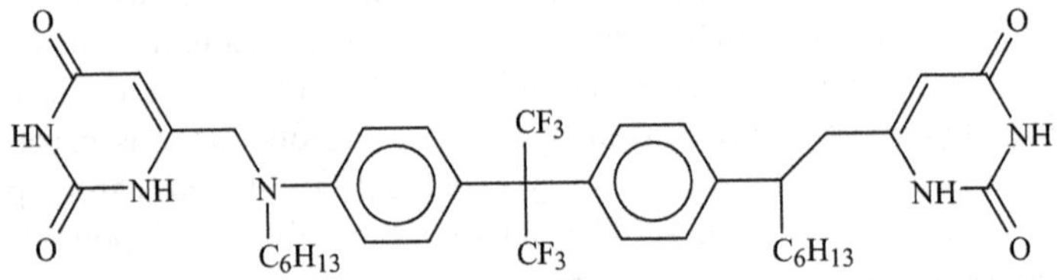

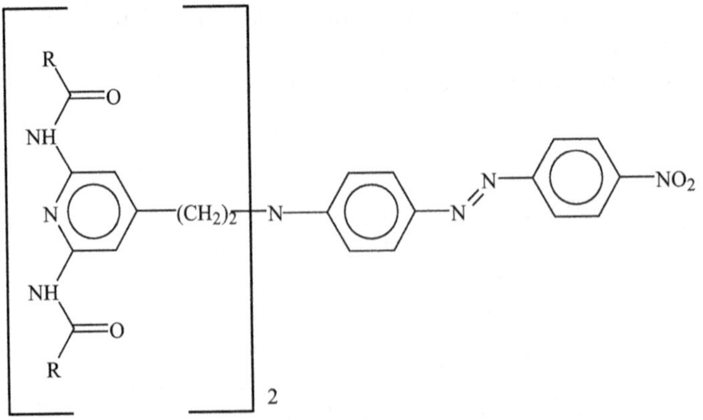

The "lock" and "key" structure in these super-molecular assemblies is provided by three hydrogen bonds per repeat unit. The nonlinear optics materials have T_g

values in the 72-100 °C range and are stable up to 120 °C, at which point the hydrogen bonds break and the structure is lost.

Kimura, et al.[92] synthesized polymers bearing oligo-aromatic esters as side chains to form second-order nonlinear optical active polymers on the basis of architecture. The cut-off wavelength (λ_{co}) of these polymers is shorter than the visible region, i.e., λ_{co} ca. 330-370 nm, which are much shorter than the typical second order nonlinear optics polymer containing chromophores like azobenzene. These polymer films exhibit good transparency in the visible region. The second-order nonlinear optical coefficient, d_{33}, is 2.2-9.8 pm.V^{-1}.

Twieg and coworkers, [93] reported preparation of poly(norbornene) copolymers functionalized with nonlinear optical chromophore side groups. Use was made of (η^6-toluene)Ni(C$_6$F$_5$)$_2$, catalyst in the polymerization of norbornene. The nickel complex used to polymerize the norbornene monomers is tolerant to many functional groups found in nonlinear optical chromophores. On the other hand, nitriles and amines other than trisubstituted amines strongly inhibit the reaction. A copolymer of hexylnorbornene and a norbornene-functionalized Disperse Red I chromophore was scaled up and studied in detail. Initial studies indicate that electric field poling is effective but that relaxation of polar order in the poly(norbornene) is faster than in a comparable methacrylate copolymer. The copolymer can be illustrated as follows:

Katz et al.[94] utilized tetracyanoethylene to form polymers with tetracyano-derivatives that are donor-acceptor molecules in a long π-conjugated system. [55] Tetracyanoethylene was reacted quantitatively with electron-rich acetylenic polymers to form polymeric materials containing repeating nonlinear optics chromophores. This [2+2] cycloaddition is followed by electrocyclic ring-opening to give tetracyanobutadiene. Presumably, this reaction is initiated by charge transfer followed by closure via a 1,4-dipolar intermediate to a cyclobutene, which opens to give tetracyanobutadiene derivative:

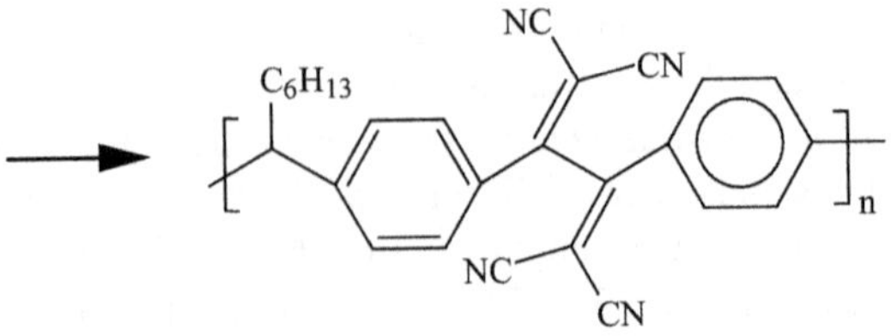

The polymer is a film former and forms by this reaction in virtually quantitative yields. The films can be cast from common organic solvents, like acetone, tetrahydrofuran, toluene, etc.

    Also, Liphardt *et al.*[95] describe photorefractive polymers with performance competitive with available crystals. The materials are a mixture of the electrooptic polymer based on bisphenol A and 4,4'-nitroaminostilbene with 29 % by weight of benzaldehyde diphenylhydrazone, a hole transporting agent:

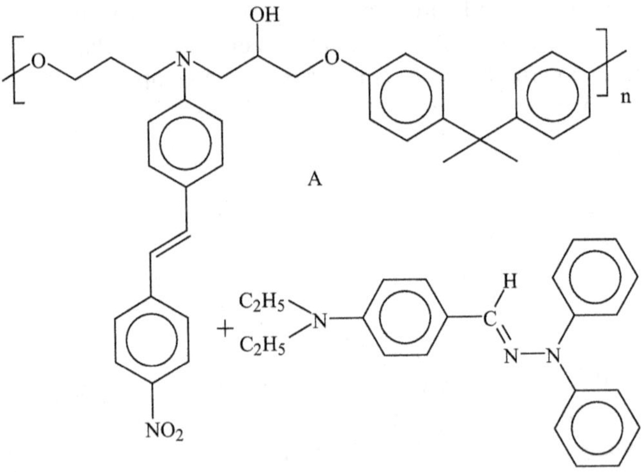

    The stilbene dye substituent plays two roles in these polymers: It is a source of charges and, when aligned in a static electric field, it produces the necessary bulk linear electrooptic response. The transport agent transfers the charge generated at the dye molecules to other parts of the material.

References

1. F. Kajzar, S. Gangadhara, S. Ponrathnam, C. Noel, and D. Reyx, *Am. Chem. Soc. Polymer Preprints*, **1997**, *38* (2), 498

2. P. Glinter, and J.-P. Huignard, In *Photorefractive Materials and Their Applications I and II,* Springer-Verlag: Berlin, Germany, 1988 (vol.l), 1989 (vol.ll).

3. Moemer, W.E.; Silence, S.M. *Chem. Rev.* **1994,** *94,* 127.

4. Solymar, L.; Webb, D.J.; Grunnet-Jepsen, A. In *The Physics and Applications of Photorefractive Material,* Oxford, U. K., 1996.

5. D.M. Burland, R.D. Miller, and C.A. Walsh, **Chem. Rev.**, **1994**, *94*, 32; P.N. Prasad, D.J. Williams, *Introduction to Nonlinear Optical Effeects in Molecules and Polymers,* Wiley, New York, N.Y. 1991

6. S. Ducharme, J.C. Scott, R.J. Twieg, and W.E. Moerner, *Phys. Rev. Lett.* **1991**, 66,1846

7. D.J. Williams, Angew. Chem. Int. Ed. Engl., **1984**, 23, 690

8. D.J. Williams, *Angew. Chem., Int. Ed. Eng.,* **1984**, *23*, 690

9. M. Sukwattansinitt, J. Ijadi-Maghsoodi, and T.J. Barton, **1995**, *Am. Chem. Soc. Polymer Preprints*, *36*,(1), 497

10. S. Duchame, R.W. Twieg, J.C. Scott, W.E. Moemer, *Phys. Rev. Lett.,* **1991**, *66*, 1846

11. C.A. Walsh and W.E. Moemer, **1992**, *J. Opt. Soc. Am.*, *B9*, 1642

12. S. Duchame, B. Jones, J.M. Takacs, L. Zhang, *Opt. Let.*, **1993**, *18*, 152

13. K. Tamura, A.B. Padias, J.H.K. Hall, N. Payghambarian, *Appl. Phys. Lett.*, **1992**, *60*, 108

14. L. Zhang, X.F. Ma, Y. Jin, T.-M. Lu, *Appl. Phys. Lett.*, **1992**, *61*, 3080

15. S. Ducharme, J.C Scott, R.J. Twieg, and W.E. Moemer, *Phys. Rev. Lett.,***1991**, *66*, 1846

16. G.P. Wiederrecht, B. Yoon, and M.R. Wasielewski, *Science,* **1995**, *270*, 1794

17. P.M. Lundquist, B. Wortman, C. Geletnecky, R.J. Twieg,M. Jurich, V.Y. Lee, C.R. Moylan, D.M. Burland, *Science*, **1996**, *274*, 1182

18. H. Saadeh, A. Gharavi, D. Yu, and L. Yu, *Macromolecules,* **1997**, *30*, 5403

19. T.J. Marks *and* M.A. Ratner, *Angew. Chem., Int. Ed. Engl.* **1995,**34,155.

20. C.J.K. Botttcher, *Theory of Electric Polarizaion*, Elsevier, Amsterdam, Netherlands, 1973

21. V.E. Moerner, D.M. Burland, C.R. Moylan , and R.J. Twieg, *Am. Chem. Soc. Polymer Preprints,* **1996**, *37* (1), 139

22. L. B. Schein, *Phil. Mag.***1992**, *t365*, 795

23. W.E. Moerner, S.M. Silence, F. Hache, and G.C. Bjorklund, *J. Opt. Sec.*

Am. **1994**, *B 11*, 320, also see W.E. Moerner and S.M. Silence, *Chem. Revs.*, **1994**, *94*, 127, and S.M. Silence, D.M. Burland. and W.E. Moerner, Chap. 5 in *Photorefractive Effects and Materials*, D.D. Nolte, ed. Kluwer Academic, Boston, 1995

24. W.E. Moerner and S.M. Silence, Chem. Rev., **1994**, *94,* 127
25. D.M. Buriand, R.D. Miller, and C.A. Walsh, *Chem. Rev.* **1994**, *94*, 31.
26. L.R. Dalton., A.W. Harper, B. Wu, R. Ghosn, J. Laquindanum, Z. Liang, A. Hubbel, and C. Xu, *Adv. Mater.* **1995**, *7*, 519. ; L.R. Dalton, A.W.Harper, R. Ghosn, W.H. Steier, M. Ziari, H. Fetterman, Y. Shi, R.V. Mustacich, A. K.-Y Jen, and K. Shea, *J. Chem. Mater.* **1995**, *7*, 1060
27. Y. Shi, C. Zhang, H. Zhang, J.H. Bechtel, L.R. Dalton, B.H. Robinson, and W.H. Steier, *Science* **2000**, *288,* 119
28. L.R. Dalton, *Opt. Eng.* **2000**, *39,* 589
29. H. Ma, B. Chen, T. Sassa, L.R. Dalton, and A.K-Y. Jen, *J. Am. Chem. Soc.* **2001**, *123,* 986
30. D.F. Eaton, **Chemtech, 1992**, *22* (5), 308
31. M. Ahlheim, M. Barzoukas, P.V. Bedworth, M. Blanchard-Desce, A. Fort, Z.Y. Hu, S.R. Marder, J.W. Ferry, C. Runser, M. Staehelin, and B. Zysset, *Science* **1996**, *271*, 335; V.K. Rao, Y.M. Cai, and A.K.-Y. Jen, *J. Chem. Soc., Cornmun.* **1994**, 1689; L.R. Dalton, *Am. Chem. Sec. Polym. Prep.* **1996**, *57*(]), 131.
32. S.R. Marder, L.-T. Cheng, B.G. Tiemann, A.C. Friedi, M. Blanchard-Desce, J.W. Perry, and J. Skinhoj, *Science*, **1994**, *263*, 511
33. A. K-Y. Jen, S.R. Marder, and C.-F. Shu, *Am. Chem. Soc. Polymer Preprints*, **1997**, *38* (2), 495
34. I. Cabrera, O. Althoff, H.-T. Man, and H.N. Yoon, *Adv. Mater.*, **1996**, *6*, 43
35. R. Dgani, **C&EN,** 4 (March 4, 1996)
36. C.R. Moylan, R.J. Twieg, V.Y. Lee, S.A. Swanson, K.M. Betterton, and R.D. Miller, *J. Am. Chem. Soc.* **1993**, *115*, 12599
37. N.J. Long, *Agew. Chem.*, **1995**, *107*, 37
38. M.C.J.M. Donckers, S.M. Silence, C.A. Walsh. F. Hache, D.M. Burland, W.E. Moerner, and R.J. Twieg, *Opt. Lett.* **1993**, *18*, 1044
39. W. E. Moerner, *Proc. Sec. Photo-Opt. Instrum. Engr.* **1993**, *1852*, 253
40. K. Meerholz, B.L. Volodin, Sandalphon, B. Kippelen, and N. Peyghambarian, *Nature*, **1994**, *371*, 497
41. W.E. Moerner, S.M. Silence, F. Hache, and G.C. Bjorklund, *J. Opt. Soc. Am. B*, **1994**, *11*, 320
42. E. Hendrickx, J.F. Wang, J.L. Maldonado, B.L. Volodin, Sandalphon, E. A. Mash, A. Persoons, B. Kippelen, and N. Peyghambarian, *Macromolecules*, **1998**, *31*, 734
43. Y. Zhang, T. Wada, T. Aoyama, L. Wang, H. Sasabe, *Mol. Cryst. Liq. Cryst. Sci. Technol., Sect. A* **1997**, *295*, 349

44. L.R. Dalton, *Am. Chem. Soc. Polymer Preprints*, **1996**, *37* (1), 131
45. M.H. Davey, Y.Y. Lee, T.J. Marks, and R.D. Miller, *Am. Chem. Soc. Polymer Preprints*, **1997**, *38* (2), 261
46. J.M.J. Frechet, T.G. Tessier, C. Grant Willson, and H. Ito, *Macromolecules*, **1985**, *8*, 317
47. Ducharme, *Science*, **1994**, *63*, 376
48. J. Luo, M. Haller, H. Li, H.-Z. Tang, A. K.-Y. Jen, K. Jakka, C.-H. Chou, and C.-F. Shu, *Macromolecules*, **2004**, 37, 248-250
49. W. You, S. Cao, Z. Hou, and L. Yu, *Macromolecules* **2003**, 36, 7014-7019
50. P.V. Bedworth, Y.M. Cai, A. Jen, S.R. Marder, *J. Org. Chem.* **1996**, *61*, 2242.
51. Y.S. Ra, S.S.H. Mao, B.Wu, L. Guo, and L.R. Dalton, A. Chen and W.H. \ Steier, *Am. Chem. Soc. Polymer Preprints*, **1997**, *38* (1), 86, 926
52. M. Haller, J. Luo, H. Li, T.-D. Kim, Y. Liao, B.H. Robinson, L.R. Dalton, and A. K.-Y. Jen *Macromolecules* **2004**,37, 688-690
53. J.E. Beecher, T. Durst, J.M.J. Frechet, A. Godt, and C.W. Willand. *Macromolecules*, **1994**, *27*, 3472-3477
54. J.C. Laquindanum, L.R. Dalton, A.K.Y. Jen, and C.W. Spangler, *Am.Chem.Soc. Polymr Preprints*, **1995**, *36* (1), 487
55. K.-J.Moon, H-K. Shim, K.-S. Lee, J. Zieba, and P.N. Prasad, *Macromolecules*, **1996**, *29*, 861
56. J. Wang, Y. Shen, Z. Shi, *J. Mater. Sci.* **2000**, 35(9), 2139-2143
57. C.Weder, P. Neuenschwander, U. W. Suter, P. Pretre, P. Kaatz, and P. Gunter, *Macromolecules*, **1995**, *28*, 2377-2382
58. M. Trollsis, *et al., J. Am. Chem. Soc.*, **1996**, *118*, 8542
59. L. Yu et al. *J. Am. Chem. Sec.*, **1994**, *116*, 6003
60. K.J. Drost, A.K.-Y. Jen, and V.P. Rao, *Chemtech*, **1995**, *25*(9), 16
61. S. Yang, Z. Peng, and L. Yv, *Macromolecules*, **1994**, *27*, 5858-5862
62. T. Matsuura, S. Ando, S. Sasaki, and F. Yamamoto, *Macromolecules*, **1994**, *27*, 6665-6670
63. T-A. Chen, A.K-Y. Jen, and Y. Cai, *Macromolecules*, **1996**, *29*, 535-539
64. T-A. Chen, A.K-Y. Jen, and Y. Cai, *J. Am. Chem. Soc.*, **1995**, *117*, 7295
65. D.M. Burland, R.D. Miller, R.J. Twieg, and V.Y. Lee *Science*, **1005**, *268*, 1604
66. R.S. Marder, M.Perry, M. Staehelin, and B. Zysset, *Science*, **1966**, *271*, 335
67. L. Bes, A. Rousseau, B. Boutevin, R. Mercier, B. Sillion, E. Toussaere, *High Perform. Polym.* **2000**, 12(1), 69-176
68. P. Zhu, P. Wang, W. Wu, and I. Ye, *Nonlinear Opt. Phys. Mater.* **1999**, *8*(4), 461-468
69. L.P. Yu and co-workers, *J. Mater. Chem.* **1999**, *9*. 1865-1873
70. J. Lu, N. Chen, S. Ji, Z. Jia, Z. Sun, X. Zhu, and W. Shi, *Chemical Journal on Internet,* **2003**, 5(4)

71. T-A. Chen, A.K-Y. Jen, and Y., Cai *Chem. Mater.* **1996**, *8*, 60

72. G. Koeckelberghs, S. Sioncke, T. Verbiest, I. Van Severen. I. Picard, A. Persoons, and C. Samyn, *Macromolecules* **2003**, 36, 9736-9741; G. Koeckelberghs, M. Vangheluwe, I. Picard, L. De Groof, T. Verbiest, A. Persoons, and C. Samyn, *Macromolecules* **2004**, 37, 8530

73. Z. Lu, P. Shao, J. Li, J. Hua, J. Qin, A. Qin, and C. Ye *Macromolecules,* **2004**, 37, 7089-7096

74. C. Sanchez , U. Pierre, and M. Curie, *Chem. Mater.*, **1997**, *9*(4), 1012 (From *Chemtech*, **1997**, *27* [8], 4)

75. R.B. Burkhardt, O. Lee, S. Boileau, and S. Boivin, *Macromolecules*, **1985**, *18*, 1277

76. N.-J. Kim, H. Chun, I. K. Moon, W.-J. Joo, and N. Kim *Bull. Korean Chem. Soc.,* **2002**, Vol. 23, No. 4 571

77. D.-W. Cheong, W-H. Kim, L.A. Samuelson, J. Kumar, and S.K. Tripathy, *Macromolecules*, **1996**, *29*, 1416

78. K. E. Huggins, S. Son, and S. I. Stupp, *Macromolecules* **1997**, *30*, 5305

79. J.-H. Wang, Y.-O. Shen, G.-X. Yu, and J. Zheng, *Synth.Met.* **2000**, 113 (1-2), 73-76

80. Y. Zhang, L. Wang, T. Wada, and H. Sasabe, *Macromolecules*, **1996**, *29*, 1569-1573

81. D.M. Walba, M.B. Ros, N.A. Clark, R. Shao, K.M. Johnson, M.G. Robinson, J.Y. Liu, and D. Doroski, *Ctyst. Liq. Cryst.*, **1991**, *198*, 51

82. H. Kapitza, R. Zentel, R.J. Twieg, C. Nguyen, , S.U. Valerien, L.F. Kremer, C.G. Wilson, *Adv. Mater.* **1990**, *11*, 539

83. P. Keller, R. Shao, D.M. Walba, and M Brunet, *Liq. Cryst.*. **1995**, *18*, 915

84. A.V. Medvedev, E.B. Barmatov, A.S. Medvedev, V.P. Shibaev, S.A. Ivanov, and J. Stumpe, *Macromolecules*, **2005**, *38*, 2223

85. D. Van Steenwinckel, C. Engels, B. Gubbelmans, E. Hendrickx, C. Samyn, and A. Persoons, *Macromolecules,* **2000,** 33(11), 4074-4079

86. M. Hirai, T. Yuzawa, Y. Haramoto, and M. Nanasawa, *React. Funct. Polym.* **2000**, 45(31, 175-181

87. H. Moon, J. Hwang, N. Kim, and S.Y. Park, *Macromolecules*, **2000**, *33*, 5116-5123

88. P. Persico, R. Centore, A. Sirigu, M. Casalboni, A. Quatela, and F. Sarcinelli, *Journal of Polymer Science, Part A: Polymer Chemistry* **2003**, *41*(12), 1841-1847

89. J. Qin, Z. Yang, Z. Li, C. Huang, *Macromolecules* **2003**, *36,* 1145

90. Z. Li, C. Huang, J. Hua, J. Qin, Z. Yang and C. Ye, *Macromolecules* **2004**, 37, 371-376

91. H. Saadeh, L. Wang, and L. Yu, J. *Am. Chem. Soc.* **2000**, *122,*546-547

92. T. Kimura, T. Fukuda, H. Matsuda, M. Kato, H. Nakanishi, *Kobunshi Ronbunshu* **2003**, 60(12), 682-692 (from Chem. Abstr.)

93. K.H. Park, R.J. Twieg, R. Ravikiran, L.F. Rhodes, R.A. Shick, D. Yankelevich and A. Knoesen, *Macromolecules,* **2004**, *37*, 5163-5178

94. H.E. Katz, K.D. Singer, J.E. Sohn, C.D, Dirk, L.A. King, and H.M. Gordon, *J. Am. Chem. Soc.*, **1987**, *109*, 6561

95. M. Liphardt, A. Goonesekera, B.E. Jones, S. Duchame, J.M. Takacs, and L. Zhang, *Science*, **1994**, *26*, 367

Index